Wave Propagation Theory

Pergamon Titles of Related Interest

Bass/Fuchs   Wave Scattering from Statistically Rough Surfaces
Chantry   Submillimetre Waves and Their Applications
Kao/Hwang   Electrical Transport in Solids
Lominadze   Cyclotron Waves in Plasma
Moruzzi et al.   Calculated Electronic Properties of Metals
Pozhela   Plasma and Current Instabilities in Semiconductors

Related Journals*

Acta Metallurgica
Electrochimica Acta
International Journal of Solids and Structures
Journal of Atmospheric and Terrestrial Physics
The Journal of Physics and Chemistry of Solids
Ocean Engineering

*Free specimen copies available upon request.

(Lectures on) Wave Propagation Theory

James R. Wait
University of Arizona
Tucson, AZ 85721

PERGAMON PRESS
New York • Oxford • Toronto • Sydney • Paris • Frankfurt

Pergamon Press Offices:

U.S.A.	Pergamon Press Inc., Maxwell House, Fairview Park, Elmsford, New York 10523, U.S.A.
U.K.	Pergamon Press Ltd., Headington Hill Hall, Oxford OX3 0BW, England
CANADA	Pergamon Canada Ltd., Suite 104, 150 Consumers Road, Willowdale, Ontario M2J 1P9, Canada
AUSTRALIA	Pergamon Press (Aust.) Pty. Ltd., P.O. Box 544, Potts Point, NSW 2011, Australia
FRANCE	Pergamon Press SARL, 24 rue des Ecoles, 75240 Paris, Cedex 05, France
FEDERAL REPUBLIC OF GERMANY	Pergamon Press GmbH, Hammerweg 6 6242 Kronberg/Taunus, Federal Republic of Germany

Copyright © 1981 Pergamon Press Inc.

Library of Congress Cataloging in Publication Data

Wait, James R
  Wave propagation theory.

  1. Electromagnetic waves—Transmission.
I. Title
QC665.T7W34   1981   530.1'43   80-23286
ISBN 0-08-026345-3
ISBN 0-08-026344-5 (pbk.)

All Rights reserved. No part of this publication may be reproduced, stored in a retrieval system or transmitted in any form or by any means: electronic, electrostatic, magnetic tape, mechanical, photocopying, recording or otherwise, without permission in writing from the publishers.

Printed in the United States of America

Contents

		Preface	vii
Note			
	1	Notation and Some Basic Ideas	1
	2	Reflection from Stratified Media	7
	3	Magneto-Telluric Fields	18
	4	General Surface Impedance	43
	5	The Zenneck Wave	56
	6	Excitation of the Layered Half-Space — The Green's Function	61
	7	On the Excitation of the Zenneck Surface Wave	67
	8	Surface Impedance of a Spherically Stratified Conductor	76
	9	Excitation of the H.F. Surface Wave by Vertical and Horizontal Apertures	83
	10	Fields of a Dipole Over an Homogeneous Anisotropic Half-Space	96
	11	Fields of an Horizontal Dipole Over a Stratified Anisotropic Half-Space	110
	12	Asymptotic Evaluation of the Field of a Vertical Dipole Over an Impedance Plane Surface	117
	13	Fields of a Circular Loop of Current Buried in a Two-Layer Earth	130
	14	Transmissions in an Idealized Earth Crust Waveguide	153
	15	Reflection from Inhomogeneous Media with Special Profiles	164

16	Approximate Methods for Inhomogeneous Media	184
17	High Frequency Electromagnetic Coupling Between Small Loops Over an Inhomogeneous Half-Space	203
18	Reflection of VLF Radio Waves from an Inhomogeneous Isotropic Ionosphere	212
19	Reflection from a Lossy Magnetoplasma Half-Space	234
20	EM Propagation in the Earth-Ionosphere Waveguide	244
21	Guiding of Microwaves by an Elevated Tropospheric Layer	310
22	Scattering from an Isolated Irregularity in a Tropospheric Duct	320
23	Coupled Mode Analysis for a Non-Uniform Tropospheric Waveguide	332
	Index	347
	About the Author	349

Preface

The subject matter for this series of lectures was taken, for the most part, from previous publications by the author over a twenty year span. To provide some continuity, considerable revision was made in the source material and various gaps in the coverage were filled. The general theme deals with the analysis of electromagnetic wave propagation in various kinds of media with special reference to terrestrial environments. The same methods can be used in acoustic waveguides either in the ocean or in the atmosphere.

These notes were used in a one semester course (three hours per week) to first and second year graduates in Electrical Engineering and Physics at the University of Colorado and, later, at the University of Arizona. The class assignments are given as exercises in these notes. It is also hoped the material will provide a useful reference to engineers and scientists who are concerned with radio propagation both above and below the earth's surface. Also, exploration geophysicists should find much of the subject matter to be relevant to electromagnetic probing.

1
Notation and Some Basic Ideas

Throughout, the rationalized MKS system of units is used. Since attention is only confined to linear phenomena, the electrical properties of the medium (or media) can also be defined in terms of the constants

ε, the dielectric constant (F/m)

σ, the conductivity (mho/m)

μ, the permeability (H/m)

In some cases these "constants" will be functions of the coordinate. Locally, however, they are always considered constant. In later chapters ε and σ are regarded as tensors to account for the anisotropic characteristics of magneto-plasmic media. Later we will also deal with related acoustic problems where analogous "constants" will be introduced.

Nearly always the time factor in this book is $\exp(+i\omega t)$ when ω is the angular frequency and t is the time. Consequently, the actual electric field $e(t)$ is related to the complex phasor E by

$$e(t) = \text{Real part of } (Ee^{i\omega t})$$

To explain some of the basic notation, a very short exposition of plane electromagnetic waves in a homogeneous medium is presented.

Ohm's law in the complex form is

$$J = (\sigma + i\varepsilon\omega)E \qquad (1)$$

where J is the current density vector and E is the electric field vector. The dimensions of J are A/m^2 and those of E are V/m. The analogous relation for magnetic quantities is

$$B = \mu H \qquad (2)$$

where B is the magnetic vector density and H is the magnetic vector intensity. The dimensions of B are wb/m^2 and those of H are A/m.

In source free media the above vector quantities are related by

$$\text{curl } E = -i\mu\omega H \qquad (3)$$

and

$$\text{curl } H = (\sigma + i\varepsilon\omega)E \qquad (4)$$

These are Maxwell's equations. (Maxwell 1888).

For a homogeneous medium

$$\text{curl curl } E = \text{grad div } E - \nabla^2 E = -i\mu\omega(\sigma + i\varepsilon\omega)E \qquad (5)$$

Since div $E = 0$, this can be reduced to

$$(\nabla^2 - \gamma^2)E = 0 \qquad (6)$$

where $\nabla^2 =$ div grad is the Laplacian operator (which operates on the rectangular components of E) and $\gamma^2 = i\mu\omega(\sigma + i\varepsilon\omega)$. The quantity γ is defined as the propagation constant.

As a simple preliminary problem, the fields are assumed not to vary in either the x or y directions with reference to a conventional coordinate system (x,y,z). Furthermore, the electric field is taken to have only an x component E_x. Therefore, Eq. (6) reduces to

$$\left(\frac{d^2}{dz^2} - \gamma^2\right)E_x = 0 \qquad (7)$$

and the solutions are $e^{+\gamma z}$ and $e^{-\gamma z}$. Therefore, the general solution is

$$E_x = Ae^{\gamma z} + Be^{-\gamma z} \qquad (8)$$

where A and B are constants. The magnetic field component then has only a y component given by

$$H_y = -\frac{1}{i\mu\omega}\frac{\partial E_x}{\partial z} = -\eta^{-1}(Ae^{\gamma z} - Be^{-\gamma z}) \qquad (9)$$

where $\eta = [i\mu\omega/(\sigma + i\varepsilon\omega)]^{1/2}$ is by definition the characteristic impedance of the medium for plane wave propagation. Remembering that the time factor is $e^{i\omega t}$, it can be seen that the term $Be^{-\gamma z}$ is a wave travelling in the positive

Notation and Some Basic Ideas

z direction with a diminishing amplitude, and the term Ae^{yz} is a wave travelling in the negative z direction with a diminishing amplitude. The quantity is thus equal to the complex ratio of the electric and magnetic field components in the x and y directions, respectively, for plane waves in an unbounded homogeneous medium.

The quantities defined by

$$\gamma = [i\mu\omega(\sigma + i\varepsilon\omega)]^{1/2} \text{ and } \eta = [i\mu\omega/(\sigma + i\varepsilon\omega)]^{1/2}$$

are sometimes called the secondary constants. In the case of free space

$$\varepsilon = \varepsilon_0 = 8.854 \times 10^{-12} \text{ F/m}$$

$$\mu = \mu_0 = 4\pi \times 10^{-7} \text{ H/m}$$

$$\sigma = 0$$

and then $\gamma = ik$ where $k = (\varepsilon_0\mu_0)^{1/2}\omega = 2\pi/\lambda$ and λ is the wavelength. Furthermore

$$\eta = \eta_0 = (\mu_0/\varepsilon_0)^{1/2} \approx 120\pi\Omega$$

GENERAL REFERENCES

The following are selected texts or review papers which deal with the general subject matter. They are recommended as supplementary reading.

AL'PERT, Ia. L., GINZBURG, V.L. and FEINBURG, E.L. (1953) Radio wave propagation, State Printing House for Technical-Theoretical Literature, Moscow.

ARORA, R.K. and WAIT, J.R., (1978) Refraction theories of radio wave propagation through the troposphere - A review, *Radio Science,* 13, No. 3, 599-600.

BANOS, A. (1966), *Dipole Radiation in the Presence of a Conducting Half-Space,* Pergamon Press, Oxford.

BARLOW, H.E.M. and BROWN, J., (1962), *Surface Waves,* Oxford University Press.

BEZRODNY, V.G., NICKOLAENKO, A.P., and SINITSIN, V.G., (1977), Radio propagation in natural waveguides, *Jour. Atmos. and Terr. Phys.,* 39, 661-688, (uses non-rationalized Gaussian units).

BREKHOVSKIKH, L.M. (1960), *Waves in Layered Media,* Academic Press, New York.

BREMMER, H., (1958), Propagation of electromagnetic waves, *Handbuch der Physik,* 16, 423-639, Springer-Verlag, Berlin.

BUDDEN, K.G., (1961), *Radio Waves in the Ionosphere,* Cambridge University Press.

BUDDEN, K.G., (1962), *The waveguide mode theory of wave propagation,* Prentice-Hall, New York.

COLLIN, R.E., (1960), *Field Theory of Guided Waves,* McGraw-Hill, New York.

COLLIN, R.E., and ZUCKER, F.J., (1969), *Antenna Theory, Parts I and II,* McGraw-Hill, New York.

EWING, M., JARDETZKY, W., and PRESS, F., (1958) *Elastic Waves in Layered Media,* McGraw-Hill, New York.

FELSEN, L.B., and MARCUVITZ, N., (1976), *Scattering and Diffraction of Waves,* Prentice Hall, New York.

Notation and Some Basic Ideas

FOCK, V.A. (1946), *The Diffraction of Radio Waves Around the Earth*, Acad. of Sciences of USSR, Moscow.

FRANZ, W., (1957), Theorie der Beugung elektromagnetischer Wellen, *Ergebnisse der angewandten Mathematik*, Pt. 4, Springer-Verlag, Berlin.

GALEJS, J., (1972), *Terrestrial Propagation of Long Electromagnetic Waves*, Pergamon Press, Oxford.

HARRINGTON, R.F., (1961) *Time Harmonic Electromagnetic Fields*, McGraw-Hill, New York.

HOLTET, J.A., editor (1974), *ELF-VLF Radio Wave Propagation*, D. Reidel Publ. Co., Dordrecht, Holland

HÖNL, H., MAVE, A.W. and WESTFALL, K., (1961), Theorie der Beugung, *Handbuch der Physik*, 25, 218-583, Springer-Verlag, Berlin.

KING, R.J. and WAIT, J.R., (1976), Electromagnetic groundwave propagation theory and experiment, *Symposia Mathematica*, 18, 107-208, Academic Press.

KONG, J.A., (1975), *Theory of Electromagnetic Waves*, John Wiley & Sons, New York.

LIGHTHILL, SIR JAMES,(1978), *Waves in Fluids*, Cambridge University Press.

LOGAN, N.A. and YEE, K.S., (1962), A mathematical model for diffraction by convex surfaces, *Proc. Symp. of Electromag. Waves*, 139-180, University of Wisconsin Press.

MAXWELL, J.C., (1888), *A Treatise on Electricity and magnetism*, Dover reprint.

MENTZER, J.R., (1955), *Scattering and Diffraction of Radio Waves*, Pergamon Press, London and New York.

MOORE, R.K. and BLAIR, W.E., (1961) Dipole radiation in a conducting half-space, *J. Res. Nat. Bur. Stand.*, 65D, (Radio Prop.) 547-563.

OFFICER, C.B.,(1958) *Introduction to the Theory of Sound Transmission*, McGraw-Hill, New York.

SCHELKUNOFF, S.A., (1943), *Electromagnetic Waves*, Van Nostrand, New York.

USLENGHI, P.L.E., editor, (1978), *Electromagnetic Scattering*, Academic Press, New York.

VAN DE HULST, H.C., (1957), *Light Scattering by Small Particles*, Wiley, New York.

WAIT, J.R., (1959), *Electromagnetic Radiation From Cylindrical Structures*, Pergamon Press, New York and London.

WAIT, J.R., (1961), The electromagnetic fields of a horizontal dipole in the presence of a conducting half-space, *Can. J. Phys.*, $\underline{39}$, 1017-1028.

WAIT, J.R., (1962), The propagation of electromagnetic waves along the earth's surface, *Proc. Symp. of Electromag. Waves*, 243-290, University of Wisconsin Press (ed. b y R.E. Langer).

WAIT, J.R., and SPIES, K.P., (1964), Characteristics of the Earth-ionosphere waveguide for VLF radio waves, National Bureau of Standards, TN-300, available from U.S. National Technical Information Service, Springfield, Va., (Accession No. PB 168 048). (An extensive compilation of numerical results for the mode equation, but restricted to equatorial transmission).

WAIT, J.R., (1966), Electromagnetic fields of dipole over an anisotropic half-space., *Can. J. Phys.*, $\underline{44}$, 2387-2401.

WAIT, J.R., (1966), Fields of a horizontal dipole over a stratified half-space, *IEEE Trans.*, $\underline{AP-14}$, 6 , 790-792.

WAIT, J.R., (1968), *Electromagnetics and Plasmas*, Holt, Rinehart and Winston, New York.

WAIT, J.R., (1977), Propagation of ELF electromagnetic waves and Project Sanguine/Seafarer, *IEEE J. of Oceanic Engrng.*, $\underline{OE-2}$, No. 2, 161-172.

Late Addition:

JONES, D.S., (1979), *Methods in Electromagnetic Wave Propagation*, Oxford University Press, (deals mostly with basic antenna theory and scattering from complex shapes).

2
Reflection from Stratified Media

Here we present a general analysis of reflection of plane waves from a parallel stratified medium consisting of M homogeneous slabs.

PARALLEL INCIDENCE

A plane wave with a time factor $\exp(i\omega t)$ is incident at an angle θ on a stratified medium composed of M homogeneous layers. The electric vector is in the plane of incidence (xz plane). The situation is illustrated in Fig. 1

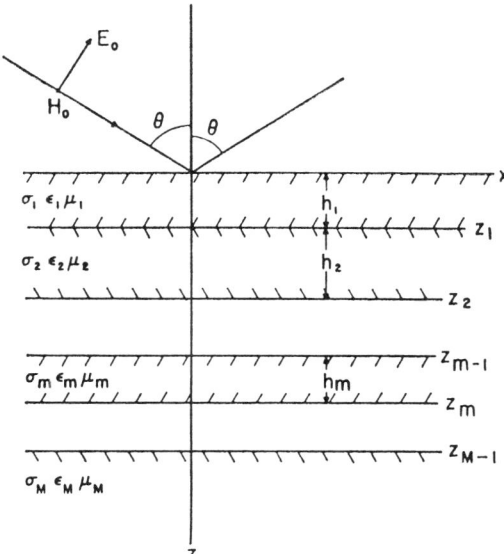

FIG. 1. A stratified medium consisting of M homogeneous layers.

where the y axis is out of the paper. The electrical constants of the layers are σ_m, ε_m and μ_m where the subscripts m indicate the mth layer below the surface.

From symmetry it can be seen that the magnetic field has only a y component and for the mth layer, it is a solution of the equation

$$(\nabla^2 - \gamma_m^2)H_{my} = 0 \qquad (1)$$

where

$$\gamma_m^2 = i\sigma_m\mu_m\omega - \varepsilon_m\mu_m\omega^2 \quad \text{with real part of } \gamma_m > 0.$$

(Reprinted in part in revised form, from: J.R. Wait, Electromagnetic Waves in Stratified Media, 2nd Edition, Pergamon Press, New York, 1970)

The general solution is of the form

$$H_{my} = [a_m e^{-u_m z} + b_m e^{u_m z}] e^{-i\lambda x} \qquad (2)$$

where $u_m^2 = \lambda^2 + \gamma_m^2$ and λ can take any value. However, real part of $u_m > 0$. The incident field H^{inc} can be written

$$H_{0y}^{inc} = H_0 e^{-\gamma_0 \cos\theta \cdot z} \times e^{-\gamma_0 \sin\theta x}$$

Therefore, in equation (2) $a_0 e^{-u_0 z} e^{-i\lambda x}$ can be identified with H_{0y}^{inc} if $a_0 = H_0$ and $i\lambda = \gamma_0 \sin\theta$. Consequently, $b_0 e^{u_0 z} e^{-i\lambda x}$ is a reflected wave and the angle of reflection is θ.

The boundary conditions at the interface $z = 0, z = z_1, \ldots, z = z_{m-1}$ are that the tangential fields should be continuous. Now since

$$E_{mx} = -(\sigma_m + i\omega\varepsilon_m)^{-1} \frac{\partial H_{my}}{\partial z} \qquad (3)$$

this means that the boundary conditions can be written

$$\left[\begin{array}{c} H_{m-1,y} = H_{m,y} \\ (\sigma_{m-1} + i\omega\varepsilon_{m-1})^{-1} \dfrac{\partial H_{m-1,y}}{\partial z} = (\sigma_m + i\omega\varepsilon_m)^{-1} \dfrac{\partial H_{m,y}}{\partial z} \end{array} \right]_{z=z_{m-1}} \qquad (4)$$

where $m = 0, 1, 2, \ldots, M-2, M-1$.

Imposing the condition that only outgoing waves are permissible in the lowest layer (which is semi-infinite) it follows that $b_M = 0$. The boundary conditions then lead to $2(M-1)$ equations which are linear in a_m and b_m to solve for $2(M-1)$ unknowns in terms of the known coefficient a_0. The solution* is

$$\frac{b_0}{a_0} = \frac{K_0 - Z_1}{K_0 + Z_1} \qquad (5)$$

where

$$Z_1 = K_1 \frac{Z_2 + K_1 \tanh u_1 h_1}{K_1 + Z_2 \tanh u_1 h_1}$$

$$Z_2 = K_2 \frac{Z_3 + K_2 \tanh u_2 h_2}{K_2 + Z_3 \tanh u_2 h_2} \qquad (6)$$

$$\ldots\ldots\ldots\ldots\ldots\ldots\ldots$$

$$Z_m = K_m \frac{Z_{m+1} + K_m \tanh u_m h_m}{K_m + Z_{m+1} \tanh u_m h_m}$$

$$\ldots\ldots\ldots\ldots\ldots\ldots\ldots$$

$$Z_{M-1} = K_{M-1} \frac{K_M + K_{M-1} \tanh u_{M-1} h_{M-1}}{K_{M-1} + K_M \tanh u_{M-1} h_{M-1}}$$

* The quantity λ should not be confused with the wavelength.

where

$$K_m = \frac{u_m}{\sigma + i\omega\varepsilon_m}, \quad \text{and} \quad u_m = (\lambda^2 + \gamma_m^2)^{1/2}$$

(Note that h_m is the thickness of the mth slab).

The quantity b_0/a_0 is the ratio of the amplitude of the reflected wave to the amplitude of the incident wave. It is denoted by $R_\parallel$ to indicate that the electric field of the incident wave (and also that of the reflected wave) is in the plane of incidence.

The present problem has a well defined analogy in transmission line theory.

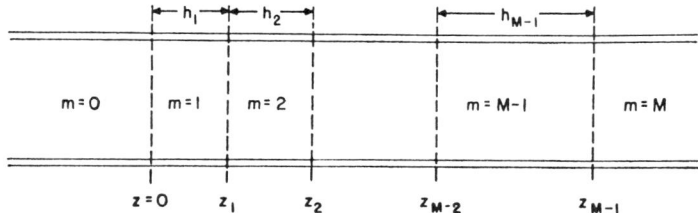

FIG. 2. Transmission line analogy for the stratified medium of M layers.

In this analogy each section of the multiple sectioned line is to correspond to a slab. The voltage across the line is E_{mx} and the current is H_{my} for nth section. The propagation constant is u_m and the characteristic or surge impedance is K_m of the mth slab. The incident wave which comes from the left in Fig. 2 is given by

$$a_0 \, e^{-u_0 z}$$

and the wave reflected at the junction, $z = 0$, is

$$b_0 \, e^{u_0 z}$$

The input impedance of the line which is the ratio of the voltage to the current at $z = 0$ is Z_1. Furthermore, the impedance at the junction $z = z_m$ is Z_{m+1}. With this analogy and a knowledge of the behavior of one dimensional transmission lines, one could write down the solution of the 2 dimensional reflection problem.

Some features of the wave problem will now be discussed. The quantity Z_1 will play an important role in the following. Here

$$Z_1 = E_{0x}/H_{0y}]_{z=0} = E_{1x}/H_{1y}]_{z=0} \; ,$$

is called the surface impedance, being the ratio of the

tangential fields at the air-ground interface. In the case of normal incidence, $\theta = 0$ and $\lambda = 0$ we see that $u_m = \gamma_m$ and $K_m = \eta_m$ where

$$\gamma_m = [i\sigma_m \mu_m \omega - \varepsilon_m \mu_m \omega^2]^{1/2}$$

and

$$\eta_m = [i\mu_m \omega/(\sigma_m + i\varepsilon_m \omega)]^{1/2}$$

For example, in the case of a homogeneous ground ($h_1 \to \infty$),

$$Z_1 = \eta_1, \quad K_0 = \eta_0$$

and the reflection coefficient becomes simply

$$R_\parallel = \frac{b_0}{a_0} = \frac{\eta_0 - \eta_1}{\eta_0 + \eta_1} \tag{7}$$

Another special case of considerable interest is when θ approaches 90 degrees corresponding to glancing incidence, then

$$u_m \cong (\gamma_m^2 - \gamma_0^2)^{1/2}, \quad K_m \cong (\gamma_m^2 - \gamma_0^2)^{1/2}/(\sigma_m + i\omega\varepsilon_m)$$

which yields for the homogeneous ground

$$Z_1 = K_1 = \eta_1(1 - \gamma_0^2/\gamma_1^2)^{1/2} \tag{8}$$

This special value of Z_1, which relates the tangential fields in the limiting case of glancing incidence, turns out to be an important quantity in further work. For this reason, it is denoted by Z^v where the superscript v indicates that the electric field in the air is nearly vertical for $|\gamma_1| \gg |\gamma_0|$. This fact can be shown by evaluating the wave tilt which is defined by

$$W = \left. \frac{E_{0x}}{E_{0z}} \right|_{z=0} \tag{9}$$

and is the complex ratio of the horizontal to the vertical electric field in the air just above the ground. It readily follows that for the general case [Norton, 1935, 1936, 1937].

$$W = \left. \frac{E_{1x}/H_{1y}}{E_{0z}/H_{0y}} \right|_{z=0} = \frac{i\omega\varepsilon_0 Z_1}{\gamma_0 \sin \theta} = \frac{Z_1}{\eta_0 \sin \theta} \tag{10}$$

In the case of a homogeneous ground

$$W = \frac{i\omega\varepsilon_0 K_1}{\gamma_0 \sin \theta} = \frac{\eta_1}{\eta_0} \frac{\left(1 - \frac{\gamma_0^2}{\gamma_1^2} \sin^2 \theta\right)^{1/2}}{\sin \theta} \tag{11}$$

For grazing incidence this becomes

$$W = \frac{\eta_1}{\eta_0}\left(1 - \frac{\gamma_0^2}{\gamma_1^2}\right)^{1/2} = \left(\frac{\mu_1}{\mu_0}\right)\frac{\gamma_0}{\gamma_1}\left(1 - \frac{\gamma_0^2}{\gamma_1^2}\right)^{1/2} \tag{12}$$

To indicate in a simple way as is possible, the influence of stratification on the reflection of waves from the ground surface, a 2 layer case will be considered. This is effected by letting $h_2 \to \infty$. Furthermore, it will be assumed that $|\gamma_1/\gamma_0|$ and $|\gamma_2/\gamma_0| \gg 1$. Then for any angle of incidence

$$u_m = \gamma_m\left(1 - \frac{\gamma_0^2}{\gamma_m^2}\sin^2\theta\right)^{1/2} \cong \gamma_m \quad (m = 1, 2)$$

and

$$K_m = \frac{u_m}{\sigma_m + i\varepsilon_m\omega} \cong \eta_m \quad (m = 1, 2)$$

This leads to the simple relation

$$Z_1 \cong QK_1 \tag{13}$$

where

$$Q = \frac{(\gamma_1/\gamma_2) + \tanh\gamma_1 h_1}{1 + (\gamma_1/\gamma_2)\tanh\gamma_1 h} \quad \text{for} \quad \mu_1 = \mu_2 = \mu_0,$$

and where Q is the correction to the characteristic impedance K_1 of the upper layer to account for the presence of the lower layer. Note then, if $|\gamma_1 h_1| \gg 1$, $Q \cong 1$. It can be said that the lower layer is not detectable when $|Q|$ is within 5 per cent of unity. Such a condition is met when $(\sigma_1\mu_0\omega)^{1/2}h > 3$.

It should also be noted that Q relates the wave tilts for a stratified (2 layer) ground and that of a homogeneous ground by

$$W \cong W_0 Q$$

where

$$W_0 \cong W]_{h_1 = \infty}$$

An example is here quoted to illustrate the order of magnitude of the quantities involved:

Frequency, $f = \omega/2\pi = 125$ kc/s
Upper layer conductivity, $\sigma_1 = 10^{-3}$ mho/m
Dielectric constant of air, $\varepsilon_0 = 8.854 \times 10^{-12}$ F/m
Dielectric constant of ground, $\varepsilon_1 = 10\cdot\varepsilon_0$
Magnetic permeability, $\mu = 4\pi \times 10^{-7}$ H/m.

For these values

$$W_0 = 0.082 \angle 41.1°$$

so that the boundary conditions now become

$$E_{m-1,y} = E_{m,y}$$

$$(i\mu_{m-1}\omega)^{-1} \frac{\partial E_{m-1,y}}{\partial z} = (i\mu_m\omega)^{-1} \frac{\partial E_{m,y}}{\partial z} \bigg]_{z=z_m} \quad (19)$$

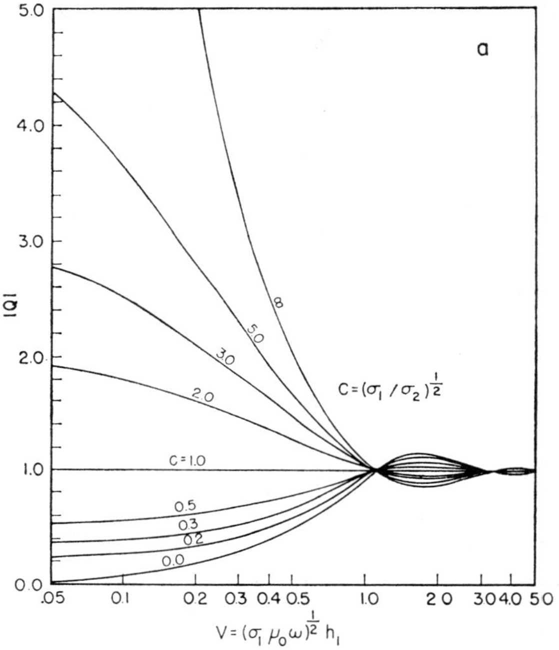

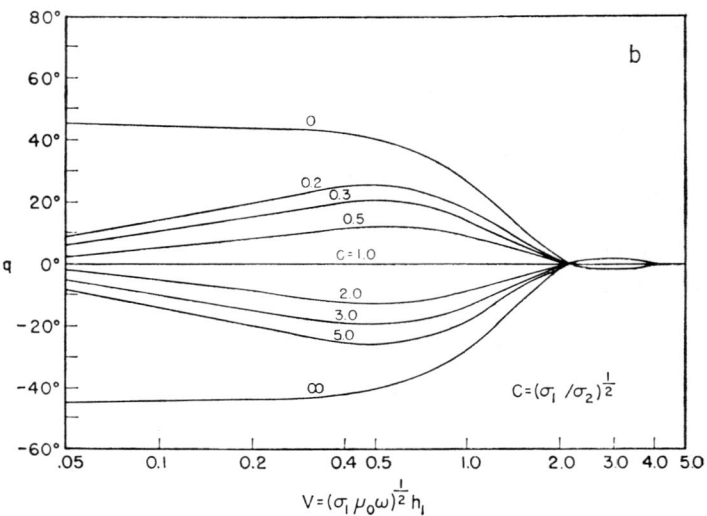

FIG. 3a. Amplitude of correction factor for a two-layer ground.

FIG. 3b. Phase of correction factor for a two-layer ground.

for a homogeneous ground. In the case of a 2 layer ground where h_1 is finite

$$W = W_0 Q$$
$$= 0.082 |Q| \angle (41.1° + q)$$

where q is the argument of Q expressed in degrees. For frequencies of this order $\varepsilon_1 \omega/\sigma_1$ and $\varepsilon_2 \omega/\sigma_2$ are small (in the above example $\varepsilon_1 \omega/\sigma_1 = 0.0069$). Therefore, $\gamma_1 \cong (i\sigma_1 \mu \omega)^{\frac{1}{2}}$ and $\gamma_2 \cong (i\sigma_2 \mu \omega)^{\frac{1}{2}}$. A formula suitable for computation of Q is then given by

$$Q = \frac{(\sigma_1/\sigma_2)^{\frac{1}{2}} + \tanh \sqrt{(i)}V}{1 + (\sigma_1/\sigma_2)^{\frac{1}{2}} \tanh \sqrt{(i)}V} \quad \text{where} \quad V = (\sigma_1 \mu \omega)^{\frac{1}{2}} h_1 \qquad (14)$$

$$= \tanh[\sqrt{(i)}V + \tanh^{-1}(\sigma_1/\sigma_2)^{\frac{1}{2}}]$$

If $\sigma_2 \gg \sigma_1$ corresponding to a highly conducting substratum,

$$Q \cong \tanh \sqrt{(i)}V \qquad (15)$$

or if $\sigma_2 \ll \sigma_1$ corresponding to an insulating substratum,

$$Q \cong \coth \sqrt{(i)}V \qquad (16)$$

In the above example, the parameter V can be replaced by $h_1/30$ where h_1 is the thickness of the upper stratum in meters. The function Q and its argument q is plotted in Fig. 3a and 3b as a function of V for various values of the ratio σ_2/σ_1.

EXTENSION TO PERPENDICULAR INCIDENCE

In the preceding problem, the incident plane wave has the electric vector contained in the plane of incidence (and the magnetic vector parallel to the interfaces). For this reason, it is called parallel incidence. The other important case is when the electric vector is perpendicular to the plane of incidence. This is termed perpendicular incidence.

Again choosing the plane of incidence to be the (xz) plane, the incident wave now has only a y component of the electric field. By analogy to Eq. (2), the general solution is of the form

$$E_{my} = [\bar{a}_m e^{-u_m z} + \bar{b}_m e^{u_m z}] e^{-i\lambda x} \qquad (17)$$

where

$u_m^2 = \lambda^2 + \gamma_m^2$, $\bar{a}_0$ is the amplitude of the incident wave, $\bar{b}_0$ is the amplitude of the reflected wave, and $\bar{b}_M = 0$. In this case

$$i\mu_m \omega H_{mx} = \frac{\partial E_{my}}{\partial z} \qquad (18)$$

These are transformable to the boundary conditions for parallel incidence by making the substitutions: E_{my} by H_{my}, $i\mu_m\omega$ by $\sigma_m + i\varepsilon_m\omega$. Then using the previous results, the solution for perpendicular incidence can be written down:

$$\frac{\bar{b}_0}{\bar{a}_0} = \frac{N_0 - Y_1}{N_0 + Y_1} \tag{20}$$

where

$$Y_m = N_m \frac{Y_{m+1} + N_m \tanh u_m h_m}{N_m + Y_{m+1} \tanh u_m h_m} \tag{21}$$

for $m = 1, 2, 3, \ldots, M - 1$ and

$$Y_M = N_M$$

In the preceding

$$N_m = \frac{u_m}{i\mu_m\omega} \tag{22}$$

where as before $u_m = (\lambda^2 + \gamma_m^2)^{1/2}$.

The quantity $\bar{b}_0/\bar{a}_0$ which is the ratio of the amplitude of the reflected wave to the incident wave is denoted by $R_\perp$. There is a similar transmission line analogy for this problem which need not be pointed out.

In analogy to the surface impedance function Z_1, the quantity Y_1 is a surface admittance and is given by

$$Y_1 = -H_{0x}/E_{0y}]_{z=0} = -H_{1x}/E_{1y}]_{z=0} \tag{23}$$

In the case of a homogeneous ground at glancing incidence ($\theta \to \pi/2$), it follows that

$$Y_1 = N_1 = (1/\eta_1)(1 - \gamma_0^2/\gamma_1^2)^{1/2} \tag{24}$$

which is denoted Y^h where the superscript h indicates that the electric field is horizontal in contrast to the near vertical electric field associated with Z^v. It is interesting to note that

$$Y^h Z^v = (1 - \gamma_0^2/\gamma_1^2) \tag{25}$$

and for $|\gamma_1^2| \gg |\gamma_0^2|$

$$Y^h Z^v \cong 1$$

Impedance Matching and Natural Oscillations in Stratified Media

In this section, some remarks will be made concerning the nature of extreme conditions where the reflection coefficient on a plane stratified media becomes zero or infinite. The discussion will be confined primarily to parallel incidence, although the results are easily carried over to perpendicular incidence.

Reflection from Stratified Media

As indicated, the reflection coefficient $R_\parallel$ is given by

$$R_\parallel = \frac{K_0 - Z_1}{K_0 + Z_1} \qquad (26)$$

where

$$K_0 = u_0/i\omega\varepsilon_0 = (\lambda^2 - k_0^2)^{1/2}/i\omega\varepsilon_0$$

and Z_1 is the normal impedance at the interface $z = 0$. In the case of a homogeneous half space ($h_1 \to \infty$)

$$Z_1 = K_1 = u_1/(\sigma_1 + i\omega\varepsilon_1) = (\lambda^2 + \gamma_1^2)^{1/2}/(\sigma_1 + i\omega\varepsilon_1) \qquad (27)$$

The condition for matching is that $R_\parallel = 0$ corresponding to the absence of a reflected wave. This requires that

$$K_0 = Z_1 \qquad (28)$$

For a homogeneous half space, the condition is simply

$$K_0 = K_1 \qquad (29)$$

which when solved for λ yields

$$\lambda = \pm \left[\frac{\mu_1^2/\gamma_1^2 - \mu_0^2/\gamma_0^2}{\mu_0^2/\gamma_0^4 - \mu_1^2/\gamma_1^4}\right]^{1/2} \qquad (30)$$

In the case of no magnetic permeability contrast ($\mu_1 = \mu_0$), the above simplifies* to

$$i\lambda = \pm \frac{\gamma_1\gamma_0}{(\gamma_1^2 + \gamma_0^2)^{1/2}} \qquad (31)$$

and if both media are perfect dielectrics $\gamma_1 = ik_1$ and $\gamma_0 = ik_0$,

$$\lambda = \pm \frac{k_1 k_0}{(k_1^2 + k_0^2)^{1/2}} \qquad (32)$$

But, since $\lambda = k_0 \sin \theta$, the preceding equations for a condition of matching can be employed to determine the angle of incidence when there shall be no reflection. In the case of the 2 dielectric half spaces, the condition is

$$\sin \theta = \pm \frac{k_1}{(k_1^2 + k_0^2)^{1/2}} \qquad (33)$$

or

$$\tan \theta = \pm(k_1/k_0)$$

This latter equation is well known from optics and there the angle θ is known as the Brewster angle and the ratio k_1/k_0 is called the relative refractive index.

* In this case the wave tilt $W = \gamma_0/\gamma_1$ as can be deduced from Eq. 11.

EXCITATION BY A DIPOLE FIELD

Up to this point we have been dealing with plane wave excitation of the layered medium. The extension to localized source excitation is really straight forward, although the evaluation of the field expressions can be tricky. Here we will just indicate in a preliminary fashion how the resultant fields for a vertical electric dipole V.E.D. can be derived.

We located the V.E.D. of current moment Ids (i.e., current × infinitesimal length ds) at $z = -h$ on the axis $\rho = 0$ of a cylindrical coordinate system (ρ, ϕ, z). The layered medium occupies the region $z > 0$ and has the same parameters as indicated in Fig. 1. In this case the fields can be derived from a single scalar wave function which is actually the z component Π_z of a Hertz vector $\vec{\Pi}$. For the region $z < 0$, $E_{o\rho} = \partial^2 \Pi/\partial\rho\partial z$, $E_{o\phi} = 0$, $E_{oz} = (k^2 + \partial^2/\partial z^2)\Pi_{oz}$, $H_{o\rho} = 0$, $H_{o\rho} = -i\varepsilon_o \omega \partial \Pi_{oz}/\partial \rho$, $H_{oz} = 0$. Now in the absence of the layered half-space we would have

$$\Pi_{oz} = \Pi_{oz}^{primary} = \frac{Ids}{4\pi i \varepsilon_o \omega} \frac{e^{-ikR_o}}{R_o}$$

where $R_o = [(z+h)^2 + \rho^2]^{1/2}$ is the distance from the source dipole to the observer. Here we now make use of the known Sommerfeld integral representation

$$\frac{e^{-ikR_o}}{R_o} = \int_0^\infty J_o(\lambda\rho) \frac{e^{\pm u_o(z+h)}}{u_o} \lambda d\lambda$$

where $u_o = (\lambda^2 - k^2)^{1/2}$. Here the sign of the exponent is chosen so that the integral converges [i.e., + for $(z+h) < 0$ and − for $(z+h) > 0$]. Remember that the source location produces a singularity at $\rho = 0$ and $z = -h$.

Using a little hindsight we can now construct the resultant Hertz potential Π_{oz} in the region $z < 0$ as follows

$$\Pi_o = \frac{Ids}{4\pi i \varepsilon_o \omega} \int_0^\infty \frac{\lambda}{u_o} [e^{\pm u_o(z+h)} + R_{\shortparallel}(\lambda) e^{u_o(z-h)}] J_o(\lambda\rho) d\lambda$$

where $R_{\shortparallel}(\lambda)$ is the "reflection coefficient" defined by (26). Here we can still identify λ with $k \sin\theta$ where θ is the angle of incidence, but θ is now a spectrum of angles that can vary from 0 to $\pi/2$ and then from $\pi/2$ to $\pi/2 + i\infty$ along a line parallel to the imaginary axis in the complex θ plane. We shall defer further discussion of this problem until later.

REFERENCES

NORTON, K.A., (1935), Propagation of radio waves over plane earth, *Nature*, 135, 954-955.

NORTON, K.A. (1936, 1937), The propagation of radio waves over the surface of the earth and in the upper atmosphere, *Proc. I.R.E.*, Pt. I, 1367-1387 and Pt. II, 1203-1236, 25.

WAIT, J.R., (1970), *Electromagnetic Waves in Stratified Media*, 2nd Edition, Pergamon Press, New York.

3
Magneto-Telluric Fields

INTRODUCTION

The temporal variations of the geomagnetic field have been studied for many years [Lahari and Price, 1939]. However, more recently have the rapid magnetic variations been studied in conjunction with the variations of the telluric (earth-current) fields. In fact, almost 30 years ago, Tikhonov [1950] in the USSR, and Kato and Kikuchi [1950] in Japan, pointed out that the electrical characteristics of the deep strata of the earth's crust could be determined from a combined analysis of geomagnetic and telluric field variations. Since then, a large number of related investigations have been carried out, particularly in the USSR.

The actual mechanism which produces the short-period variations (i.e., frequencies of the order of 1 to 10^{-3} Hz) is not yet well understood. In recent years it has been suggested on numerous occasions that the phenomena are related to magneto-hydrodynamic (i.e., MHD) waves in the exosphere or the ionosphere of the earth. [See Wait, 1962, and Matushita and Campbell 1967 for a review of early suggestions.]

SOME HISTORY

Most investigations of magneto-telluric fields boil down to a study of the interrelation between the tangential components of the horizontal electric and magnetic fields at the surface of the earth. As far as this writer is able to ascertain the first definitive paper dealing with this subject appeared in 1950, and was authored by Tikhonov [1950]. He realized that rapid geomagnetic variations and earth currents, observed at the surface of the earth, must be connected by some definite relationship. He showed that, at low frequencies, the amplitude of the derivative of the component H_x of the magnetic variations is proportional to the (orthogonal) component of the electric field E_y.[2] This was in agreement with the experimentally established fact that there is a proportionality between these quantities. Tikhonov's model of the earth's crust is a planar layer $0 \leq z \leq l$ of finite conductivity σ lying upon an ideally conducting substrate. Implicitly in his analysis, it was assumed that horizontal gradients of the fields could be neglected. Thus, for a spectral component of frequency ω he found that, at $z=0$,

$$i\mu_0 \omega H_x \cong E_y \gamma \coth(\gamma l)$$

where

$$\gamma = (i\sigma\mu_0\omega)^{\frac{1}{2}} \text{ and } \mu_0 \cong 4 \times 10^{-7}.$$

Using previously published data on the observed diurnal variations at Tucson (Arizona) and Zui (USSR), Tikhonov computed the value of σ and l which best fitted the first four harmonics. For Tucson, the values were about 4×10^{-3} mhos/m and 1,000 km, respectively. For Zui, the corresponding values were about 3×10^{-1} mhos/m and 100 km.

In a later paper [Tikhonov and Lipskaya, 1952], the horizontal gradient of the fields was allowed to be finite (although the earth's crust was still regarded as a horizontally stratified medium). The authors postulated that the field components of long period may be represented as a wave which is propagated from east to west with the velocity of the earth's rotation. Using the same data as mentioned above, they obtained revised estimates for the conductivity σ of the upper stratum of thickness l under the assumption of an ideally conducting substrate. For Tucson, the values were approximately 10^{-2} mhos/m and 1,100 km, where for Zui the corresponding values were about 7×10^{-1} mhos/m and 110 km. In this paper they also showed

(Reprinted, in part, from J.R. Wait, Jour. Res. N.B.S., 66D, 509-541, 1962).

that the measured values of the *vertical* magnetic field variations were consistent with the model and the postulation of linear east-to-west motion of the fields.

In the third paper of this sequence [Lipskaya, 1953], the electromagnetic equations for the model described above were cast in a form to clearly demonstrate the various relationships between the field components.

Subsequent to the Russian work mentioned above, Cagniard [1953] published a paper which has been extensively referenced since. His analysis which assumed plane wave incidence, developed formulas which related H_x and E_y on the surface of a stratified conducting medium. A discussion of the limitations in Cagniard's results appeared shortly thereafter [Wait, 1954]. The essential point made by this writer is that the proportionality between H_x and E_y is only valid if the fields themselves do not vary appreciably in a horizontal distance of the order of a "skin depth" in the ground. In the case of a homogeneous flat earth it was indicated that this distance was of the order of $|\gamma^{-1}|$. For example, at a frequency of 10^{-3} Hz ; $\sigma \sim 10^{-3}$ mhos/m, $|\gamma^{-1}|$ is about 350 km. Consequently, the field should be uniform over a considerably broad area to permit the Cagniard interpretive procedure to be applied. This limitation is, of course, in addition to the requirement that the crustal layers themselves are uniform. Very recently Price [1962] has indicated that the limitation mentioned [Wait, 1954] becomes much more stringent when the magneto-telluric method is applied to a stratified earth in certain important instances.

In a later paper, Tikhonov and Shakhsuvarov [1956] discussed the methods for calculating the admittance H_x/E_y at the surface of a horizontally stratified earth of any number of layers. Again, horizontal field gradients were neglected. Actually, the formulas given were a special case of an earlier general analysis [Wait, 1953a and b] where the incidence was oblique. Tikhonov and Shakhsuvarov [1956] give a few curves of the ratio H_x/E_y for both two- and three-layer structures. Some asymptotic approximations of the impedance formula for parallel layers were discussed by Berdichevsky and Brunelli [1959]. The derivation of the basic formulas is very similar to that found in earlier papers [e.g., Wait, 1953a].

The problem has also been investigated by Scholte and Veldkamp [1955] in Holland. Their analysis of the variations of H_x and E_y is essentially the same as those of Tikhonov, Cagniard, and others. By a relatively simple method of data analysis they estimate the amplitude and phase of H_x/E_y without actually performing spectral analyses. They used the experimental data from the magnetic observatory at Witteveen for frequencies in the range from 10^{-4} to 10^{-1} sec. The resulting curves of amplitude and phase versus frequency fitted a two-layer earth model with the upper conductivity $\sigma_1=2$ mhos/m and the lower conductivity $\sigma_2=10^{-1}$ mhos/m. The thickness of the upper stratum was $d_1=600$ m. In the same paper Scholte and Veldkamp also discuss the influence of earth curvature on the ratio H_y/E_x. They conclude that the effect is very small.

An analysis of the continuously varying conductivity profile was carried out by Bossy and De Vuyst [1959] in Belgium. They showed that for normal incidence the ratio H_x/E_y on the surface could be expressed in closed form when the conductivity $\sigma(z)$ varied with depth in the manner

$$\sigma(z)=\sigma_0\left(1+\frac{z}{a}\right)^{-\beta} \qquad (1)$$

where σ_0 and a are constants. They used such a model to interpret experimental data of the phase of E_x/H_y for frequencies from about 3×10^{-3} to 1 c/s at Dourbes. However, they found it necessary to modify the model by having a (well conducting) homogeneous surface layer of thickness about 500 m overlying the inhomogeneous substrate of poor conductivity (with β about 9).

In all the work mentioned above, the natural magneto-telluric field has been studied only at frequencies less than about 1 Hz. Vladimirov [1960] has described an investigation of the use of higher frequencies in the range from 0.3 to 1,000 c/s. Field experiments were made in the Ryl'sk district of the Kursk region in the USSR. The area consists of sand and clay deposits underlain at a depth of approximately 500 m by a crystalline base of practically infinite resistivity. The measured values of the ratios $|E_y/H_x|^2$ and $|E_x/H_y|^2$ on the surface were consistent with an assumed two-layer structure where the upper conductivity $\sigma_1 \simeq 5 \times 10^{-2}$ mhos/m and the lower conductivity $\sigma_2 \simeq 0$. From the experimental curve, the thickness of the upper stratum was deduced to be 450 m. In a subsequent paper Vladimirov and Kolmakov [1960] discuss the resolving power of the magneto-telluric method. They confine their discussion to a three-layer model with an infinite resistive basement. The principle of equivalence stated by them is that for a given conductivity σ_1 and thickness h_1 of the uppermost layer, the theoretical three-layer curves of $|H_x/E_y|$ practically preserve their shape when the conductivity σ_2 and thickness h_2 of the intermediate layer vary over specified limits. (This question is discussed again in the present paper.) Vladimirov and Kolmakov then conclude that the same curve may characterize different geoelectric sections and can be interpreted only with known values of σ_2 and h_2. Some further comments on the same subject are given by Vladimirov and Nikiforova [1961] and Vladimirov and An [1961].

In an interesting paper, Chetaev [1960], also from the USSR, illustrates a procedure to obtain information about the anisotropy of a homogeneous stratum by measuring the surface impedance in orthogonal directions. The model he considers is a half-space with a medium whose longitudinal conductivity is σ_l and the transverse conductivity is σ_t. The angle of inclination of the anisotropy is α (i.e., $\alpha=0$ corresponds to a flat lying anisotropic medium with σ_t the horizontally directed conductivity). Chetaev shows that if the electric component E_y is transverse to the strike of the structure,

$$Z = E_y/H_x = -(i\mu\omega/\sigma_l)^{\frac{1}{2}}[1 + [(\sigma_t/\sigma_l) - 1]\sin^2\alpha]^{\frac{1}{2}} \qquad (2)$$

and if the electric component E_x is along the strike,

$$Z = E_x/H_y = (i\mu\omega/\sigma_l)^{\frac{1}{2}} \qquad (3)$$

which is not affected by σ_t. Since σ_t/σ_l is usually greater than one, the minimum value of the surface impedance $|Z|$, as a function of azimuth, is $(\mu\omega/\sigma_l)^{\frac{1}{2}}$ which occurs when the electric vector is along the strike.

It has been pointed out by Pokityanski [1961] that the interpretation of field results should take account of the anisotropy of the structure. In fact, in general, the tangential fields are related by

$$E_x = Z_{xx}H_y - Z_{xy}H_x \qquad (4)$$

$$E_y = Z_{yx}H_y - Z_{yy}H_x \qquad (5)$$

where the coefficients are the components of a surface impedance tensor.

In the case of parallel stratified or flat-lying media Z_{xy} and Z_{yx} vanish and $Z_{xx} = Z_{yy}$. Of course, the components of the impedance tensor depend on the choice of the (x,y) axes. For an anisotropic homogeneous half-space or for an inhomogeneous structure where the conductivity does not vary along one of the horizontal directions, it is possible to find special directions of x and y (denoted u and v, respectively), such that

$$E_u = Z_1 H_v \qquad (6)$$

and

$$E_v = -Z_2 H_u \qquad (7)$$

where Z_1 and Z_2 are the principal values of the tensor impedance. Consequently, if the u and v components of the tangential fields are measured, Z_1 and Z_2 may be simply calculated. Unfortunately, those principal directions are not always known beforehand. Pokityanski [1961] has described an ingenious graphical scheme to determine the u and v directions from measurements of the nonorthogonality of the tangential E and H vectors. Strictly speaking, the method is only valid for source fields which are linearly polarized. A more straightforward procedure would be to return to the surface impedance relations and write them in the form

$$Z_z = E_z/H_y = [Z_{xx} - Z_{xy}\alpha] \qquad (8)$$
and
$$Z_y = E_y/H_x = -[Z_{yy} - Z_{yx}/\alpha] \qquad (9)$$

where $\alpha = H_x/H_y$ and where Z_x and Z_y are the impedances as measured in the x and y directions, respectively.

It is clear that Z_x and Z_y depend on α and, consequently, they will be a function of the source field. However, the elements of the surface impedance tensor can be calculated if at least two independent measurements are made. Thus, for a particular frequency, these will provide the following sets of values, $Z_{x,1}$, $Z_{y,1}$, α_1 and $Z_{x,2}$, $Z_{y,2}$, and α_2 where the subscript 1 or 2 is used to distinguish between the two measurements. Then, provided $\alpha_1 \neq \alpha_2$, it easily follows that

$$Z_{xy} = \frac{Z_{x,1} - Z_{x,2}}{\alpha_1 - \alpha_2}, \qquad (10)$$

$$Z_{yx} = \frac{Z_{y,1} - Z_{y,2}}{(1/\alpha_2) - (1/\alpha_1)}; \qquad (11)$$

$$Z_{xx} = Z_{x,1} + (Z_{xy}/\alpha_1)$$
$$= Z_{x,2} + (Z_{xy}/\alpha_2); \qquad (12)$$

$$Z_{yy} = -Z_{y,1} + (Z_{yx}/\alpha_1)$$
$$= -Z_{y,2} + (Z_{yx}/\alpha_2). \qquad (13)$$

In principle these equations could be used to determine the elements of the surface impedance tensor for an elliptically polarized field. It is probable that the apparent scatter in the magneto-telluric data, as measured by Garland and Webster [1960] and Watt et al. [1962], is a consequence of the anisotropy or of tilting the structure.

In another paper, Kovtun [1961] has discussed, in a general way, the nature of the magneto-telluric fields for two-dimensional inhomogeneous structures. It is assumed that electrical properties vary only in the (x,z) plane; the surface of the earth is the (x,y) plane. It is stated incorrectly that the field can be decomposed into an *independent* set of TM (transverse magnetic) and TE (transverse electric) waves. While this may be true for the incident or primary field, it is overlooked that the boundary conditions couple these waves together [Wait, 1959] in the general case. Consequently, his subsequent discussion can only be considered approximate. The most recent Soviet writings on the subject have fully accounted for this type of coupling [e.g., see Dmitriev and Bevdichevsky, 1979].

NUMERICAL RESULTS FOR HORIZONTALLY STRATIFIED STRUCTURE

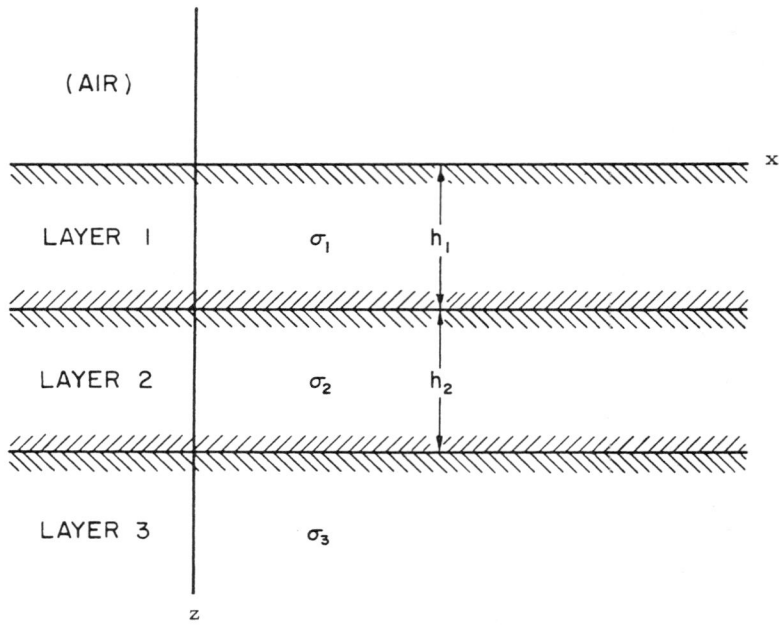

FIGURE 1. *Stratified model of the earth's crust.*

The model is quite simple, the earth is assumed to be horizontally stratified and consists of three homogeneous layers. The upper layer is of thickness h_1 with conductivity σ_1, the middle or intervening layer is of thickness h_2 with conductivity σ_2, and the bottom layer is of infinite thickness and has a conductivity σ_3. In terms of a Cartesian coordinate system (x,y,z) the earth's surface is $z=0$ and the interfaces between layers[4] are at $z=h_1$ and $z=h_1+h_2$. The magnetic permeability is assumed to be constant throughout and equal to μ_0. Furthermore, displacement currents are neglected in all three conducting layers. The problem is now to find an expression for the ratios of the tangential fields E and H at the boundary between free space and the earth. If the horizontal gradients of the fields are negligible the result can be written (for a time factor $e^{i\omega t}$)

$$Z = \frac{-E_y}{H_x} = \frac{E_x}{H_y} = \left(\frac{i\mu_0\omega}{\sigma_1}\right)^{\frac{1}{2}} Q\left(\sqrt{\sigma_1\mu_0\omega}h_1, \frac{\sigma_2}{\sigma_1}, \frac{\sigma_3}{\sigma_1}, \frac{h_1}{h_2}\right) \tag{24}$$

where Q is a function of the four variables indicated. Explicitly,

$$Q = \frac{\hat{Q} + (\sigma_2/\sigma_1)^{\frac{1}{2}} \tanh\left[(i\sigma_1\mu_0\omega)^{\frac{1}{2}}h_1\right]}{(\sigma_2/\sigma_1)^{\frac{1}{2}} + \hat{Q} \tanh\left[(i\sigma_1\mu_0\omega)^{\frac{1}{2}}h_2\right]} \tag{25}$$

where

$$\hat{Q} = \frac{1 + (\sigma_3/\sigma_2)^{\frac{1}{2}} \tanh\left[(i\sigma_2\mu_0\omega)^{\frac{1}{2}}h_2\right]}{(\sigma_3/\sigma_2)^{\frac{1}{2}} + \tanh\left[(i\sigma_2\mu_0\omega)^{\frac{1}{2}}h_2\right]}. \tag{26}$$

This result follows directly from a previously derived general formula applicable to any number of layers [Wait, 1953a and b].

---
[4] See figure 1.

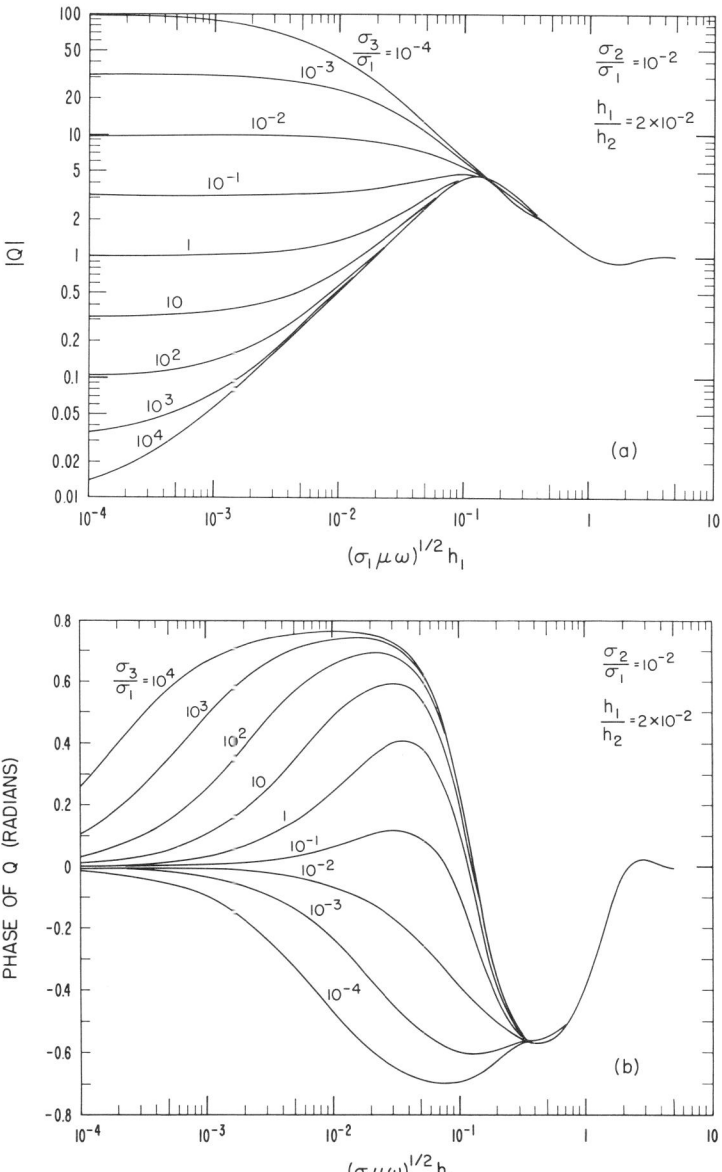

FIGURE 2. Amplitude and phase of the complex quantity Q for a three-layer model plotted as a function of the dimensional frequency/depth factor $(\sigma_1\mu\omega)^{\frac{1}{2}}h_1$.

[For this set of curves the exciting field is assumed to be uniform (see text).]

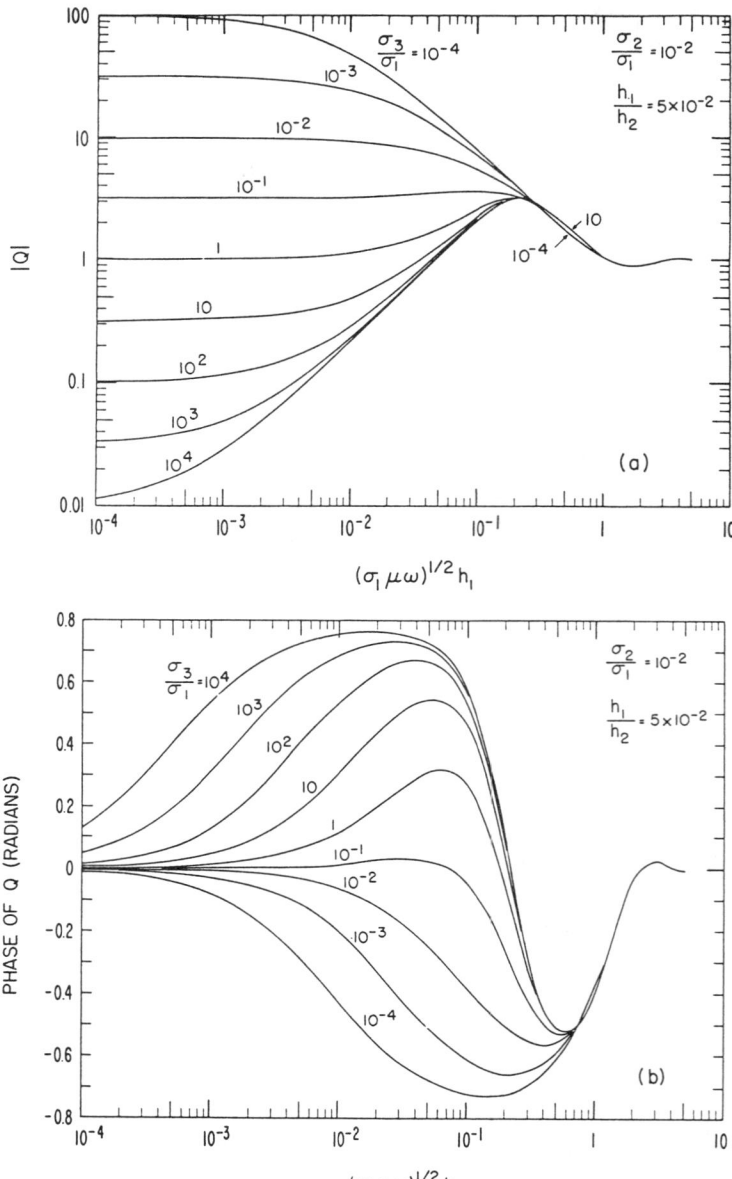

FIGURE 3. Amplitude and phase of the complex quantity Q for a three-layer model plotted as a function of the dimensional frequency/depth factor $(\sigma_1\mu\omega)^{\frac{1}{2}}h_1$.

[For this set of curves the exciting field is assumed to be uniform (see text).]

Magneto-Telluric Fields

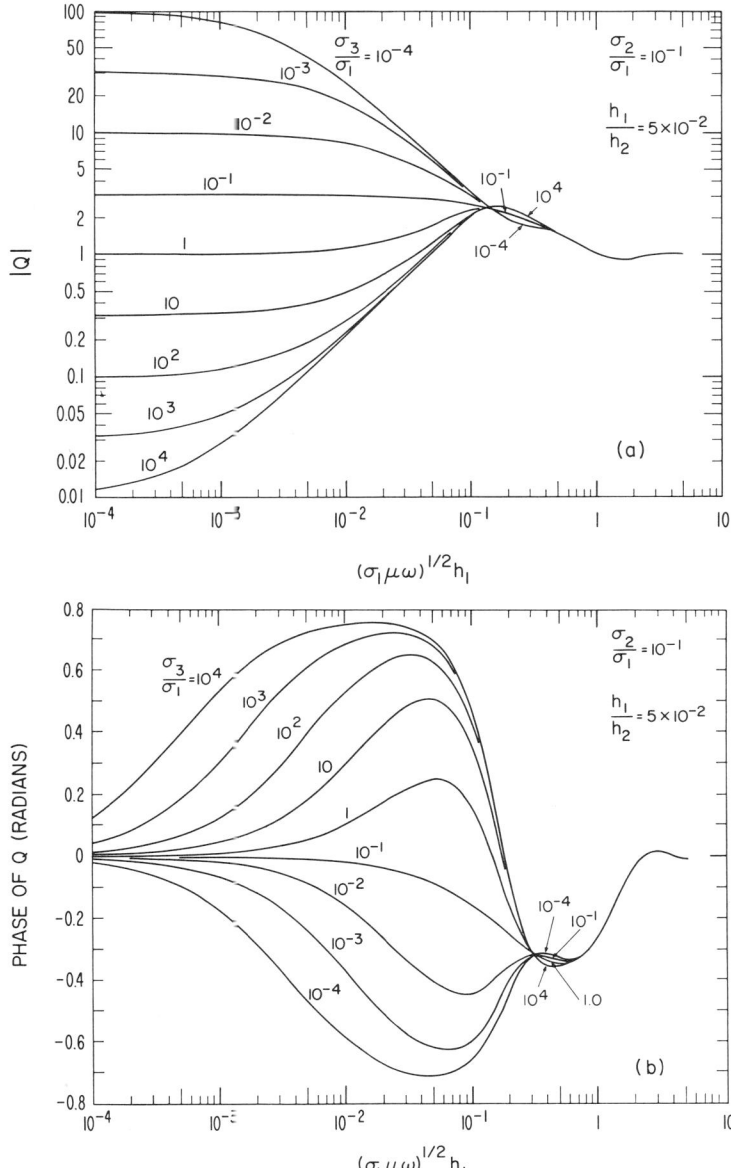

FIGURE 4. Amplitude and phase of the complex quantity Q for a three-layer model plotted as a function of the dimensional frequency/depth factor $(\sigma_1\mu\omega)^{\frac{1}{2}}h_1$.

[For this set of curves the exciting field is assumed to be uniform (see text).]

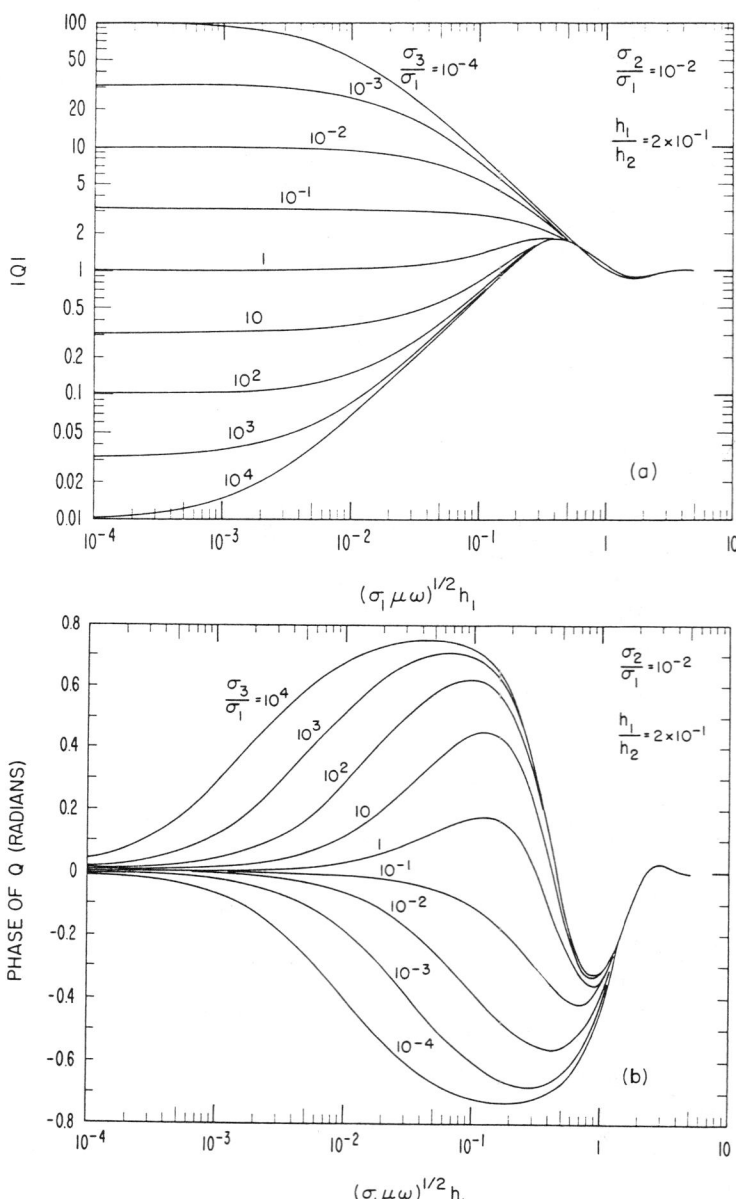

FIGURE 5. Amplitude and phase of the complex quantity Q for a three-layer model plotted as a function of the dimensional frequency/depth factor $(\sigma_1\mu\omega)^{1/2}h_1$.
[For this set of curves the exciting field is assumed to be uniform (see text).]

Magneto-Telluric Fields

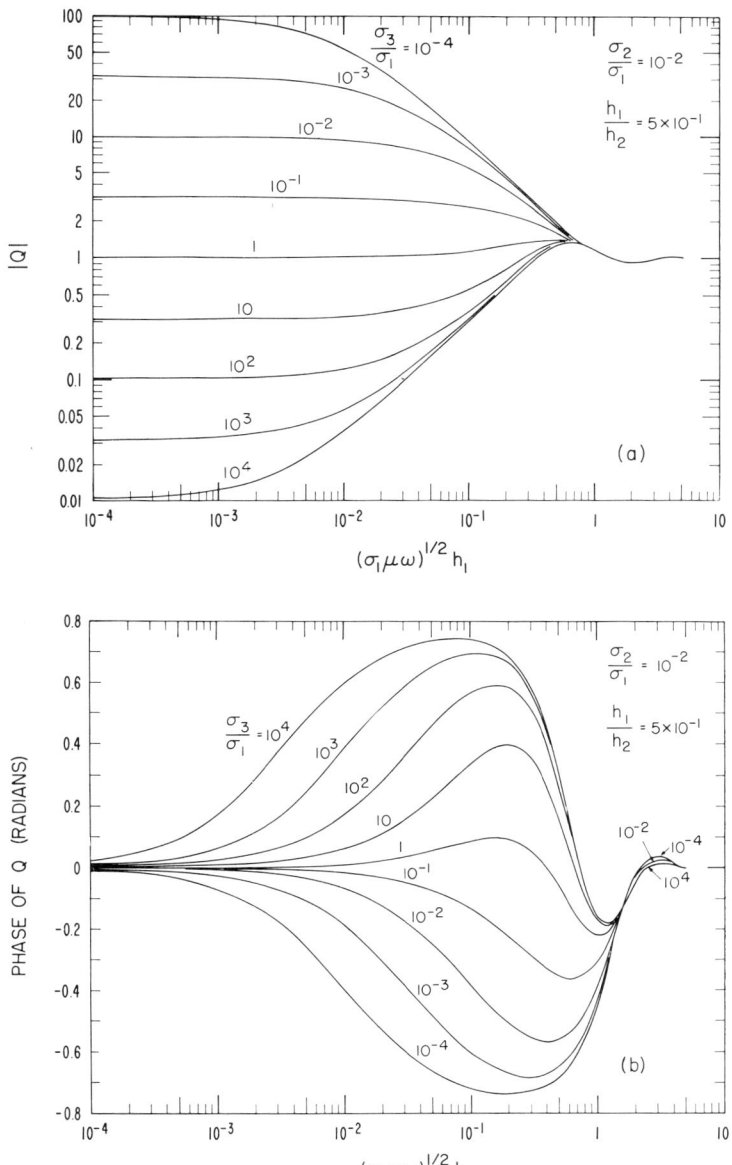

FIGURE 6. Amplitude and phase of the complex quantity Q for a three-layer model plotted as a function of the dimensional frequency/depth factor $(\sigma_1 \mu \omega)^{\frac{1}{2}} h_1$.

[For this set of curves the exciting field is assumed to be uniform (see text).]

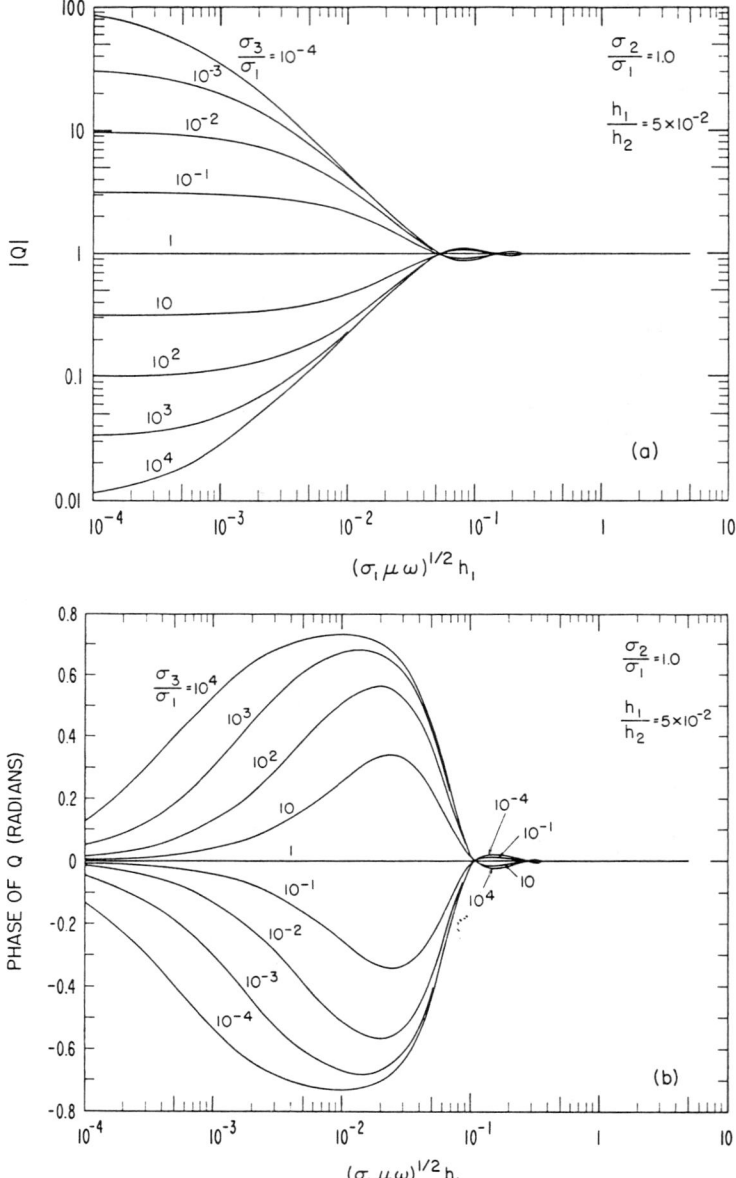

FIGURE 7. Amplitude and phase of the complex quantity Q for a three-layer model plotted as a function of the dimensional frequency/depth factor $(\mu_1 \sigma \omega)^{\frac{1}{2}} h_1$.

[For this set of curves the exciting field is assumed to be uniform (see text).]

Magneto-Telluric Fields

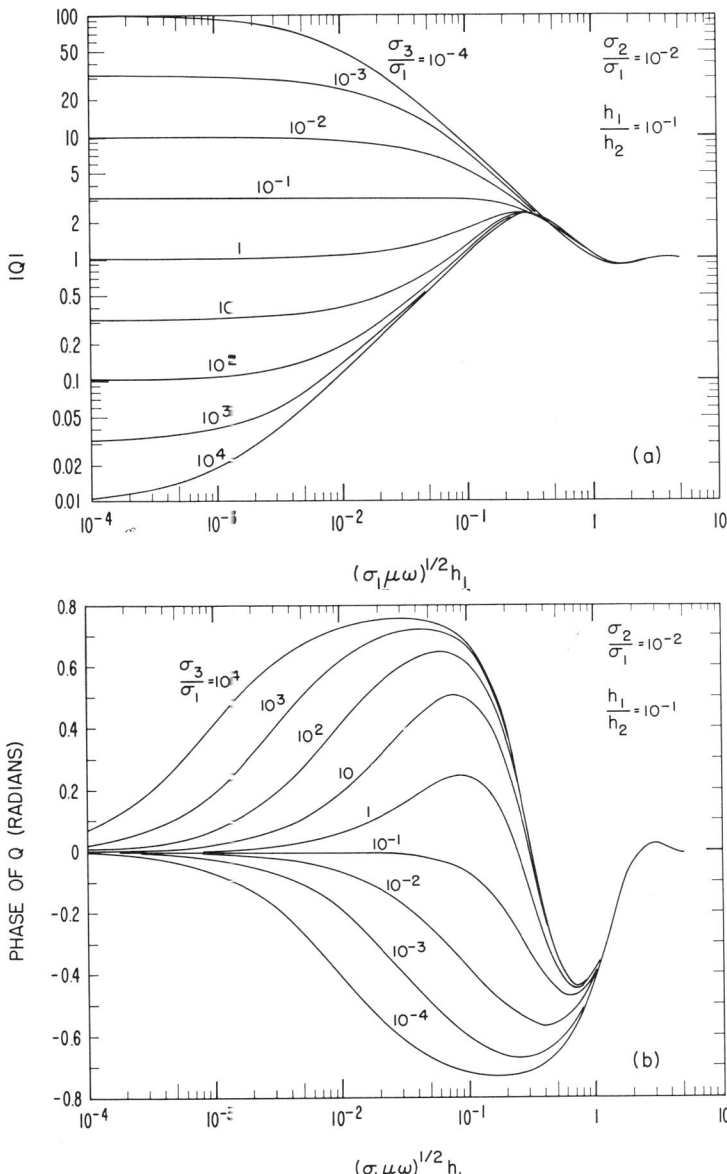

FIGURE 8. Amplitude and phase of the complex quantity Q for a three-layer model plotted as a function of the dimensional frequency/depth factor $(\sigma_1 \mu \omega)^{\frac{1}{2}} h_1$.

[For this set of curves the exciting field is assumed to be uniform (see text).]

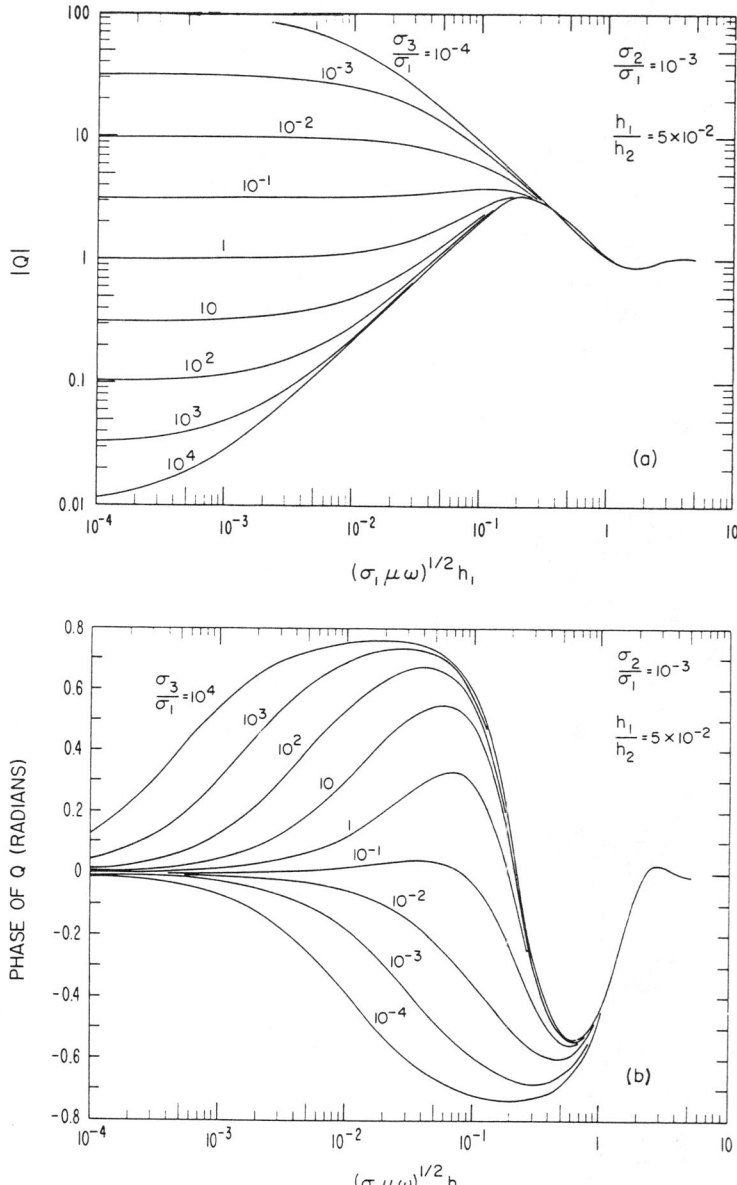

FIGURE 9. *Amplitude and phase of the complex quantity Q for a three-layer model plotted as a function of the dimensional frequency/depth factor* $(\sigma_1\mu\omega)^{\frac{1}{2}}h_1$.

[For this set of curves the exciting field is assumed to be uniform (see text).]

It can be seen from eq (25) that if $h_1 \to \infty$, Q tends to unity. This suggests that, for the general case, one defines an apparent conductivity σ_a such that

$$Z = \left(\frac{i\mu_0\omega}{\sigma_a}\right)^{\frac{1}{2}} = \left(\frac{i\mu_0\omega}{\sigma_1}\right)^{\frac{1}{2}} Q. \tag{17}$$

Thus
$$Q = (\sigma_1/\sigma_a)^{\frac{1}{2}} \text{ or } \sigma_a/\sigma_1 = Q^{-2}.$$

Clearly, σ_a must be complex in order to admit such an equivalence with a homogeneous half-space. The concept of an equivalent or complex conductivity has been discussed previously in connection with radio propagation over stratified media [Wait, 1953b, 1958].

Extensive numerical results for Q over a wide range of the four parameters are available [Jackson et al., 1962]. These greatly extend the set of curves of $|\sigma_a/\sigma_1|$, for $\sigma_3 = 0$ only, published by Yungul [1961]. A sample of some of these numerical data is given in this paper in graphical form. In the first set, shown in figures 2a to 6b, the ratio σ_2/σ_1 is fixed at $1/100$. This corresponds to an intervening layer of very poor conductivity. In each case, the abscissa is the dimensionless ratio $(\sigma_1\mu_0\omega)^{\frac{1}{2}}h_1$. For these five sets of curves, the values of h_1/h_2 are $1/50$, $1/20$, $1/10$, $1/5$, and $1/2$, respectively. Thus, in all cases, the upper layer is thin compared with the middle or intervening layer.

It is interesting to note that for the low frequencies [corresponding to small values of $(\sigma_1\mu_0\omega)^{\frac{1}{2}}h_1$], σ_a always approaches σ_3 the conductivity of the bottom layer. At the high frequencies, σ_a approaches σ_1 as noted above. In the intermediate region an interesting transition takes place. The right portion of the curves is roughly characteristic of a two-layer structure and the shape is determined mainly by the characteristics of the upper layer.

In the second set of curves shown in figures 7a to 9b, the ratio h_1/h_2 is fixed at $1/20$ and the conductivity ratio σ_2/σ_1 takes the values 1, 0.1, and 10^{-3}. In the first set shown in figures 7a and 7b, the upper two layers combined in a single layer of thickness (h_1+h_2). For this case, the results are identical to the standard two-layer curves [Cagniard, 1953; Wait, 1953a]. The interesting thing about the three sets of curves, in figures 8a to 9b, is their similarity. Here it appears that the conductivity of the intervening layer σ_2 plays a very small role. Actually, this is a general characteristic of all three-layer magneto-telluric curves provided σ_2 is somewhat less than σ_1.

The relative insensitivity of the factor Q to the conductivity and thickness of a poorly conducting intervening layer can be demonstrated directly from the basic equations given above. For example, under the conditions that

$$(\sigma_2\mu_0\omega)^{\frac{1}{2}}h_2 \ll 1$$

$$\hat{Q} \cong (\sigma_2/\sigma_3)^{\frac{1}{2}}$$

and thus eq (25) becomes

$$Q \cong \frac{1+(\sigma_3/\sigma_1)^{\frac{1}{2}}\tanh[(i\sigma_1\mu\omega)^{\frac{1}{2}}h_1]}{(\sigma_3/\sigma_1)^{\frac{1}{2}}+\tanh[(i\sigma_1\mu\omega)^{\frac{1}{2}}h_1]}. \tag{18}$$

Therefore, under the restriction stated above, the three-layer structure is equivalent to a two-layer model whose parameters are σ_1, σ_3, and h_1. The result is essentially independent of σ_2 and h_2.

It is worth pointing out that a dual three-layer model exists which broadens the application of these numerical results. The parameters of the dual model are indicated by primed symbols and they are related to the original problem as follows:

$$\sigma_2'/\sigma_1' = \sigma_1/\sigma_2$$
$$\sigma_3'/\sigma_1' = \sigma_1/\sigma_3$$
$$(\sigma_1'\mu_0\omega)^{\frac{1}{2}}h_1' = (\sigma_1\mu_0\omega)^{\frac{1}{2}}h_1$$
$$(\sigma_2'\mu_0\omega)^{\frac{1}{2}}h_2' = (\sigma_2\mu_0\omega)^{\frac{1}{2}}h_2.$$

It then immediately follows from eqs (25) and (26) that

$$Q'\left(\sqrt{\sigma_1\mu_0\omega}h_1', \frac{\sigma_2'}{\sigma_1'}, \frac{\sigma_3'}{\sigma_1'}, \frac{h_1'}{h_2'}\right) = \frac{1}{Q\left(\sqrt{\sigma_1\mu_0\omega}h_1, \frac{\sigma_2}{\sigma_1}, \frac{\sigma_3}{\sigma_1}, \frac{h_1}{h_2}\right)}. \tag{19}$$

The apparent conductivity σ_a' in the dual or transformed problem is given by

$$\sigma_a'/\sigma_1' = (Q')^{-2} = Q^2.$$

Consequently, the ordinate in figures 2a, 3a, to 9a can be regarded as the quantity $|(\sigma_a'/\sigma_1')^{\frac{1}{2}}|$ in the transformed problem. The conductivity ratios *are all inverted* and

$$\frac{h_1'}{h_2'} = \frac{\sigma_1}{\sigma_2}\frac{h_1}{h_2}.$$

As is obvious, the ordinate in the phase curves in figures 2b, 3b, to 9b is the *negative* of the phase of Q'.

In the five sets of curves from figures 2a to 6b, the values of h_1'/h_2' are thus 2, 5, 10, 20, and 50, respectively, and the conductivity ratio σ_2'/σ_1' is fixed at 100.

The dual property of magneto-telluric curves for horizontally stratified structures has been discussed recently by Kolmakov [1961]. His analysis seems to be unnecessarily involved. Actually, using our correction factors (i.e., amplitude and phase of Q), the duality is almost self-evident.

The curves shown in the preceding figures are based on the important and often overlooked assumption that the exciting field is effectively uniform. Stated in another way, the fields should not vary appreciably in the horizontal direction. To indicate the significance of this assumption for a horizontally stratified medium, it is desirable to consider a two-dimensional spatially periodic field. The coordinate system may then be chosen so that $\partial/\partial y = 0$ and $\partial/\partial x = -ikS$ where $k = 2\pi/$free-space wavelength $= \omega/c$ and S is a dimensionless quantity which may be complex. Therefore, the fields are of the form.

$$u(x,z) \sim f(z) \exp(-ikSx) \tag{20}$$

where $f(z)$ is some function of z. The usual assumption in magneto-telluric studies is to set $S = 0$ insofar as the fields in the earth are concerned. Cagniard [1953, 1954] justifies the assumption by saying that kx is always small compared with unity; consequently, $\exp(-ikSx)$ may be replaced by unity. The fallacy in his argument is that S may be very large when the source of the field is in the near zone. Cagniard does not recognize this important fact and, rather surprisingly, he imagines the source field to be a uniform plane wave at a real angle of incidence. In this trivial case S is then the sine of the (real) angle of incidence and, of course, it could never exceed unity.

To demonstrate that S may be greater than one, we need only choose a very simple model for the source. It is a uniform line current I at some height H above the surface of the earth. For simplicity, it will be located at $z = -H$ and runs parallel to the y axis of the coordinate system (x,y,z). The primary field (neglecting the influence of the earth) of this line current has only y component of the electric field. It is well known and is given, in terms of a modified Bessel function, by [Wait, 1959]

$$E_y = \frac{i\mu_0\omega I}{2\pi} K_0[ikR] \tag{21}$$

where $R = [(z+H)^2 + x^2]^{\frac{1}{2}}$ is the distance to the current line from the observer at (x,z). If, rather hypothetically, $kR \gg 1$, the Bessel function may be approximated by the first term of its asymptotic expansion. Then

$$E_y = -\frac{i\mu_0\omega I}{2\pi}\left(\frac{\pi}{2ikR}\right)^{\frac{1}{2}} e^{-ikR}. \tag{22}$$

Magneto-Telluric Fields

Thus the magnitude of the field is proportional to $(kR)^{-\frac{1}{2}}$ and the wave front is nearly plane. However, unfortunate as it may be, this limiting form is never achieved in the micropulsation region. For example, at $\omega/2\pi = 10^{-2}$ the wavelength $2\pi/k = 3 \times 10^7$ km, which is rather large. In nearly all cases of practical interest the observer is in the near zone where $kR \ll 1$. Thus

$$E_y \cong -\frac{i\mu_0 \omega I}{2\pi} [\log (kR/2) + 0.5772 \ldots], \tag{23}$$

which bears little similarity to a plane wave.

To demonstrate the existence of S values greater than unity, it is desirable to write the equation in the form of an integral [Wait, 1953b], thus

$$E_y = \frac{i\mu_0 \omega I}{\pi} \int_{-\infty}^{+\infty} \frac{\exp[-ikC|z+H|]}{C} e^{-ikSx} dS \tag{24}$$

where $C = (1-S^2)^{\frac{1}{2}}$. In the far field, where $(kR) \gg 1$, the important values of S are in the range between 0 and 1 and the integral can be evaluated by the method of stationary phase to yield eq (22). However, if kR is small the important contributions in the integral are for large values of S.

To focus attention on the physical aspects of the problem, a single harmonic component of the spectrum is considered. Also, initially the earth is assumed to be a homogeneous half-space of conductivity σ_1 and permeability μ_0. The earth curvature is neglected and the assumption is justified later on. Thus, for a source field which varies as $\exp(-ikSx)$ and when the electric field has only x component, the surface impedance is easily found to be [Wait, 1953b, 1958]

$$Z_1 = -E_y/H_x]_{z=0} = \left(\frac{i\mu_0\omega}{\sigma_1}\right)^{\frac{1}{2}} (1-i\beta)^{-\frac{1}{2}} \tag{25a}$$

where

$$\beta = k^2 S^2 / \sigma_1 \mu_0 \omega.$$

If $\beta \ll 1$

$$Z_1 \cong \left(\frac{i\mu_0\omega}{\sigma_1}\right)^{\frac{1}{2}} \tag{25b}$$

which is the value appropriate for a homogeneous half-space under the assumption of negligible horizontal gradients. Since kS can be identified physically with $\partial/\partial x$ it follows that the condition $\beta \ll 1$ is equivalent to saying that the surface fields vary in the x direction in a distance small compared with δ where

$$\delta = [2/(\sigma_1 \mu \omega)]^{\frac{1}{2}}$$

is the skin depth of the conducting medium. At a frequency of 0.1 c/s and for a conductivity $\sigma_1 \cong 10^{-3}$ mhos/m, δ is approximately 50 km. Consequently, if the fields vary appreciably in a distance of the order of 50 km over the surface of a homogeneous earth, some departure from eq (25b) is to be expected at frequencies less than 0.1 c/s. This limitation was pointed out previously [Wait, 1954] using a somewhat different argument.

When the earth becomes horizontally stratified, the limitation that β is small remains. However, the stringency of this condition varies in a manner depending on the stratification as pointed out by Price [1962]. To demonstrate this interesting phenomenon, the surface impedance for a two-layer earth is considered. The model consists of a homogeneous surface layer of conductivity σ_1 and thickness h_1 and semi-infinite lower layer of conductivity σ_2. Then on the assumption that $\frac{\partial}{\partial x} = -ikS$ and $\frac{\partial}{\partial y} = 0$, it follows from previous work [Wait, 1953b] that

$$Z_1 = \left(\frac{i\mu\omega}{\sigma_1}\right)^{\frac{1}{2}} (1-i\beta)^{-\frac{1}{2}} Q \tag{26}$$

where
$$Q = \frac{G + \tanh \chi}{1 + G \tanh \chi} \qquad (27)$$

where
$$G = \left[\frac{1 - i\beta}{(\sigma_2/\sigma_1) - i\beta}\right]^{\frac{1}{2}} \qquad (28)$$

and
$$\chi = (i\sigma_1\mu\omega)^{\frac{1}{2}} h_1 (1 - i\beta)^{\frac{1}{2}}. \qquad (29)$$

If h_1 tends to infinity, Q approaches unity and eq (6) reduces to (5a). Thus Q can be regarded again as a correction factor which accounts for the stratification in the earth's crust. If β is replaced by zero, the correction factor becomes

$$Q = \frac{(\sigma_1/\sigma_2)^{\frac{1}{2}} + \tanh[(i\sigma_1\mu\omega)^{\frac{1}{2}} h_1]}{1 + (\sigma_1/\sigma_2)^{\frac{1}{2}} \tanh[(i\sigma_1\mu\omega)^{\frac{1}{2}} h_1]}, \qquad (30)$$

which is a special case of eq (15) when $h_3 = \infty$.

To illustrate the influence of finite β on the behavior of Q, a number of calculations were carried out on a digital computer. Extensive tabulations of this quantity for a range of values of the parameters are available [Jackson et al., 1962]. Some of the results are shown graphically in figures 10a to 12b. In the first pair, σ_2 is effectively zero corresponding to a conducting layer (of thickness h) lying on an insulating substratum. It is apparent from these curves that a finite value of β leads to a major change in the shape of the curves. The same behavior is evident when the lower layer is finitely conducting as can be evidenced in figures 11a and 11b where $\sigma_1/\sigma_2 = 25$. However, in this case, the effect is not so pronounced. In fact, when the lower layer is relatively highly conducting, the influence of finite β is relatively small as can be seen from figures 12a and 12b.

To discuss the significance of the results shown in figures 10a to 12b, it is desirable to introduce a scale distance L defined by

$$\left|\frac{\partial}{\partial x}\right| = kS = \frac{1}{L}.$$

It is a measure of the horizontal distance in which the field changes by an appreciable amount. For a periodic disturbance, L is the (spatial) wavelength. Thus

$$\beta = \delta^2/(2L^2)$$

where δ is the skin depth in the upper layer. For a highly insulating substratum, it appears that β must be very small or that δ must be much less than L if the standard magneto-telluric interpretation is to be applied. For example, if

$$h_1 = 1.4 \text{ km}, \sigma_1 = 10^{-3} \text{ mhos/m}, f = 0.1 \text{ Hz and } \sigma_2 \ll \sigma_1,$$

it follows that

$$\delta = 50 \text{ km and } (\sigma_1\mu\omega)^{\frac{1}{2}} h_1 = \sqrt{2} h_1/\delta \cong 0.02.$$

Then from figure 10a it can be seen that if $\beta < 10^{-3}$, $|Q|$ is within 10 percent of its value for $\beta = 0$. This condition is equivalent to

$$L > \delta/(\beta\sqrt{2})$$

or

$$L > 1{,}100 \text{ km}.$$

The condition is even more stringent for the phase.

Uniformity of the exciting fields over distances of the order of 1,000 km would not be very common, particularly in higher latitudes where the currents could be quite localized. Also, at low latitudes the presence of equatorial electrojet [Vestine, 1960] at heights of the order of 100

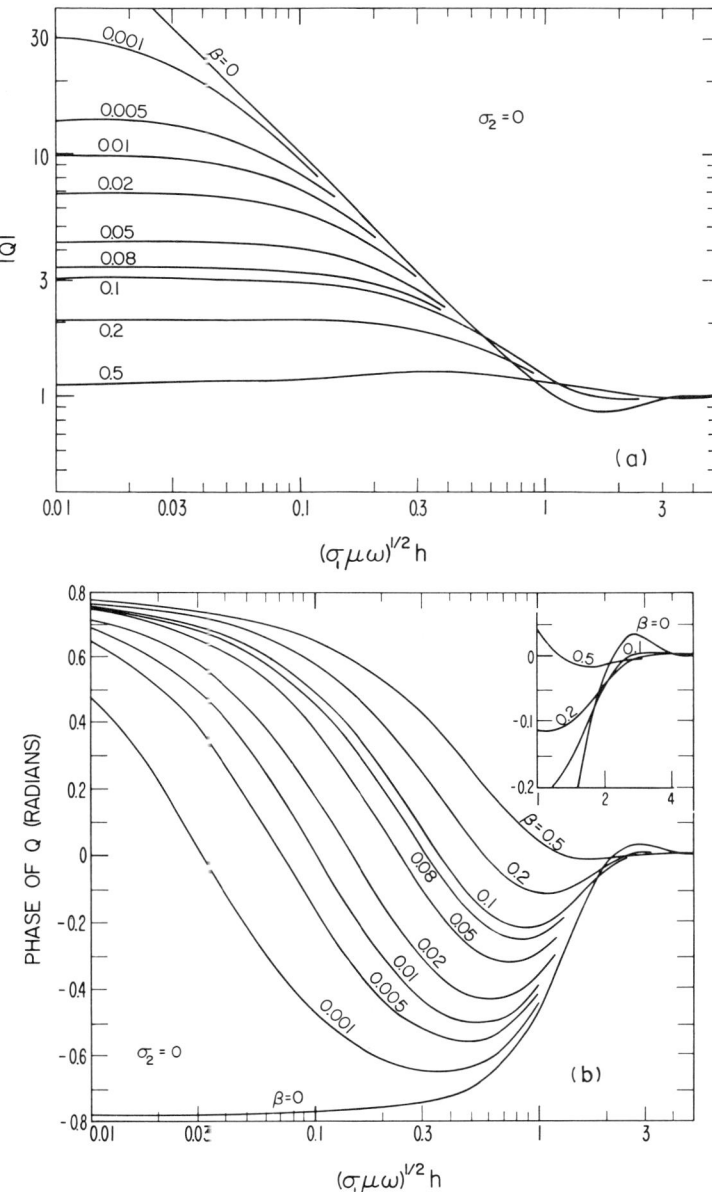

FIGURE 10. Amplitude and phase of Q for a two-layer model plotted as a function of $(\sigma_1\mu\omega)^{\frac{1}{2}}h_1$.
[For this set of curves the exciting field is non-uniform (i.e., $\beta > 0$, see text).]

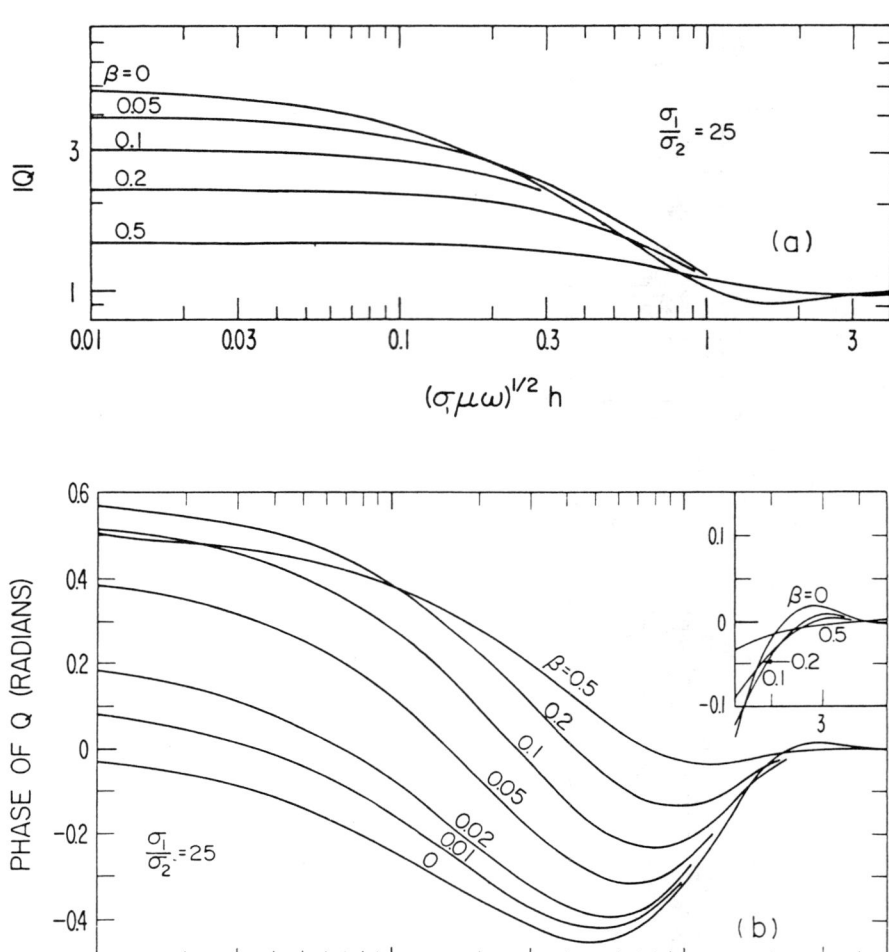

FIGURE 11. *Amplitude and phase of Q for a two-layer model plotted as a function of $(\sigma_1\mu\omega)^{\frac{1}{2}}h_1$.*
[For this set of curves the exciting field is non-uniform (i.e., $\beta>0$, see text).]

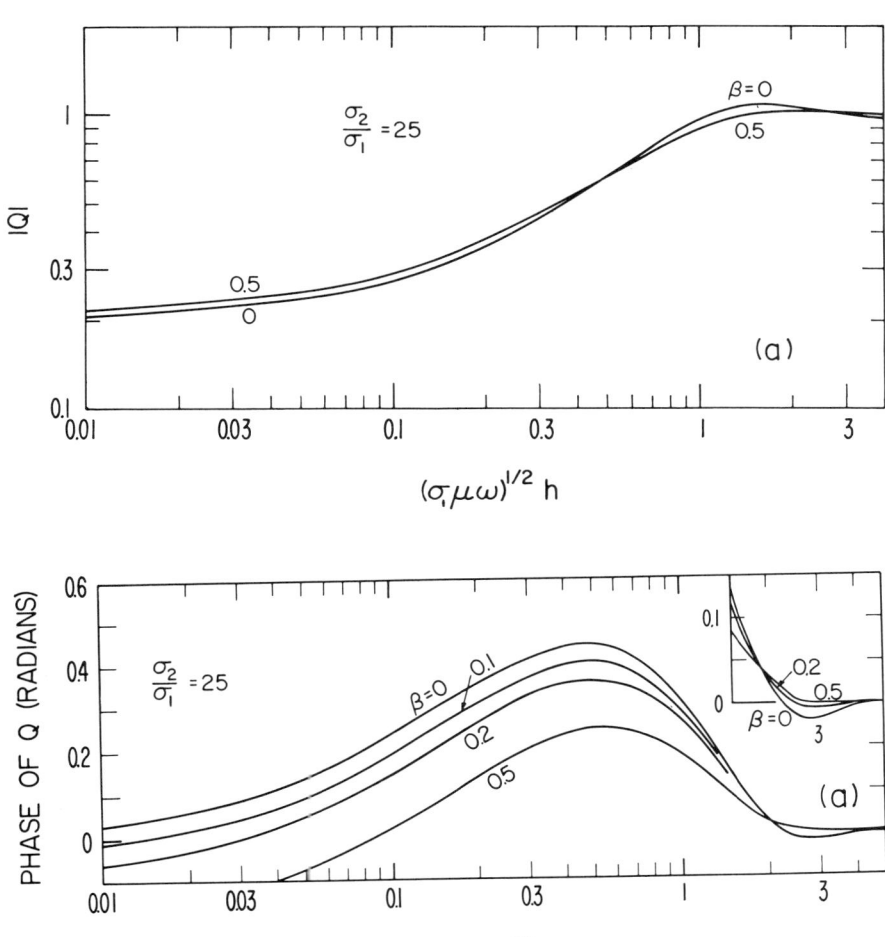

FIGURE 12. *Amplitude and phase of Q for a two-layer model plotted as a function of* $(\sigma_1\mu\omega)^{\frac{1}{2}}h_1$.
[For this set of curves the exciting field is non-uniform (i.e., $\beta > 0$, see text).]

km would be expected to produce considerable nonuniformity. Actually, the two-layer model with a highly insulating substratum is a most unfavorable circumstance. Furthermore, it is somewhat hypothetical since the poorly conducting crystalline rock will also be of finite depth. Consequently, a three-layer model should again be adopted for a more realistic appraisal. In this case, eq (36) may be generalized [Wait, 1953b] to

$$Q = \frac{G\hat{Q} + \tanh x}{1 + G\hat{Q} \tanh x} \qquad (31)$$

where

$$\hat{Q} = \frac{\hat{G} + \tanh \hat{x}}{1 + \hat{G} \tanh \hat{x}} \qquad (32)$$

$$\hat{G} = \left[\frac{\sigma_2/\sigma_1 - i\beta}{\sigma_3/\sigma_1 - i\beta}\right]^{\frac{1}{2}} \text{ and } \hat{x} = (i\sigma_2\mu_0\omega)^{\frac{1}{2}} h_2 (1 - i\beta\sigma_1/\sigma_2)^{\frac{1}{2}}.$$

Now, if the intervening stratum is poorly conducting such that

$$(\sigma_2\mu_0\omega)^{\frac{1}{2}} h_2 \ll 1 \text{ and } \beta \sqrt{\sigma_1\mu_0\omega} h_2 \ll 1$$

it readily follows that

$$Q \cong \frac{\hat{G}G + \tanh x}{1 + \hat{G}G \tanh x} \qquad (33)$$

where

$$\hat{G}G = \left[\frac{1 - i\beta}{\sigma_3/\sigma_1 - i\beta}\right]^{\frac{1}{2}}. \qquad (34)$$

The latter equation for Q does not depend on σ_2 or h_2 and thus the three-layer structure is equivalent to a two-layer model whose constants are σ_1, σ_3, and h_1. If, *in addition*, $\beta \ll 1$ this formula for Q reduces to eq (29).

In summary, the restrictions on the use of the curves of Q given in figures 2a to 9b can be applied to a three-layer structure provided the single condition

$$|\beta Q^2| \ll 1$$

is met. Thus, for highly insulating substrata, the condition can become quite stringent since $|Q^2|$ is then large compared with unity. On the other hand, for highly conducting substrata, $|Q^2|$ may be small and the condition is not at all stringent.

REFERENCES

BERDICHEVSKY, M.N., and BRUNELLI, (1959), Theoretical premises of magneto-telluric profiling, *Bull. (Izv.) Acad. Sci. USSR*, Geophys. Ser. No. 7, 1061-1069.

BOSSY, L., and De Vuyst, (1959/III), Relations entre les champs électrique et magnétique d'une onde de periode tres longue induts dans un milieu de conductivité variable, *Geofis. pura e Applicata* (Milano) 44, 119-134.

CAGNIARD, L., (1953), Basic theory of the magneto-telluric method of geophysical prospecting, *Geophys. XVIII*, 3, 605-635.

CAMPBELL, W.H., (1959), Studies of magnetic field micropulsations with periods of 5 to 30 seconds, *J. Geophys. Res.*, 64, 1819-1826.

CAMPBELL, W.H., (1960), Natural electromagnetic energy below the ELF range, *J. Res. NBS*, 64D (Radio Prop.), No. 4, 409-411.

CHETAEV, D.N., (1960), The determination of the anisotropy coefficient and the angle of inclination of a homogeneous anisotropic medium, by measuring the impedance of the natural electromagnetic field, *Bull. (Izv.) Acad. Sci. USSR*, Geophys. Series No. 4, 617-619.

DMITRIEV, V.I. and BERDICHEVSKY, M.N., (1979), The fundamental model of magneto-telluric sounding, *Proc. IEEE*, 67, No. 7, 1034-1044.

GARLAND, G.D., and WEBSTER, T.F., (1960), Studies of natural electric and magnetic fields, *J. Res. NBS*, 64D (Radio Prop.), 405-408.

JACKSON, C.M., WAIT, J.R., and WALTERS, L.C., (1962), Numerical results for the surface impedance of a stratified conductor, *NBS Tech. Note No. 143* (PB161644).

KATO, Y., and KIBUCHI, T., (1950), On the phase difference of earth current induced by changes of the earth's magnetic field, Part I, Science

Reports of Tohoku Univ., Series 5, *Geophys.*, 2, 139-141, Part II, 142-145.

KOLMAKOV, M.V., (1961), An interesting property of theoretical magneto-telluric sounding curves, *Bull. (Izv.) Acad. Sci. USSR*, Geophys. Series No. 4 583-587.

KOLMAKOV, M.V., and VLADIMIROV, N.P., (1961), On the equivalence of magneto-telluric sounding curves, *Bull. (Izv.) Acad. Sci. USSR*, Geophys. Series No. 4, 544-552.

KOVTUN, A.A., (1961), The magneto-telluric investigation of structures inhomogeneous in layers, *Bull. (Izv.) Acad. Sci. USSR*, Geophys. Series No. 11, 1663-1667.

KOZULIN, YU. N., (1961), On the theory of electromagnetic frequency sounding of multilayered structures, *Bull. (Izv.) Acad. Sci. USSR*, Geophys. Series No. 8, 1204-1212.

LAHIRI, B.N., and PRICE, A.T., (1939), Electromagnetic induction in non-uniform conductors and the determination of the conductivity of the earth from terrestrial magnetic variations, *Phil. Trans. Roy. Soc.*, 237A, 509-540.

LIPSKAYA, N.V., (1953), On certain relationships between harmonics of the periodic variations of the terrestrial electric and magneto fields, *Izv. Akad. Nauk, USSR*, Geophys. Series No. 1, 41-47

MATUSHITA, S., and CAMPBELL, W.H., (1967), *Physics of Geomagnetic Phenomena*, Vol. I and II, Academic Press, New York.

NIBLETT, E.R., and SAYN-WITTGENSTEIN, C., (1960), Variation of electrical conductivity with depth by the magneto-telluric method, *Geophys.*, XXV, No. 5, 998-1008.

POKITYANSKI, I.I., (1961), On the application of the magneto-telluric method to anisotropic and inhomogeneous masses, *Bull. (Izv.) Acad. Sci. USSR*, Geophys. Series No. 11, 1607-1613.

PRICE, A.T., (1962), The theory of magneto-telluric methods when the source field is considered, *J. Geophys. Res.*, 67, No. 5, 1907-1918.

SCHOLTE, J.G., and VELDKAMP, J., (1955), Geomagnetic and Geoelectric variations, *J. Atmos. Terrest. Phys.*, 6, 33-45.

TIKHONOV, A.N., (1950), Determination of the electrical characteristics of the deep strate of the earth's crust, *Dok. Akad. Nauk, USSR*, 73, 2, 295-297.

TIKHONOV, A.N., and LIPSKAYA, N.V., (1952), Terrestrial electric field variations, *Dok. Akad. Nauk*, 87, 4, 547-550.

TIKHONOV, A.N., and SHAKHSUVAROV, D.N., (1956), Concerning the possibility of using the impedance of the earth's natural electromagnetic field for investigating its upper layers, *Bull. (Izv.) Acad. Sci. USSR*, Geophys. Series No. 4, 410-418.

VESTINE, E.H., (1960), The upper atmosphere and geomagnetism, *Physics of the Upper Atmosphere* (ed. J.A. Ratcliffe), Academic Press, New York and London.

VLADIMIROV, N.P., (1960), The feasibility of using the earth's natural electromagnetic field for geological surveying, *Bull. (Izv.) Acad. Sci. USSR*, Geophys. Series No. 1, 139-141.

VLADIMIROV, N.P., and AN, V.A., (1961), A method of processing magneto-telluric oscillograms, *Bull. (Izv.) Acad. Sci. USSR*, Geophys. Series No. 11, 1649-1654.

VLADIMIROV, N.P., and KOLMAKOV, M.V., (1960), The resolving power of the magneto-telluric method, *Bull. (Izv.) Acad. Sci. USSR*, Geophys. Series.

No. 11, 1598-1600.

VLADIMIROV, N.P., and NIKIFOROVA, N.N., (1961), On the interpretation of magneto-telluric sounding curves, *Bull. (Izv.) Acad. Sci. USSR, Geophys. Series* No. 1, 111-113.

WAIT, J.R., (1953a), Propagation of radio waves over a stratified ground, *Geophys.*, 18, 416-422.

WAIT, J.R., (1953b), The fields of a line source of current over a stratified conductor, *Appl. Sc. Res.*, Sec. B, 3, 279-292.

WAIT, J.R., (1954), On the relation between telluric currents and the earth's magnetic field, *Geophys.*, XIX, No. 2, 281-289.

WAIT, J.R., (1959), *Electromagnetic radiation from cylindrical structures*, Pergamon Press, New York.

WAIT, J.R., (1962), Theory of magneto-telluric fields, *J. Res. of NBS*, 66D, No. 5, 509-541.

WATT, A.D., MAXWELL, E.L. and MATHEWS, F.S., (1962), Some electrical properties of the earth's crust, *DECO Electronics, Inc. Boulder, Colo.* Report No. 30-S-1.

YUNGUL, S.H., (1961), Magneto-telluric sounding three-layer interpretation curves, *Geophys.*, XXVI, No. 4, 465-473.

4
General Surface Impedance

Here we outline a fairly general formulation of the surface impedance for a uniformly stratified model of the earth when the source can be described as an overhead current system with any specified variation in the horizontal plane. We believe this development gives much needed insight to the general electromagnetic induction problem. An earlier attempt [Berdichevsky, Vanyan and Faynberg, 1969] to provide such an overall theoretical framework was very limited in scope, since all displacement currents were neglected and the vertical current accross the air/earth interface was ignored.

FORMULATION

We shall formulate the problem in as simple manner as possible, but without restricting it's utility. We neglect earth curvature and choose a rectangular coordinate system (x,y,z) with the earth's surface to be the plane $z = 0$ and with the z axis pointing downward as indicated in Fig. 1. The region $z < 0$ is assumed to be free space with permittivity ε_o and permeability μ_o except for a source region at the level $z = -z_o$ (i.e., at a height z_o <u>above</u> the earth's surface). In this region we assume that a spectral component of the current in the x direction is $j_x e^{-i\beta x} e^{-i\lambda y} e^{i\omega t}$ amps/m where ω is the angular frequency. Similarly, in the y direction the spectral component of the current is $j_y e^{-i\beta x} e^{-i\lambda y} e^{i\omega t}$ amps/m. Here β and λ are wave numbers that describe the spatial variation of the source field. Thus, all field components vary as $\exp[-i\beta x - i\lambda y + i\omega t]$. The objective then is to deduce the corresponding surface impedance relation at the air/earth interface $z = 0$. This is postulated to have the form

$$E_x = Z_{xx}H_x + Z_{xy}H_y \qquad (1)$$

$$E_y = Z_{yx}H_x + Z_{yy}H_y \qquad (2)$$

where as indicated Z_{xx}, etc., are the elements of the surface impedance matrix [Z] that relates the tangential electric field $\vec{E}_t$ and the tangential magnetic field $\vec{H}_t$.

THE HERTZ POTENTIAL

For the present problem it is convenient, but certainly not mandatory, to describe the fields in terms of electric and magnetic Hertz vectors that have only vertical or z components. These are denoted by Π and Π^*, respectively. Within any of the homogeneous regions (with say a conductivity σ, permittivity ε and permeability μ) we can thus write

$$E_z = (-\gamma^2 + \partial^2/\partial z^2)\Pi \qquad (3)$$

and

$$H_z = (-\gamma^2 + \partial^2/\partial z^2)\Pi^* \qquad (4)$$

where $\gamma^2 = i\mu\omega(\sigma + i\varepsilon\omega)$. Except at the source itself Π and Π^* satisfy the Helmholz equation $(\nabla^2 - \gamma^2)\Pi = 0$. From Maxwell's equations it is now a simple matter to deduce that the other field components are to be obtained from

$$E_x = -i\beta \frac{\partial \Pi}{\partial z} - \mu\omega\lambda\Pi^* \qquad (5)$$

$$H_x = -i\beta \frac{\partial \Pi^*}{\partial z} - i\lambda(\sigma + i\varepsilon\omega)\Pi \qquad (6)$$

$$E_y = -i\lambda \frac{\partial \Pi}{\partial z} + \mu\omega\beta\Pi^* \qquad (7)$$

$$H_y = -i\lambda \frac{\partial \Pi^*}{\partial z} + i\beta(\sigma + i\varepsilon\omega)\Pi \qquad (8)$$

The partial fields associated with the Π potential are usually designated TM (transverse magnetic) and those with the Π^* potential as TE (transverse electric).

Note that in general, at least for a homogeneous region:
$$\vec{E} = (-\gamma^2 + \text{grad div})\Pi + (\sigma + i\varepsilon\omega)\text{curl } \Pi^*$$
$$\vec{H} = (-\gamma^2 + \text{grad div})\Pi^* - i\mu\omega \text{ curl } \vec{\Pi}$$
but we are free here to assume that $\vec{\Pi}$ and $\vec{\Pi}^*$ have only z components Π and Π^*

General Surface Impedance

We will first deal with the source region $z < 0$ and add a subscript zero to pertinent quantities. The required forms for the Hertz potentials are easily seen to be

$$\Pi_o = ae^{u_o z} + be^{-u_o z} \quad (9)$$
$$\Pi_o^* = a^* e^{u_o z} + b^* e^{-u_o z} \quad \text{for } 0 > z > -z_o \quad (10)$$

and

$$\Pi_{\circ} = ce^{u_o z} \quad (11)$$
$$\Pi_{\circ}^* = c^* e^{u_o z} \quad \text{for } z < -z_o \quad (12)$$

where $u_o = (\lambda^2 + \beta^2 + \gamma_o^2)^{1/2}$, $\gamma_o^2 = -\varepsilon_o \mu_o \omega^2$, and a, b, c, a^*, b^* and c^* are coefficients, as yet, undetermined. In writing (9) to (12) and for subsequent field quantities, we omit the common factor $\exp[-i\beta x - i\lambda y + i\omega t]$.

Ampere's law immediately tells us that the source conditions are

$$H_{oy}(-z_o - 0) - H_{oy}(-z_o + 0) = j_x \quad (13)$$
$$H_{ox}(-z_o - 0) - H_{ox}(-z_o + 0) = -j_y \quad (14)$$
$$E_{oy}(-z_o - 0) - E_{oy}(-z_o + 0) = 0 \quad (15)$$
$$E_{ox}(-z_o - 0) - E_{ox}(-z_o + 0) = 0 \quad (16)$$

Subsequent application gives

$$b = \frac{(\beta j_x + \lambda j_y)}{2\varepsilon_o \omega (\lambda^2 + \beta^2)} e^{-u_o z_o} \quad (17)$$

$$b^* = -\frac{(\lambda j_x - \beta j_y)}{2iu_o(\lambda^2 + \beta^2)} e^{-u_o z_o} \quad (18)$$

The other coefficients are related to the "source" coefficients b and b^* as follows

$$a = Rb, \quad a^* = R^* b^* \quad (19)$$
$$c = (R - e^{2u_o z_o})b, \quad c^* = (R^* + e^{2u_o z_o})b^* \quad (20)$$

where R and R^* are reflection coefficients that we discuss below.

For the region $(-z_o < z < 0)$, explicit expressions for the Hertz potentials are thus

$$\Pi_o = \frac{(\beta j_x + \lambda j_y)}{2\varepsilon_o \omega (\lambda^2 + \beta^2)} [e^{-u_o(z+z_o)} + R e^{u_o(z-z_o)}] \tag{21}$$

and

$$\Pi_o^* = -\frac{(\lambda j_x - \beta j_y)}{2i u_o (\lambda^2 + \beta^2)} [e^{-u_o(z+z_o)} + R^* e^{u_o(z-z_o)}] \tag{22}$$

It is now a simple matter to show that

$$R = (K_o - Z)/(K_o + Z) \tag{23}$$

and

$$R^* = (N_o - Y)/(N_o + Y) \tag{24}$$

when $K_o = u_o/(i\varepsilon_o \omega)$ and $N_o = u_o/(i\mu_o \omega)$ are the TM wave impedance and TE wave admittance, respectively. If the earth were homogeneous with electrical constants σ, ε, and μ we would have simply that $Z = u/(\sigma + i\varepsilon\omega)$ and $Y = u/(i\mu\omega)$ where $u = (\lambda^2 + \beta^2 + \gamma^2)^{1/2}$. But, in general, we can define

$$Z = E_x/H_y\Big|_{TM} = -E_y/H_x\Big|_{TM} = -\frac{\partial \Pi_o/\partial z}{i\varepsilon_o \omega \Pi_o}\Bigg]_{z=0} \tag{25}$$

and

$$Y = -H_x/E_y\Big|_{TE} = H_y/E_x\Big|_{TE} = -\frac{\partial \Pi_o^*/\partial z}{i\mu_o \omega \Pi_o^*}\Bigg]_{z=0} \tag{26}$$

Actually, explicit expressions for Z and Y are given in the Appendix.

It is useful here to give further insight to the basic wave immittances Z and Y by considering the so-called "wave tilts". To this end we first set $\lambda = 0$ so that the excitation is uniform in the y direction (i.e., $\partial/\partial y = 0$ everywhere). It is then a simple exercise to show that the electric wave tilt W, as defined, is given by

$$W = -\frac{E_{ox}}{E_{oz}}\Bigg|_{\substack{\lambda=0 \\ z=0}} = \frac{u_o}{i\beta} \frac{(1 - R)}{(1 + R)} = \frac{\varepsilon_o \omega}{\beta} Z \tag{27}$$

General Surface Impedance

Similarly the magnetic wave tilt W^* is

$$W^* = -\left.\frac{H_{ox}}{H_{oz}}\right|_{\substack{\lambda=0 \\ z=0}} = \frac{u_o}{i\beta}\frac{(1-R^*)}{(1+R^*)} = \frac{\mu_o\omega}{\beta}Y \quad (28)$$

These wave tilt ratios are measureable quantities and they are simply related to the basic wave immitances at least for the two dimensional field configuration. However, in general, the field ratios E_{ox}/E_{oz} and H_{ox}/H_{oz} do not bear any simple relationship to Z and Y.

As a further iterim step we may exhibit the fields components in the region $(0 > z > -z_o)$ as follows

$$\begin{Bmatrix} E_{ox} \\ E_{oy} \end{Bmatrix} = \frac{(\beta j_x + \lambda j_y)i u_o}{2\epsilon_o\omega(\lambda^2 + \beta^2)}\begin{Bmatrix} \beta \\ \lambda \end{Bmatrix}[e^{-u_o(z+z_o)} - Re^{u_o(z-z_o)}]$$

$$+ \frac{(\lambda j_x - \beta\lambda_y)\mu_o\omega}{2iu_o(\lambda^2+\beta^2)}\begin{Bmatrix} \lambda \\ -\beta \end{Bmatrix}[e^{-u_o(z+z_o)} + R^*e^{u_o(z-z_o)}] \quad (29)$$

and

$$\begin{Bmatrix} H_{ox} \\ H_{oy} \end{Bmatrix} = -\frac{(\lambda j_x - \beta j_y)}{2(\lambda^2+\beta^2)}\begin{Bmatrix} \beta \\ \lambda \end{Bmatrix}[e^{-u_o(z+z_o)} - R^*e^{u_o(z-z_o)}]$$

$$+ \frac{(\beta j_x + \lambda j_y)}{2(\lambda^2+\beta^2)}\begin{Bmatrix} \lambda \\ -\beta \end{Bmatrix}[e^{-u_o(z+z_o)} + Re^{u_o(z-z_o)}] \quad (30)$$

Now we insist that

$$E_{ox}(z=0) = Z_{xx}H_{ox}(z=0) + Z_{xy}H_{oy}(z=0) \quad (31)$$

and

$$E_{oy}(z=0) = Z_{yx}H_{ox}(z=0) + Z_{yy}H_{oy}(z=0) \quad (32)$$

Since these must hold for any linear combination of j_x and j_y we are lead to the following set of linear equations:

$$\frac{i\beta u_o}{\varepsilon_o \omega} (1 - R) = Z_{xx}\lambda(1 + R) - Z_{xy}\beta(1 + R) \tag{33}$$

$$\frac{\mu_o \omega \lambda}{i u_o} (1 + R^*) = -Z_{xx}\beta(1 - R^*) - Z_{xy}\lambda(1 - R^*) \tag{34}$$

$$\frac{i\lambda u_o}{\varepsilon_o \omega} (1 - R) = Z_{yx}\lambda(1 + R) - Z_{yy}\beta(1 + R) \tag{35}$$

$$\frac{-\mu_o \omega \beta}{i u_o} (1 + R^*) = -Z_{yx}\beta(1 - R^*) - Z_{yy}\lambda(1 - R^*) \tag{36}$$

Noting that
$$(1 - R)/(1 + R) = i\varepsilon_o \omega Z/u_o \tag{37}$$
and
$$(1 - R^*)/(1 + R^*) = i\mu_o \omega Y/u_o \tag{38}$$

(33) - (36) are easily solved to obtain

$$Z_{xy} = \frac{\beta^2 Z + \lambda^2 Y^{-1}}{\beta^2 + \lambda^2} \tag{39}$$

$$Z_{yx} = -\frac{\lambda^2 Z + \beta^2 Y^{-1}}{\beta^2 + \lambda^2} \tag{40}$$

and

$$Z_{xx} = -Z_{yy} = \frac{\lambda \beta}{\beta^2 + \lambda^2} (Y^{-1} - Z) \tag{41}$$

In the case of a homogeneous earth these reduce to

$$Z_{xy} = \frac{\beta^2 + \gamma^2}{(\sigma + i\varepsilon\omega)u} \tag{42}$$

$$Z_{yx} = -\frac{\lambda^2 + \gamma^2}{(\sigma + i\varepsilon\omega)u} \tag{43}$$

and

$$Z_{xx} = -Z_{yy} = -\frac{\lambda \beta}{(\sigma + i\varepsilon\omega)u} \tag{44}$$

where $u = (\lambda^2 + \beta^2 + \gamma^2)^{1/2}$.

Then we may assert that $Z_{xy} \simeq -Z_{yx} \simeq \eta$ where $\eta = i\mu\omega/(\sigma + i\varepsilon\omega)$ provided $|\beta/\gamma|^2$ and $|\lambda/\gamma|^2 \ll 1$. To within this same approximation Z_{xx}/η and Z_{yy}/η are

General Surface Impedance

of the order of $|\lambda\beta/\gamma^2|$ and may thus be neglected. An alternative approach is to recall that $-i\beta = \partial/\partial x$ and $-i\lambda = \partial/\partial y$ which suggests that $\beta^2 = -\partial^2/\partial x^2$, $\lambda^2 = -\partial^2/\partial y^2$, $\lambda\beta = -\partial^2/\partial x \partial y$ at least in an operational sense.

Thus, if $|\gamma^2|$ is sufficiently large

$$\frac{\gamma^2 + \beta^2}{\gamma u} = \left(1 + \frac{\beta^2}{\gamma^2}\right)\left(1 + \frac{\lambda^2}{\gamma^2} + \frac{\beta^2}{\gamma^2}\right)^{-1/2}$$

$$\simeq 1 + \frac{1}{2\gamma^2}(\beta^2 - \lambda^2) + \text{terms in } \frac{1}{\gamma^4}, \text{ etc.} \qquad (45)$$

$$\simeq 1 - \frac{1}{2\gamma^2}\left(\frac{\partial^2}{\partial x^2} - \frac{\partial^2}{\partial y^2}\right) + \text{terms in } \frac{1}{\gamma^4}, \text{ etc.}$$

Using this type of argument led Monteath [1973] to write the following surface impedance condition for the tangential fields at the boundary of a homogeneous half-space:

$$E_x \simeq \eta H_y - \frac{\eta}{2\gamma^2}\left[\left(\frac{\partial^2 H_y}{\partial x^2} - \frac{\partial^2 H_y}{\partial y^2}\right) - 2\frac{\partial^2 H_x}{\partial x \partial y}\right]$$

$$+ \text{terms in } \gamma^{-4}, \text{ etc.} \qquad (46)$$

and

$$E_y = -\eta H_x + \frac{\eta}{2\gamma^2}\left[\left(\frac{\partial^2 H_x}{\partial y^2} - \frac{\partial^2 H_x}{\partial x^2}\right) - 2\frac{\partial^2 H_y}{\partial y \partial x}\right]$$

$$+ \text{terms in } \gamma^{-4}, \text{ etc.} \qquad (47)$$

These show in a very clear manner that the leading terms (i.e., $E_x \simeq \eta H_y$ and $E_y \simeq -\eta H_x$) are only valid if the horizontal variation of the tangential fields are sufficiently slow. It is significant that the first correction terms involve second derivatives only. Thus linear variations of the fields do not have a major effect on upsetting the simple proportionality of the $E_x(E_y)$ and $H_z(-H_x)$ tangential field components. This point has been made in a rather emphatic manner by Dmitriev and Berdichevsky [1979]. But their argument does not obviate the fact that the simple source-independent assumption is difficult to justify in cases where the causative ionosphere currents are localized such as in an electrojet. This is particularly the concern when applying the magnetotelluric method to highly resistive sub-layers where the excitation periods are greater than 60 seconds [Wait and Spies, 1973].

The utility of the surface impedance description, of course, is enhanced when the dependence on the spatial wave numbers β and λ is minimized. In the general case of an N-layered half-space model the desired simplification is as follows

$$Z_{xy} \simeq -Z_{yx} \simeq \eta_1 Q \tag{48}$$

where

$$\eta_1 Q = Z]_{\lambda=\beta=0} = Y^{-1}]_{\lambda=\beta=0} \tag{49}$$

where $\eta_1 = \gamma_1/(\sigma_1 + i\varepsilon_1\omega) = i\mu_1\omega/\gamma_1$ and $Z_{xx} \simeq -Z_{yy} \simeq 0$. Here Q is a normalized surface impedance for an N-layered half-space when the incidence is perpendicular. Tabulations and graphical plots have been given elsewhere [Jackson, Wait and Walter, 1967; Wait, 1970] for the case N = 2 and 3. Of course, if the half space is homogeneous, we have merely that Q = 1. The condition that (48) is a valid approximation can be stated

$$|\beta/\gamma_n|^2 \text{ and } |\lambda/\gamma_n|^2 << 1$$

for every layer n = 1,2,3.... where an appreciable fraction of the energy propagates.

Concluding Remarks

We have considered the general theory of the surface impedance of an N-layered half-space for non-uniform excitation. For convenience in the formulation, only a single spectral component was considered. Actually, there is no loss in generality here. For example, by Fourier synthesis the actual x-directed time dependent source current could be written

$$j_x(x,y,t) = \frac{1}{(2\pi)^3} \iiint_{-\infty}^{+\infty} j_x(\beta,\lambda,\omega) e^{-i\beta x} e^{-i\lambda y} e^{i\omega t} \, d\beta d\lambda d\omega$$

General Surface Impedance

and similarly for $j_y(x,y,t)$. The field components are synthesized in like fashion. For example the actual time dependent x-directed electric field is given by

$$E_{ox}(x,y,z,t) = \frac{1}{(2\pi)^3} \int\!\!\!\int\!\!\!\int_{-\infty}^{+\infty} E_{ox}(\beta,\lambda,z,\omega) e^{-i\beta x} e^{-i\lambda y} e^{i\omega t} \times d\beta d\lambda d\omega$$

where $E_{ox}(\beta,\lambda,z,\omega)$ is given by (29) and (30). It is now rather obvious when we say that the matrix surface impedance relationship given by (1) and (2) applies only to the Fourier spectral component with spatial wave numbers β and λ and angular frequency ω. It is possible this restriction on the surface impedance description is not always appreciated.

Appendix

Expressions for Y and Z for an N layered half-space are given here. The method of solution is indicated only briefly since the derivation for closely allied problems are given elsewhere [Strutt, 1912; Wait, 1953; Wait, 1970].

As indicated in Fig. 1, the half-space region $z > 0$ is represented by N homogeneous regions. These are defined in such a fashion that the electrical properties σ_n, ε_n, μ_n and the thicknesses are:

$$0 < z < z_1 \; ; \; \sigma_1, \varepsilon_1, \mu_1 \quad (h_1 = z_1)$$

$$z_1 < z < z_2 \; ; \; \sigma_2, \varepsilon_2, \mu_2 \quad (h_2 = z_2 - z_1)$$

- -

$$z_{n-1} < z < z_n \; ; \; \sigma_n, \varepsilon_n, \mu_n \quad (h_n = z_n - z_{n-1})$$

- -

$$z_{N-2} < z < z_{N-1} \; ; \; \sigma_{N-1}, \varepsilon_{N-1}, \mu_{N-1} \quad (h_{N-1} = z_{N-1} - z_{N-2})$$

$$z_{N-1} < z < \infty \; ; \; \sigma_N, \varepsilon_N, \mu_N \quad (h_N = \infty)$$

In the n'th layer the Hertz potentials have the forms

$$\Pi_n = a_n e^{u_n z} + b_n^* e^{-u_n z}$$

and

$$\Pi_n^* = a_n^* e^{u_n z} + b_n^* e^{-u_n z}$$

where the common factor is $\exp(-i\beta x - i\lambda y + i\omega t)$ and $u_n = (\lambda^2 + \beta^2 + \gamma_n^2)^{1/2}$, $\gamma_n^2 = i\mu_n\omega(\sigma_n + i\varepsilon_n\omega)$.

The unknown coefficients a_n, b_n, a_n^* and b_n^* are found from imposing the boundary conditions at each of the interfaces. This requires the individual continuity of the functions $(\sigma_n + i\varepsilon_n\omega)\Pi_n$, $\partial\Pi_n/\partial z$, $i\mu_n\omega\Pi_n^*$ and $\partial\Pi_n^*/\partial z$. Thus we have two uncoupled sets of $(2N - 1)$ linear equations for the $(2N - 1)$ unknown coefficients. Of course, a_N and a_N^* are identically zero because only outgoing waves are permissible in the bottom semi-infinite region.

We deduce that

$$Z = Z_1 = K_1 \frac{Z_2 + K_1 \tanh u_1 h_1}{K_1 + Z_2 \tanh u_1 h_1}$$

General Surface Impedance

where
$$Z_2 = K_2 \frac{Z_3 + K_2 \tanh u_2 h_2}{K_2 + Z_3 \tanh u_2 h_2}$$

- - - - - - - - - - - - - - - -

$$Z_n = K_n \frac{Z_{n+1} + K_n \tanh u_n h_n}{K_n + Z_{n+1} \tanh u_n h_n}$$

- - - - - - - - - - - - - - - -

$$Z_{N-1} = K_{N-1} \frac{K_N + K_{N-1} \tanh u_{N-1} h_{N-1}}{K_{N-1} + K_N \tanh u_{N-1} h_{N-1}}$$

where $K_n = u_n/(\sigma_n + i\varepsilon_n \omega)$ for $n = 1,2,3\ldots N$). The corresponding expression for Y (or Y_1) is identical in form if we merely replace Z_n by Y_n and K_n by N_n where $N_n = u_n/(i\mu_n \omega)$ for $n = 1,2,3\ldots N$.

A great simplification ensues when the excitation is constant (i.e. $\beta = \lambda = 0$). Then both expressions for Z and Y^{-1} are identical and given by the above sequence where $K_n = \eta_n = i\mu_n \omega/\gamma_n$ and $u_n = \gamma_n$ for $n = 1,2,3\ldots N$.

REFERENCES

BERDICHEVSKY, M.N., VANYAN, L.L. and FAYNBERG, E.B., (1969), Theoretical principles in using electromagnetic variations to study the electric conductivity of the earth, *Geomagnetism and Aeronomy*, Vol. 9, No. 3, 465-467.

DMITRIEV, V.I. and BERDICHEVSKY, M.N., (1979), The fundamental model of magneto-telluric sounding, *Proc. IEEE*, Vol. 67, No. 7, 1034-1044.

JACKSON, C.M., WAIT, J.R., WALTERS,L.C., (1962), Numerical results for the surface impedance of a stratified conductor, *NBS Tech. Note*, No. 143, March, (available from National Technical Information Service, Alexandria, Va., 22151, accession No. PB161644).

MONTEATH, G.D., (1973), *Applications of the Electromagnetic Reciprocity Principle*, p. 72, Pergamon Press, Oxford.

SRIVASTAVA, S.P., (1966), Method of interpretation of magneto-telluric data when (the) source field is considered, *Jour. Geophys. Res.*, Vol. 70, No. 4, 945-954.

STRUTT, J.W. (Lord Rayleigh), (1912), On the propagation of waves through a stratified medium, with special reference to the question of reflection, *Proc. of the Royal Society*, A, Vol. LXXXVI, pp. 207-266,

WAIT, J.R., (1953), Propagation of radio waves over a stratified ground, *Geophys.*, Vol. 18, 418-422.

WAIT, J.R., (1970), *Electromagnetic Waves in Stratified Media*, 2nd Ed., Chapt. 2 Pergamon Press, Oxford.

WAIT, J.R., and SPIES, K.P., (1973), Range dependence of the surface impedance and wave tilt for a time-source excited two layer earth, *IEEE Trans.*, Vol. AP-21, No. 6, 905-907.

General Surface Impedance

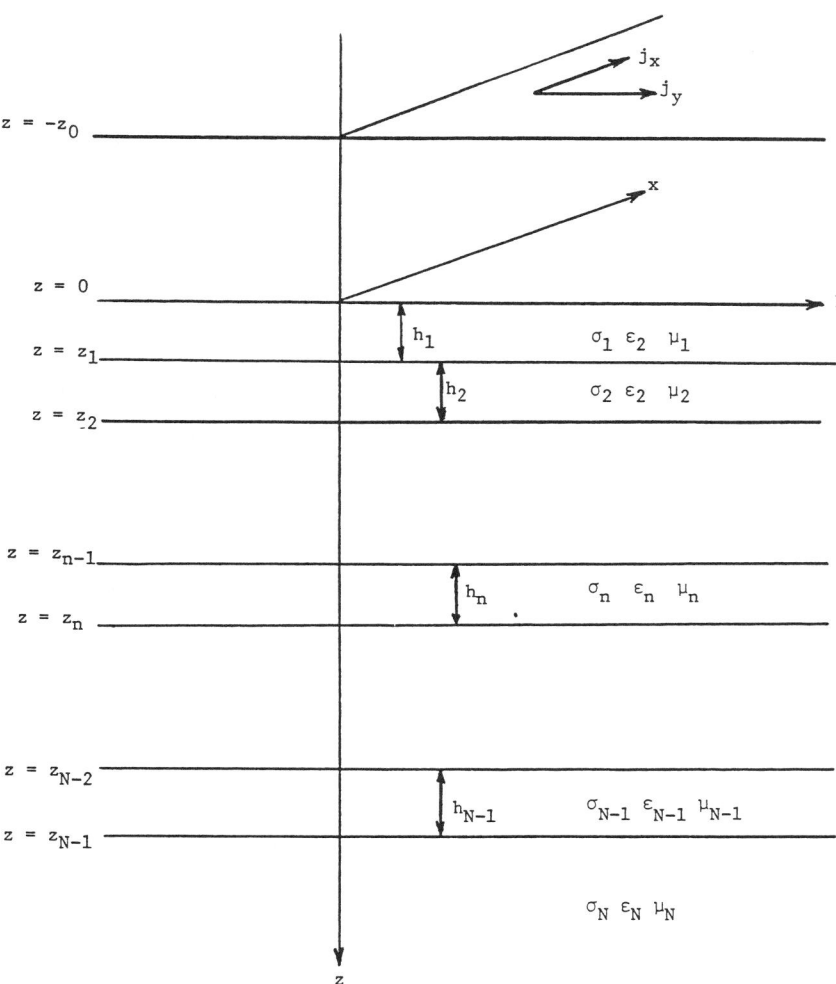

Fig. 1   Overhead current excitation of an N-layered half-space.

5
The Zenneck Wave

BASIC FORMULATION

Early in this century, J. Zenneck published a wave solution to Maxwell's equations for a flat earth that was modelled as a homogeneous half-space. This has been the source of much controversy, which is an understatement. [Wait, 1957]. Here we will present a concise development of Zenneck's solution from the standpoint of reflection of a plane wave from the air/earth interface.

We consider a rectangular coordinate system (x,y,z) with the lower conducting half-space of electric constants $(\sigma_g, \varepsilon_g, \mu_g)$ to be described by $z < 0$. The upper half-space is assumed to be free space with constants (ε_o, μ_o). If a plane wave is taken to be incident at angle of incidence θ and polarized such that the magnetic field has <u>only</u> a y component H_y. The resulting solution is

$$H_y = H_o(e^{u_o z} + R_{\parallel} e^{-u_o z}) e^{-i\lambda x} e^{i\omega t} \tag{1}$$

for $z > 0$, while

$$H_y = H_o(1 + R_{\parallel}) e^{-uz} e^{-i\lambda x} e^{i\omega t} \tag{2}$$

for $z > 0$. Here $\lambda = k \sin \theta$, $k = (\varepsilon_o \mu_o)^{1/2} \omega$, $u_o = ik \cos \theta = (\lambda^2 - k^2)^{1/2}$, $u = (\lambda^2 + \gamma_g^2)^{1/2}$, $\gamma_g = [i\mu_g \omega (\sigma_g + i\varepsilon_g \omega)]^{1/2}$ and ω is the angular frequency.

Now the electric field component E_x is obtained from

$$E_x = -(i\varepsilon_o \omega)^{-1} \partial H_y / \partial z \text{ for } z > 0 \tag{3}$$

and

$$E_x = -(\sigma_g + i\varepsilon_g \omega)^{-1} \partial H_y / \partial z \text{ for } z < 0 \tag{4}$$

Now on matching E_x and H_y at the interface $z = 0$, we readily deduce that

$$R_{\parallel} = (K_o - K)/(K_o + K). \tag{5}$$

where $K_o = u_o/(i\varepsilon_o \omega)$ and $K = u/(\sigma_g + i\varepsilon_g \omega)$. As we indicated before, a condition of wave matching occurs where $K_o = K$. The field in the upper region is then of the form $\exp(+u_o z)\exp(-i\lambda x)$. This matching condition is explicitly

The Zenneck Wave

written

$$\frac{u_o}{i\varepsilon_o\omega} = \frac{u}{\sigma_g + i\varepsilon_g\omega} \qquad (6)$$

To solve this equation, we set $u_o = -\alpha + i\beta$ and $u = \alpha_g + i\beta_g$ where α, β, α_g and β_g are real. To simplify our discussion, we consider $\varepsilon_g\omega/\sigma_g \ll 1$. Thus, we require that

$$i\frac{\alpha}{\varepsilon_o\omega} + \beta\frac{1}{\varepsilon_o\omega} = \frac{\alpha_g}{\sigma_g} + i\frac{\beta_g}{\sigma_g} \qquad (7)$$

Then
$$\alpha = (\varepsilon_o\omega/\sigma_g)\beta_g \text{ and } \beta = (\varepsilon_o\omega/\sigma_g)\alpha_g$$

Furthermore
$$u_g = \alpha_g + i\beta_g \simeq \gamma_g \simeq (1+i)/\delta_g$$

where
$$\delta_g = [2/(\sigma_g\mu_g\omega)]^{1/2}$$

is the "skin depth." Thus,

$$\alpha \simeq \beta \simeq (\varepsilon_o\omega/\sigma_g)\delta_g^{-1} = \varepsilon_o\omega[\mu_o\omega/(2\sigma_g)]^{1/2} \qquad (8)$$

Here we see that α and β are positive real. This corresponds to an amplitude decaying in the positive direction above the interface, while the phase velocity is directed downwards (i.e., towards the interface). On the other hand, within the conductor, the wave is attenuated downward and the phase velocity is also directed downwards.

This is compatible with the physical view of the Zenneck wave as envisaged by Barlow and Brown [1962]. However, there is an alternative point of view that we will now mention.

Imagine that the reflection coefficient has a pole (i.e., a resonance). Then the field in the upper free-space region must have the form $\exp(-u_o z)\exp(-i\lambda x)$. The pole condition is clearly

$$K_o = -K$$

or (9)

$$\frac{-u_o}{i\varepsilon_o\omega} = \frac{u}{\sigma_g + i\varepsilon_g\omega}$$

If we now set

$$-u_o = -\alpha + i\beta \quad \text{and} \quad u = \alpha_g + i\beta_g$$

we find precisely the same value for α and β, including the fact that they should both be positive real.

We thus conclude that the same Zenneck wave corresponds to a zero or a pole of the reflection coefficient depending on how we choose to define the vertical propagation constant of the wave function in the earth.

It is now a simple matter to solve either (6) or (9) for horizontal wavenumber λ_o of the Zenneck wave. Assuming the permeability of the earth is μ_o, we readily deduce that

$$\frac{1}{\lambda_o^2} = \frac{1}{k^2} - \frac{1}{\gamma_g^2} \tag{10}$$

or on setting $i\lambda = \alpha_o + i\beta_o$ we see that

$$\frac{\lambda_o}{k} \doteq \frac{\beta_o - i\alpha_o}{k} = \left[1 - \frac{k^2}{\gamma_g^2}\right]^{-1/2} \tag{11}$$

Here α_o is the attenuation rate in nepers/m and β_o is the phase in radians/meter. If $|k/\gamma_g|^4 \ll 1$ we can write

$$\frac{\beta_o - i\alpha_o}{k} \simeq 1 + \frac{k^2}{2\gamma_g^2} + \frac{3k^4}{8\gamma_g^4} \tag{12}$$

In this case $\gamma_g^2 \simeq i2/\delta_g^2$ or $\gamma_g \simeq (1+i)/\delta_g$. Then we find that the normalized attenuation rate is

$$\alpha_o/k \simeq k^2\delta_g^2/4 \tag{13}$$

while the normalized phase is

$$\beta_o/k \simeq 1 - \frac{3k^4\delta^4}{32} \tag{14}$$

Thus the phase velocity ω/β_o is slightly greater than c. That is, the Zenneck wave is a fast wave!

The Zenneck Wave

SURFACE IMPEDANCE BOUNDARY

In the context of the above discussion it is useful to consider the type of wave that can be supported by a boundary with a fixed surface impedance Z_s. The "matching condition" is clearly

$$Z_s = u_o/(i\varepsilon_o\omega) \tag{15}$$

Again we write $u_o = -\alpha + i\beta$ where α and β are real. Furthermore, we stipulate that $Z_s = R_s + iX_s$ where R_x and X_s are the specified surface resistance and reactance of the boundary. Thus, we require that

$$(-\alpha + i\beta)/(i\varepsilon_o\omega) = R_s + iX_s \tag{16}$$

so that

$$\alpha = \varepsilon_o\omega X_s \quad \text{and} \quad \beta = \varepsilon_o\omega R_s \tag{17}$$

It is now a simple matter to solve for the transverse wave number λ_o from

$$\lambda_o^2 = u_o^2 + k^2 = (-\alpha + i\beta)^2 + k^2 \tag{18}$$

Indeed, if $i\lambda_o = \alpha_o + i\beta_o$ where α_o and β_o are real, it is algebraically simple to deduce that

$$2\begin{pmatrix}\alpha_o^2\\ \beta_o^2\end{pmatrix} = \mp(\alpha^2 - \beta^2 + k^2) + [(\alpha^2+\beta^2)^2 + 2k^2(\alpha^2 - \beta^2) + k^4]^{1/2} \tag{19}$$

where the $-$ sign is to be used for α_o and the $+$ sign for β_o. As a check, we can confirm that $\alpha_o\beta_o = \alpha\beta$. For the special case $\alpha = \beta$, (i.e., $R_s = X_s$), we see that

$$\begin{pmatrix}\alpha_o^2\\ \beta_o^2\end{pmatrix} = \frac{\mp k^2 + (4\alpha^4 + k^4)^{1/2}}{2} = \frac{k^2}{2}\left[\mp 1 + \left(1 + \frac{4\alpha^4}{k^4}\right)^{1/2}\right] \tag{20}$$

which is still formally exact! But now if $\alpha \ll k$,

$$\alpha_o/k \simeq \alpha^2/k^2 \tag{21}$$

and

$$\beta_o/k \simeq 1 + \alpha^4/(2k^4). \tag{22}$$

If we now consider a conducting half-space in this context, we would set

$$R_s \simeq X_s \simeq [\mu_o\omega/(2\sigma_g)]^{1/2} \tag{23}$$

and then

$$\alpha \simeq \beta \simeq \varepsilon_o\omega[\mu_o\omega/(2\sigma_g)]^{1/2} \simeq k^2\delta_g/2 \tag{24}$$

This result shows that attenuation rate as given by (21) is the same as that predicted by (13) based on the "more exact" mode equation for the same problem. Also, to the same order, the phase factor $\beta_o \simeq k$ according to both (22) and (14) but the higher order corrections do not agree. This is expected because in writing down (16) the dependence of R_s and X_s on the propagation constant of the wave has been neglected.

THE INDUCTIVE BOUNDARY

An important limiting case of the preceding development is the case where the surface impedance is highly inductive (i.e., $R_s = 0$ and $X_s > 0$). Then clearly $\alpha = \varepsilon_o \omega X_s$ and $\beta = 0$. Then from (18) we see that

$$\lambda_o = (\alpha^2 + k^2)^{1/2} = \beta_o$$

$\alpha_o = 0$ and $u_o = -\alpha = -\varepsilon_o \omega X_s$. Thus, for $z > 0$, the wave has the form

$$H_y \simeq \text{const.} \times \exp(-\varepsilon_o \omega X_s z) \exp[-i(\alpha^2 + k^2)^{1/2} x].$$

Clearly this is a trapped surface wave; it is confined to a region just above the ground plane; it has a phase velocity *less* than c, and it is unattenuated in the horizontal direction.

REFERENCES

BARLOW, H.E.M., and BROWN, J., (1962), *Surface Waves*, Oxford University Press.

WAIT, J.R., (1957), Excitation of surface waves on conducting, stratified and corrugated surfaces, *Jour. Res. NBS*, **59**, 365-377.

EXERCISE: Solve equ. (10) for λ_o when displacement currents are non negligible. Compare your result with that predicted by equ. (18) when the value of Z_s is either the normal incidence or the appropriate grazing incidence value.

6
Excitation of the Layered Half-Space —The Green's Function

In dealing with the interaction of sources with stratified media, we need to incorporate a spectrum of plane waves. The simplest model here is a two-dimensional current distribution. The latter may be electric or magnetic; here we will deal with the magnetic type because this is a convenient model to consider the excitation of vertically polarized surface waves over the earth [e.g., see King and Wait, 1976].

We begin with a postulated aperture distribution at $z = h$ over a planar layered media that occupies the space $z < 0$ as indicated in Fig. 1. The magnetic current $m(x)$ is defined as follows

$$H_y(z = h - 0) - H_y(z = h + 0) = 0$$
$$E_x(z = h - 0) - E_x(z = h + 0) = m(x) \quad (1)$$

Now we assume the following Fourier transform exists:

$$M(\lambda) = \int_{-\infty}^{+\infty} m(x) e^{+i\lambda x} \, dx \quad (2)$$

then the spectral representation for the source current distribution is given by the inverse transform

$$m(x) = \frac{1}{2\pi} \int_{-\infty}^{+\infty} M(\lambda) e^{-i\lambda x} \, d\lambda \quad (3)$$

Here λ, the transform variable, can be envisaged as the transverse wave number.

Using some hindsight, this suggests we use the following spectral form for the resultant magnetic field in the free space region:

$$H_y = \int_{-\infty}^{+\infty} f(\lambda) [e^{u_0 z} + R e^{-u_0 z}] e^{-i\lambda x} \, d\lambda \quad ; \quad 0 < z < h \quad (4)$$

$$= \int_{-\infty}^{+\infty} g(\lambda) e^{-u_0 z} e^{-i\lambda x} \, d\lambda \quad ; \quad z > h \quad (5)$$

where $u_0 = (\lambda^2 - k^2)^{1/2}$, $k = \omega/c = (\varepsilon_0 \mu_0)^{1/2} \omega$. Here R is the appropriate reflection coefficient for TM(transverse magnetic) or parallel incidence; thus,

$$R = (K_o - Z)/(K_o + Z) \tag{6}$$

where $K_o = u_o/i\varepsilon_o\omega$ and Z is the surface impedance defined by

$$Z = -E_x/H_y\big|_{z=0} \tag{7}$$

In general, Z may be a function of λ but in any case it is considered to be specified in the present context.

The electric field components corresponding to (4) and (5) are obtained from

$$E_x = -\frac{1}{i\varepsilon_o\omega}\frac{\partial H_y}{\partial z} \tag{8} \quad \text{and} \quad E_z = \frac{1}{i\varepsilon_o\omega}\frac{\partial H_y}{\partial x} \tag{9}$$

The aperture conditions (1) can be applied to yield the algebraic pair

$$f[e^{u_o h} + Re^{-u_o h}] - ge^{-u_o h} = 0 \tag{10}$$

and

$$-K_o f[e^{u_o h} - Re^{-u_o h}] - Kge^{-u_o h} = M(\lambda)/2 \tag{11}$$

These are solved to give

$$f(\lambda) = -\frac{M(\lambda)}{4\pi K_o} e^{-u_o h} \tag{12}$$

and

$$g(\lambda) = -\frac{M(\lambda)}{4\pi K_o}[e^{u_o h} + Re^{-u_o h}] \tag{13}$$

This is the formal solution of the problem as specified.

A Green's function for the present problem can be obtained by considering that the source is

$$m(x) = m_o \delta(x - x') \tag{14}$$

where $\delta(x)$ is the unit impulse function at $x = x'$. Then clearly $M(\lambda) = m_o e^{i\lambda x'}$. The corresponding field is

$$H_y = -\frac{i\varepsilon_o\omega}{2\pi} G(z,z',x - x') m_o \tag{15}$$

where

$$G(z,z',x - x') = \frac{1}{2}\int_{-\infty}^{+\infty} \frac{e^{u_o z'} + Re^{-u_o z'}}{u_o} e^{-u_o z} e^{-i\lambda(x-x')} d\lambda \tag{16}$$

for $z > z'$. Here we have replaced h by z'. Also we note that

$$G(z,z',x - x') = \frac{1}{2}\int_{-\infty}^{+\infty} \frac{e^{-u_o z'}}{u_o}[e^{u_o z} + Re^{-u_o z}] e^{-i\lambda(x-x')} d\lambda \tag{17}$$

Excitation of the Layered Half-Space

for $z' > z$. Of course (17) is obtained from (16) by merely interchanging z' and z.

The Green's function G is the basic building block for a general two-dimensional magnetic current distribution $m(x',z')$. Then the resultant magnetic field is obtained by superposition

$$H_y = -i\varepsilon_0\omega \int_{-\infty}^{+\infty}\int_{-\infty}^{+\infty} m(x',z')G(z,z';x-x')dx'dz' \tag{18}$$

To provide insight to the problem, we assume that the ground is perfectly conducting (i.e., $Z = 0$) and thus $R = +1$. The Green's function given by (16) for $z > z'$ then becomes

$$G = \frac{1}{2}\int_{-\infty}^{+\infty} \frac{e^{u_0 z'} + e^{-u_0 z'}}{u_0} e^{-u_0 z} e^{-i\lambda(x-x')} d\lambda \tag{19}$$

The integral is now a standard form [Campbell and Foster, 1949] and may be written

$$G = K_0\left[ik[(x-x')^2 + (z-z')^2]^{1/2}\right] + \left[K_0\ ik[(x-x')^2 + (z+z')^2]^{1/2}\right] \tag{20}$$

where K_0 is the MacDonald function or modified Bessel function of the second kind. When $k|x - x'| \gg 1$, i.e., large electrical range, we can use the leading term for the asymptotic series for the MacDonald function. Thus

$$K_0\left[ik[(x-x')^2 + (z \mp z')^2]^{1/2}\right] \tag{21}$$
$$\simeq (\pi/2)^{1/2}[x-x')^2 + (z \mp z')^2]^{-1/4} \exp\ -ik[(x-x')^2 + (z \mp z')^2]^{1/2}$$

corresponding to the source $(-)$ and image $(+)$ contributions respectively. Furthermore, if we set

$$[(x-x')^2 + (z \mp z')^2]^{1/2} = R_{\mp}$$

then R_- and R_+ are the linear distances from the source at (x',z') and the image at $(x',-z')$ to the observer at x,z.

In the case of the "image" term we note that

$$ikR_+ = i\bar{\lambda}(x-x') + \bar{u}_0(z+z') \tag{22}$$

where $\bar{\lambda} = k \sin \bar{\theta}$ and $\bar{u}_o = ik \cos \bar{\theta}$ where $\sin \bar{\theta} = z + z'/R_+$ and $\cos \bar{\theta} = x - x'/R_+$. Here $\bar{\theta}$ is the local angle of reflection. Consequently we can think of $\bar{\lambda}$ as the predominant wave number for the reflected wave. The situation is illustrated in Fig. 2.

Using the above as a basis, we can now obtain (non-rigorously) the following asymptotic corrections for the reflected wave to account for the finite conductivity of the ground plane

Thus we write
$$G(z,z';x - x') = K_o[ikR_-] + \hat{G} \qquad (23)$$
where
$$G = \frac{1}{2} \int_{-\infty}^{+\infty} R(\lambda) e^{-u_o(z+z')} e^{-i\lambda(x-x')} d\lambda \qquad (24)$$

We now express $R(\lambda)$ as a Taylor series about the predominant value $\bar{\lambda}$ in the manner
$$R(\lambda) = R(\bar{\lambda}) + (\lambda - \bar{\lambda}) R'(\bar{\lambda}) + (\lambda - \bar{\lambda})^2 2^{-1} R''(\bar{\lambda}) + \ldots \qquad (25)$$

Here we note that in an operational sense,
$$\lambda - \bar{\lambda} = i(\partial/\partial x) - \bar{\lambda} \qquad (26)$$

This suggests that
$$G \simeq R(\bar{\lambda}) \left\{ K_o[ikR_+] + \left[i\frac{\partial}{\partial x} - \bar{\lambda} \right] K_o[ikR_+] \right.$$
$$\left. + 2^{-1} \left[i\frac{\partial}{\partial x} - \bar{\lambda} \right]^2 K_o[ikR_+] + \ldots \right\} \qquad (27)$$

This is an asymptotic development that is strictly valid only when $R_+ \to \infty$ (i.e., far field). It presupposes that $R(\lambda)$ is reasonably well behaved in the narrow cone of angles in the vicinity of $\bar{\theta}$.

The above development suggests that the leading term for G is $(x - x')^{-1/2}$ for large horizontal ranges. However, near grazing angles (when $Z \neq 0$) the reflection coefficient $R(\bar{\lambda})$ will approach -1 and then the source and direct image term would clearly cancel. The first asymptotic correction term then behaves as $(x - x')^{-3/2}$. This leads to the expected "ground wave" behavior for H.F. vertically polarized waves over an imperfectly conducting flat earth.

Obviously this approximate method cannot be applied to the case where Z is purely inductive, nor is the result valid in the near field of an aperture. But the general Green's function form can be used as a starting point for a more refined asymptotic evaluation.

REFERENCES

CAMPBELL, G. and FOSTER, R.M., (1949), *Fourier Integrals for Practical Application*, Van Nostrand, New York.

KING, R.J. and WAIT, J.R., (1976), Electromagnetic ground wave propagation - theory and experiment, *Symposia Mathematica*, 18, 107-208 (Academic Press).

EXERCISE: Obtain an expression analogous to (18) that would be applicable to an electric current source of density $J_y(x',z')$ located over the ground surface. Show that the Green's function for the case $Z = 0$ is the same as eq. (20) with the + sign preceding the second K_o function becoming a - sign.

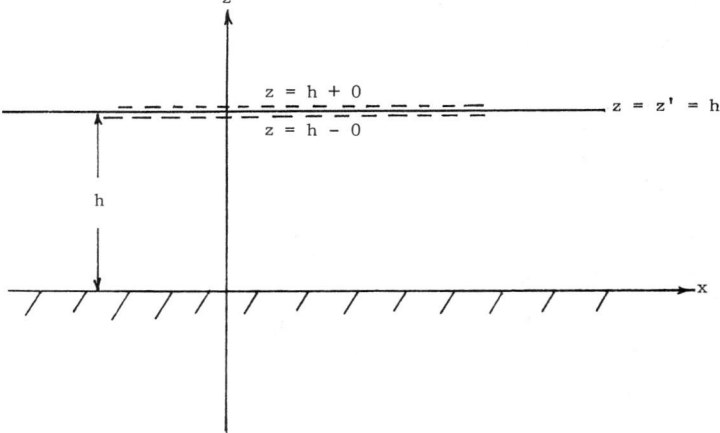

Fig. 1. A y directed magnetic current is located at height h over the layered half-space.

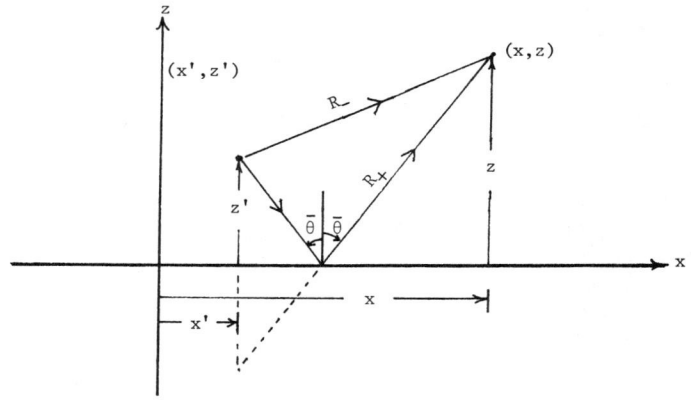

Fig. 2. Illustrating the reflected wave for the line magnetic source.

7
On the Excitation of the Zenneck Surface Wave

Zenneck [1907] showed many years ago that a surface wave, which is a solution to Maxwell's equations, could be supported by a planar interface separating two homogeneous media. When the upper medium is air and the lower medium is a dissipative ground, this Zenneck surface wave is characterized by a phase velocity greater than that of light and an attenuation in the direction of propagation along the interface. Furthermore, the wave is attenuated with height above the surface. For several decades, a controversy ensued regarding the significance of the radial Zenneck wave and surface wave in the solution given by Sommerfeld [1926] for the field of a vertical dipole over the earth's surface. We will not discuss the controversy here since extensive reviews are given elsewhere [Wait 1964, Banos 1966, King and Wait 1976]. Nonetheless, as pointed out by Goubau [1951], the Zenneck wave is orthogonal to the radiation field and therefore it should be possible to excite it.

In order to examine the excitation problem quantitatively [Hill and Wait, 1978], we have derived the fields of a vertical aperture (of either finite or infinite extent) which is modelled by a magnetic current sheet over a flat, homogeneous earth. A two-dimensional model was assumed for simplicity, but the results are relevant to the practical three dimensional case. Here we briefly review these results and present a further calculation at 10 MHz of the total field for a finite aperture.

The geometry of the vertical aperture over a lossy earth is shown in Figure 1. The region $z > 0$ is free space characterized by permittivity ε_o and permeability μ_o. The y-directed magnetic sheet current $m_y(z')$ is located in the plane $x = 0$ and is uniform in the y direction. The TM

(Transverse Magnetic) fields are also independent of y, and the nonzero components are H_y, E_x, and E_z. The time dependence is $\exp(i\omega t)$. E_x and E_z can be derived from H_y by

$$E_x = \frac{-1}{i\omega\varepsilon_o} \frac{\partial H_y}{\partial z} \quad \text{and} \quad E_z = \frac{1}{i\omega\varepsilon_o} \frac{\partial H_y}{\partial x} \qquad (1)$$

For simplicity, the earth is characterized by a surface impedance Z such that

$$E_x\big|_{z=0} = ZH_y\big|_{z=0} \qquad (2)$$

A normalized surface impedance Δ can also be defined as

$$\Delta = Z/(\mu_o/\varepsilon_o)^{\frac{1}{2}} \simeq [(\varepsilon_g - i\sigma_g/\omega)/\varepsilon_o]^{-\frac{1}{2}} \qquad (3)$$

where ε_g and σ_g are the permittivity and conductivity of the ground, respectively.

The magnetic sheet current can be viewed as a superposition of magnetic line sources, each with a strength of $m_y(z')dz'$. The fields of a y-directed magnetic line source over an impedance boundary are well known[8] and the magnetic field produced by the current sheet is thus given by the following integral:

$$H_y = \int_0^\infty m_y(z')G(z,z')dz' , \qquad (4)$$

where

$$G(z,z') = \frac{-i\omega\varepsilon_o}{4\pi} \int_{-\infty}^{\infty} \{\exp(-u|z-z'|) + R_{\shortparallel}\exp[-u(z+z')]\} u^{-1}\exp(-i\lambda x)d\lambda, \qquad (5)$$

and where

$$R_{\shortparallel} = \frac{u - ik_o\Delta}{u + ik_o\Delta}, \quad u = (\lambda^2 - k_o^2)^{\frac{1}{2}}, \text{ and } k_o = \omega(\mu_o\varepsilon_o)^{\frac{1}{2}} .$$

The z variation of the Zenneck wave is given by $\exp(-\Gamma_z z)$ where $\Gamma_z = -ik_o\Delta$. Thus we choose the magnetic sheet current $m_y(z')$ to have the same z' variation. That is

On the Excitation of the Zenneck Surface Wave

$$m_y(z') = M_o \exp(-\Gamma_z z'), \qquad (6)$$

where M_o is a constant with units of volts/m. Note from (3) that Γ_z is complex but always has a positive real part which results in an exponential decay of m_y with height. By substituting (6) into (4), we may obtain the following[7]:

$$H_y = \frac{-i\omega\varepsilon_o M_o}{2\pi} \exp(-\Gamma_z z) \int_{-\infty}^{\infty} \frac{\exp(-i\lambda x)}{\lambda^2 - k_o^2 - \Gamma_z^2} d\lambda \qquad (7)$$

The only singularities of the integrand are simple poles at $\pm\lambda_z$ where

$$\lambda_z = (k_o^2 + \Gamma_z^2)^{\frac{1}{2}} = k_o(1 - \Delta^2)^{\frac{1}{2}} \qquad (8)$$

In fact, we can identify λ_z as the horizontal wavenumber of the Zenneck surface wave. From (3) and (8), we can see that the real part of λ_z is less than k_o (fast wave behavior) and the imaginary part is negative for $\sigma_g \neq 0$. The integral in (7) can now be evaluated using the residue theorem. For positive x the contour is closed in the negative λ plane and, for negative x, the contour is closed in the positive λ plane. The result for H_y is

$$H_y = \frac{-\omega\varepsilon_o M_o}{2\lambda_z} \exp(-\Gamma_z z - i\lambda_z |x|). \qquad (9)$$

Thus the infinite Zenneck aperture excites a pure Zenneck wave with no radiation field. This is in accord with Goubau's [1951] results.

We now consider the case of a finite vertical aperture which is of more practical interest. The formulation here is analogous to earlier work by Cullen [1954] and Brown [1959] on the excitation of surface waves over purely reactive surfaces. In their case, only a small aperture is needed to efficiently launch the trapped surface wave on the structure.

For an aperture of finite height z_o, we set

$$m_y(z') = \begin{cases} M_o \exp(-\Gamma_z z'), & z' < z_o \\ 0, & z' > z_o \end{cases} \tag{10}$$

The resultant expression for H_y, for $z < z_o$, is

$$H_y = \frac{-i\omega\epsilon_o M_o}{4\pi} I_f, \tag{11}$$

where

$$I_f = \frac{-2\pi i}{\lambda_z} \exp(-\Gamma_z z - i\lambda_z |x|)$$

$$- \exp(-\Gamma_z z_o) \{ \int_{-\infty}^{\infty} u^{-1}(u+\Gamma_z)^{-1} \exp[-u(z_o-z) - i\lambda x] d\lambda \tag{12}$$

$$+ \int_{-\infty}^{\infty} u^{-1}(u-\Gamma_z)^{-1} \exp[-u(z_o+z) - i\lambda x] d\lambda \}$$

From (12) we see that the limit of the infinite aperture is easily obtained by letting z_o approach ∞. Then the coefficient of the integral terms vanishes, and the first term, which is the Zenneck surface wave, remains. For arbitrary values of z_o, z, and x, the integral (12) must be evaluated. This may be performed by using the modified saddle point method to yield[7]

$$I_f \simeq \frac{i\pi \exp(-\Gamma_z z_o)}{k_o} [\exp(-ik_o R_+ - w_+) \mathrm{erfc}(iw_+^{\frac{1}{2}}) \tag{13}$$

$$- \exp(-ik_o R_- - w_-) \mathrm{erfc}(iw_-^{\frac{1}{2}})] .$$

where

$$w_- = p_- \left(1 - \frac{z_o - z}{\Delta R_-}\right)^2, \quad w_+ = p_+ \left(1 + \frac{z_o + z}{\Delta R_+}\right)^2$$

and

$$p_- = -ik_o R_- \Delta^2/2, \quad p_+ = -ik_o R_+ \Delta^2/2$$

$$R_- = [(z_o - z)^2 + x^2]^{\frac{1}{2}}, \quad R_+ = [(z_o + z)^2 + x^2]^{\frac{1}{2}}$$

On the Excitation of the Zenneck Surface Wave

where erfc denotes the error function complement. The only restrictions on (13) are that $k_o x$ must be large and $|\Delta|$ and $(z_o + z)/x$ must be small.

To deal with the small aperture, we consider that z_o is small and z equals zero (i.e., observer on the interface). Then (13) reduces to

$$I_f \simeq 2z_o \left(\frac{2\pi}{ik_c x}\right)^{1/2} \exp(-ik_o x) F(p) \qquad (14)$$

where

$$F(p) = 1 - i(\pi p)^{1/2} \exp(-p) \operatorname{erfc}(ip^{1/2}) \qquad (15)$$
$$\text{and } p = -ik_o x \Delta^2/2$$

Here $F(p)$ is the well-known Sommerfeld attenuation function which accounts for earth conduction [Norton 1936, 1937]. For $p = 0$, we note that $F(p)$ equals unity.

As indicated by (9), an infinite vertical aperture will excite a pure Zenneck wave with exponential attenuation in x. However, a small aperture excites the usual ground wave as indicated by (14) with an $x^{-3/2}$ variation for large distances. Here we present numerical results for various aperture heights using the expression for H_y given by (11) and (13) for average ground at 10 MHz. In this case Δ is approximately $0.186 + i0.134$. The 1/e height decay of the Zenneck wave is 35.5m, and the 1/e horizontal propagation decay is $x \simeq 189$m. Thus the Zenneck wave is not significant for long distance propagation over land at 10 MHz even if it could be excited. The quantity plotted is actually $|H_y/H_{yo}|$, where

$$H_{yo} = \frac{-i\omega\epsilon_o M_o}{2\pi} \cdot 2 K_o(ik_o x) \int_o^{z_o} \exp(-\Gamma_z z') dz'$$

$$\simeq \frac{-i\omega\epsilon_o M_o}{2\pi} \left(\frac{2\pi}{ik_o x}\right)^{\frac{1}{2}} \exp(-ik_o x) \left[\frac{1 - \exp(-\Gamma_z z_o)}{\Gamma_z}\right] \quad (16)$$

H_{yo} can be considered the field of a line source located at the surface of a perfectly conducting earth. The strength of the line source is given by the integrated aperture current. Since this normalization removes the $x^{-\frac{1}{2}}$ cylindrical nature of the field, the results are approximately applicable to the three dimensional problem of an axial Zenneck wave excitation [Wait, 1964]. The curve for the small value of $z_o(=1m)$ is approximately that predicted by the Sommerfeld attenuation function $F(p)$ in (15). The phase is not shown, but it also agrees with Norton's [1936, 1937] phase data. With this type of normalization, the Zenneck wave is seen to be most effective for ranges near its 1/e value. For larger values of x, it suffers from exponential attenuation. For smaller values, a simple localized source (small z_o) is adequate. Note that for large values of x, curves for finite z_o approach the x^{-1} behavior. This is consistent with the asymptotic behavior that $F(p) \to -1/(2p)$ as $|p| \to \infty$.

As indicated in Figure 2, it is apparently in the intermediate range (numerical distance $|p| = k_o x |\Delta^2|/2 \simeq 1$) where the Zenneck wave is most significant. For larger distances, the ultimate exponential decay of the Zenneck wave takes it out of contention. However, it should be noted that the intermediate range can be quite large for frequencies in the vicinity of 1 MHz or for even higher frequencies for propagation over sea water. The primary disadvantage for the vertical aperture excitation is that the slow vertical decay of the Zenneck wave dictates the necessity of an aperture of very great vertical extent.

The corresponding theory for the spherical earth has been used to perform similar calculations for both vertical and horizontal transmitting antennas [Wait and Hill, 1979]. The quantitative results are different, but the general conclusions remain unchanged.

REFERENCES

BANOS, A., (1966), *Dipole Radiation in the Presence of a Conducting Half-Space*, Section 4.10, Pergamon Press, Oxford.

BROWN, J., (1959), Some theoretical results for surface wave launchers, *EEE Trans. Antennas Propagat.*, AP-7, S169-S174.

CULLEN, A.L., (1954), The excitation of plane surface waves, *Proc. IEE*, 101, pp. 271-277.

GOUBAU, G., (1951), Ueber die Zenneck Bodenwelle, *Zeits. ange. Physik*, 3, 103-107.

HILL, D.A. and WAIT, J.R., (1978), Excitation of the Zenneck surface wave by a vertical aperture, *Radio Science*, 13, No. 6, 969-977.

KING, R.J., and WAIT, J.R., (1976), Electromagnetic groundwave propagation - theory and experiment, *Symposia Mathematica*, 18, 107-208, (Academic Press).

NORTON, K.A., (1936, 1937), The propagation of radio waves over the surface of the earth and in the upper atmosphere, *Proc. IRE*, 24, 1367-1387 and 1203-1236.

SOMMERFELD, A.N., (1926), The propagation of waves in wireless telegraphy, *Ann. Phys.*, 81, 1135-1153.

WAIT, J.R., (1964), Electromagnetic surface waves, *Advances in Radio Research*, (Academic Press, London), 1, 157-217.

WAIT, J.R., (1970), *Electromagnetic Waves in Stratified Media*, Chapter II, Pergamon Press, Oxford, (2nd Edition).

WAIT, J.R., and HILL, D.A., (1979), Excitation of the H.F. surface wave by vertical and horizontal antennas, *Radio Science*, 14, No. 3.

ZENNECK, J., (1907), Ueber die Fortpflanzung ebener electromagnetischer Wellen einer ebenen Leiterflache und ihre Beziehung zur drahtlosen Gelegraphie, *Ann. Phys.*, 23, 846-866.

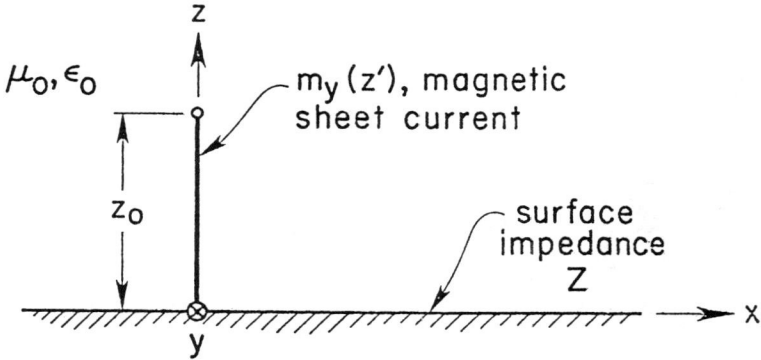

Fig. 1. Geometry for a vertical magnetic current sheet over a half space of surface impedance Z. The aperture height z_o can be either finite or infinite.

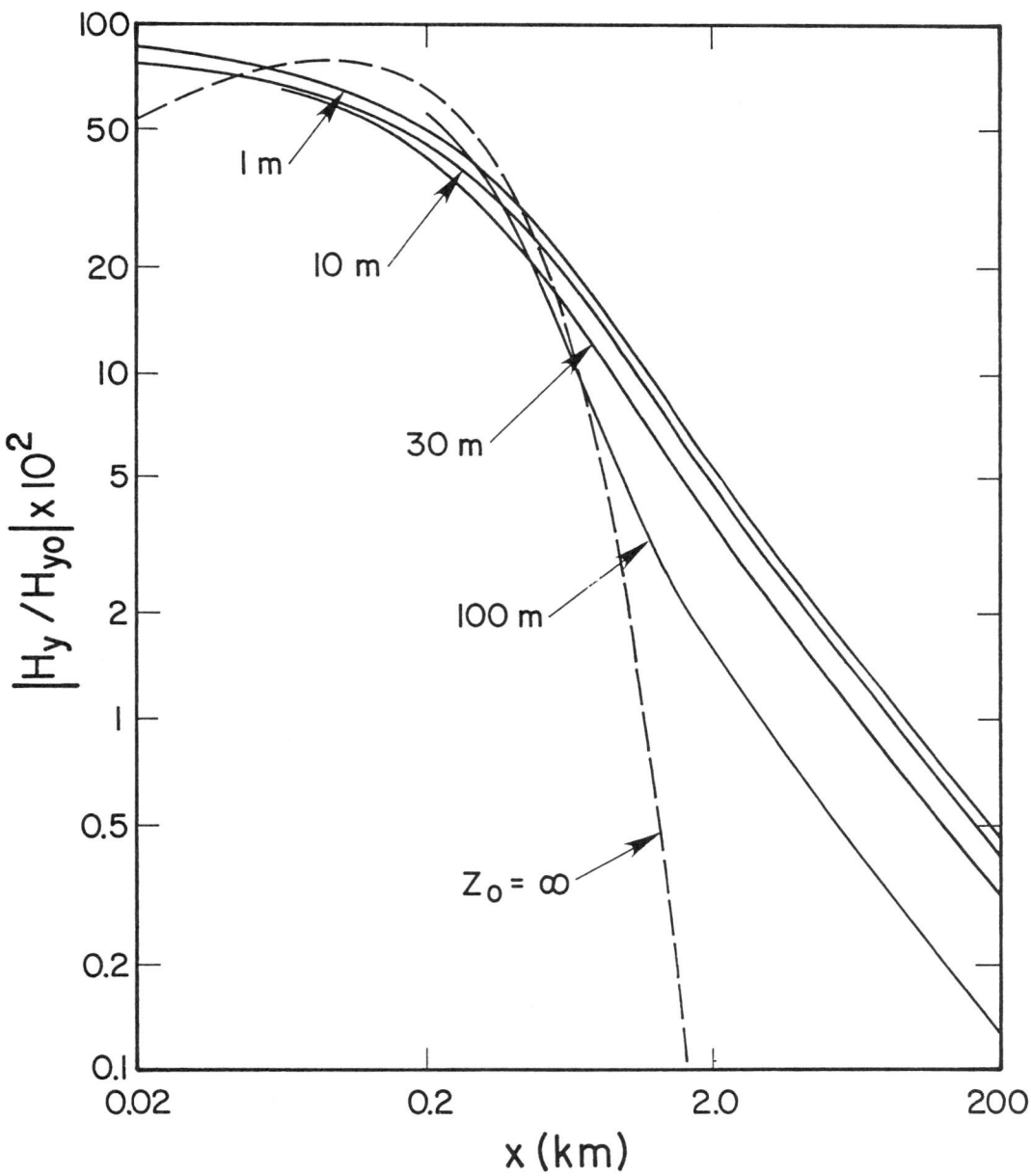

Fig. 2. The magnitude of the magnetic field normalized to the field of an equivalent line source over a perfect conductor. [$\sigma_g = 10^{-2}$ mhos/m, $\varepsilon_g/\varepsilon_o = 6$, and f = 10 MHz].

8
Surface Impedance of a Spherically Stratified Conductor

In many boundary-value problems involving waves in stratified media the solutions may be quickly obtained if the analogies with transmission line theory are exploited. In general, the analogous transmission line is nonuniform in the sense that the characteristic impedances may be different in the two directions on the line. Although Schelkunoff [1943] has discussed nonuniform transmission line theory, it seems worth while to include a short exposition of its essential features.

The starting point is the equations which connect the transverse voltage V between two parallel wires and the longitudinal current I in the lower wire. In terms of distance x down the line, these are

$$\frac{dV}{dx}=-ZI \quad \text{and} \quad \frac{dI}{dx}=-YV \qquad (1),\ (2$$

where Z is the distributed series impedance per unit length and Y is the distributed series admittance per unit length. Allowing Z and Y to be both functions of x one readily finds that

$$\frac{d^2V}{dx^2}-\frac{Z'}{Z}\frac{dV}{dx}-YZV=0. \qquad (3)$$

$$\frac{d^2I}{dx^2}-\frac{Y'}{Y}\frac{dI}{dx}-YZI=0. \qquad (4)$$

Second-order differential equations of this type possess two linearly independent solutions. The general solution is a linear combination of these solutions. For example

$$V(x)=AV^+(x)+BV^-(x) \qquad (5)$$

where A and B are independent of x and V^+ and V^- are the fundamental wave functions. Similarly,

$$I(x)=AI^+(x)+BI^-(x) \qquad (6)$$

where I^+ and I^- are the corresponding fundamental current wave functions. It is evident that the wave functions individually satisfy eqs (1) and (2).

A characteristic wave impedance may now be associated with each pair of wave functions; thus

$$K^+(x)=\frac{V^+(x)}{I^+(x)}=-\frac{1}{YI^+}\frac{dI^+}{dx}=-\frac{ZV^+}{dV^+/dx} \qquad (7)$$

and

$$K^-(x)=-\frac{V^-(x)}{I^-(x)}=\frac{1}{YI^-}\frac{dI^-}{dx}=\frac{ZV^-}{dV^-/dx}. \qquad (8)$$

It is a convenient physical artifice to consider the waves associated with K^+ as propagating in the positive x direction and those associated with K^- as waves propagating in the negative x direction. In the case of a uniform line where Z and Y are constant, the meaning of these wave functions is quite clear. In the presence of any local nonuniformity, a reflected wave would be generated and the individual wave functions no longer are purely propagating. Nevertheless, in many cases, the elementary wave functions bear considerable resemblance to "traveling" waves. For example, this happens when Z and Y are slowly varying functions of x.

(Reprinted, in part in revised form, from J.R. Wait, N.B.S. Jour. Res., 61, 205-232, 1958)

Surface Impedance of A Spherically Stratified Conductor

For present purposes the label progressive is used to describe a fundamental wave on a nonuniform line. In the limiting case of a uniform line the progressive waves become traveling waves.

It is important to note that there is some arbitrariness in the selection of the wave functions on nonuniform lines. The choice is usually made on the basis of convenience.

A useful quantity is the ratio of the wave functions at two points x_1 and x_2 on the line. Thus, by definition,

$$\chi_V^+(x_1, x_2) = \frac{V^+(x_2)}{V^+(x_1)}, \qquad \chi_V^-(x_1, x_2) = \frac{V^-(x_2)}{V^-(x_1)} \qquad (9)$$

$$\chi_I^+(x_1, x_2) = \frac{I^+(x_2)}{I^+(x_1)}, \qquad \chi_I^-(x_1, x_2) = \frac{I^-(x_2)}{I^-(x_1)}. \qquad (10)$$

We are now in the position to study the reflection in nonuniform lines. For example, a semi-infinite nonuniform line is terminated in an impedance Z. If $Z = K^+$, the voltage associated with this incident wave is exactly equal to the voltage across the terminal impedance and thus the entire incident current wave flows through this impedance. Hence, no reflection occurs. However, if $Z \neq K^+$, the incident current wave is not completely absorbed and reflection occurs. If the incident "progressive" waves are characterized by the wave functions $V^+(x)$ and $I^+(x)$, the problem is to calculate the reflection coefficient at the terminal impedance. These are defined by

$$q_I = \frac{I^-}{I^+} \quad \text{and} \quad q_V = \frac{V^-}{V^+} \qquad (11)$$

for current and voltage, respectively. Now,

$$V_t = V^+ + V^- \quad \text{and} \quad I_t = I^+ + I^- \qquad (12)$$

where V_t and I_t are the voltage current at the terminals of Z_t. Furthermore,

$$V^+ = K^+ I^+, \; V^- = K^- I^-, \; V_t = Z_t I_t. \qquad (13)$$

The latter two sets of equations are readily solved to give

$$q_V = \frac{M^- - Y_t}{M^- + Y_t} \quad \text{where} \quad M^\pm = \frac{1}{K^\pm} \quad \text{and} \quad Y_t = \frac{1}{Z_t} \qquad (14)$$

and

$$q_I = \frac{K^+ - Z_t}{K^- + Z_t}. \qquad (15)$$

A basic problem in nonuniform lines is to form wave functions $V(x)$ and $I(x)$ at $x = x_1$ in terms of the specified impedance at $x = x_2$. For convenience, $x_2 > x$. Using the preceding conventions one may easily write

$$V(x) = V^+(x) + V^+(x_2) q_V(x_2) \frac{V^-(x)}{V^-(x_2)} \qquad (16)$$

and

$$I(x) = I^+(x) + I^+(x_2) q_I(x_2) \frac{I^-(x)}{I^-(x_2)}. \qquad (17)$$

Here V^+ and I^+ are regarded as the "incident" progressive wave and V^- and I^- are the "reflected" progressive wave.

The impedance at the point $x = x_1$ is then given by

$$Z(x_1)=\frac{V(x_1)}{I(x_1)}=K^+(x_1)\frac{1+q_V(x_2)\chi_V^+(x_1,x_2)\chi_V^-(x_2,x_1)}{1+q_i(x_2)\chi_I^+(x_1,x_2)\chi_I^-(x_2,x_1)}. \qquad (18)$$

This result is most useful in reflection-type problems.

To complete this very brief survey of nonuniform transmission line theory, the corresponding formulas for transmission coefficients are also given.

The transmission coefficients at impedance discontinuity are defined by

$$p_V=\frac{V_t}{V^+} \text{ and } p_I=\frac{I_t}{I^+} \qquad (19)$$

being analogous to the previous definitions of q_V and q_I. Then from eqs (12) and (13),

$$p_V=\frac{M^-+M^+}{M^-+Y_t} \qquad (20)$$

and

$$p_I=\frac{K^-+K^+}{K^-+Z_t}. \qquad (21)$$

As expected, these coefficients become unity when $Z_t=K^+$ (or $Y_t=M^+$).

The transmission coefficient across a section (x_1,x_2) of *another* line inserted between the original line and impedance Z_t is readily obtained. The result, given by Schelkunoff [1943], is

$$T=p(1+q\chi+q^2\chi^2+\ldots)\chi^+(x_1,x_2)$$

$$=\frac{p}{1-q\chi}\chi^+(x_1,x_2) \qquad (22)$$

where

$$p=p^+(x_1)p^+(x_2),\ q=q^-(x_1)q^+(x_2) \qquad (23)$$

and $\chi=\chi^+(x_1,x_2)\chi^-(x_2,x_1)$. The result holds for both current and voltage waves which explains the absence of subscripts on I and V. The physical meaning of the results is very clear: p is the product of the two transmission coefficients at the discontinuities x_1 and x_2, q is the product of the reflection coefficient for a progressive wave incident from the right on the junction x_1, and the reflection coefficient for a progressive wave incident from the left. (Here x increases towards the right.) In the absence of multiple reflections T would be equated to $p\chi^+(x_1,x_2)$. Thus, the factor $(1+q\chi+q^2\chi^2+\ldots)$ represents the influence of multiple reflections.

We now proceed to apply nonuniform transmission line theory to a spherically stratified medium. In particular, let us consider the earth (of outer radius a_1) as consisting of a homogeneous core of radius a_2 of electrical constants σ_1, ϵ_1, and μ_1 surrounded by a homogeneous mantle of electrical constants σ_2, ϵ_2, and μ_2. It is assumed that the sources of the field are completely exterior to the earth. Thus, within the concentric homogeneous regions the fields may be derived from two scalar potential functions U and V. The first set which are TM (transverse magnetic) are derivable from U and the second set are TE (transverse electric), derivable from V.

For TM waves

$$E_r=\frac{1}{\sigma+i\epsilon\omega}\left(\frac{\partial^2}{\partial r^2}-\gamma^2\right)(rU)\quad H_r=0 \qquad (24)$$

while for the TE waves

$$H_r=\frac{1}{i\mu\omega}\left(\frac{\partial^2}{\partial r^2}-\gamma^2\right)(rV)\quad E_r=0. \qquad (25)$$

Surface Impedance of A Spherically Stratified Conductor

Both U and V satisfy

$$r^2\frac{\partial^2 U}{\partial r^2}+\frac{1}{\sin\theta}\frac{\partial}{\partial\theta}\left(\sin\theta\frac{\partial U}{\partial\theta}\right)+\frac{1}{\sin^2\theta}\frac{\partial^2 U}{\partial\phi^2}=\gamma^2 r^2 U \qquad (26)$$

as can be readily ascertained from Maxwell's equations.

Since any field may be expressed as a superposition of these two sets, it is sufficient to discuss them separately.

Using the standard separation-of-variables technique, it is assumed that

$$U=u(\theta,\phi)\hat{u}(r). \qquad (27)$$

On substituting this into eq (26) one finds that u and $\hat{u}$ must satisfy

$$\sin\theta\frac{\partial}{\partial\theta}\left(\sin\theta\frac{\partial u}{\partial\theta}\right)+\frac{\partial^2 u}{\partial\phi^2}+\nu(\nu+1)\sin^2\theta\, u=0 \qquad (28)$$

and

$$\frac{d^2\hat{u}}{dr^2}-\left[\gamma^2+\frac{\nu(\nu+1)}{r^2}\right]\hat{u}=0 \qquad (29)$$

where ν is a constant.

The u's are Legendre functions and the general form may be expressed as

$$u(\theta,\phi)=\sum_m\int[F_m(\nu)P_\nu^m(\cos\theta)+G_m(\nu)P_\nu^m(-\cos\theta)]\,d\nu\times[F_m\cos m\phi+g_m\sin m\phi] \qquad (30)$$

where the integration contour is suitably chosen in the complex ν plane and m is an integer. The constants $F_m(\nu)$, $G_m(\nu)$, f_m, and g_m depend on the nature of the source field. For example, if the source is a vertical electric dipole at $\theta=0$, $F_m(\nu)=0$, and $g_m=0$. Also, for $m\neq 0$, $f_m=0$. Thus

$$u=\int G_0(\nu)P_\nu(-\cos\theta)\,d\nu. \qquad (31)$$

It is known that integrals of this type can be deformed so as to enclose the poles of $G_0(\nu)$ and thus u can be represented as a series of residues. Therefore,

$$u\simeq 2\pi i\sum_{\nu_s}[\text{Residues of }G_0(\nu_s)]\,P_{\nu_s}(-\cos\theta) \qquad (32)$$

where ν_s are the solutions of $[G_0(\nu)]^{-1}=0$. Provided $|\nu_s|\gg 1$ and θ is not near 0 or π

$$P_{\nu_s}(-\cos\theta)\simeq\frac{\text{const}}{(\sin\theta)^{\frac{1}{2}}}\left[e^{-ika_1\theta S_s}+e^{i\pi/2}e^{-ika_1(2\pi-\theta)S_s}\right] \qquad (33)$$

where $\nu_s(\nu_s+1)\simeq\nu_s^2\simeq ka_1 S_s$. Here the angular function u has the physical character of two waves traveling in opposite directions around the cylinder. In analogy to waves on a flat surface, S_s can be interpreted as the sine of a complex angle of incidence where k is the wave number in free space.

It is clearly apparent from this simple example that ν is related to the azimuthal variation of the source field. In the general case, ν depends both on the longitudinal and latitudinal variations.

Solutions of the radial equation are conveniently expressed in the form

$$\hat{u}=A\hat{I}_\nu(\gamma r)+B\hat{K}_\nu(\gamma r) \qquad (34)$$

where $\hat{I}_\nu$ and $\hat{K}_\nu$ could be any two independent solutions of eq (29). Here $\gamma = [i\mu\omega(\sigma+i\epsilon\omega)]^{\frac{1}{2}}$ where Re $\gamma > 0$. For convenience $\hat{I}_\nu(\gamma r)$ is chosen to be a solution which is finite or zero at $r=0$ and $\hat{K}_\nu(\gamma r)$ is chosen to vanish as r tends to infinity. In terms of modified (cylindrical) Bessel functions

$$\hat{I}_\nu(z) = \left(\frac{\pi z}{2}\right)^{\frac{1}{2}} I_{\nu+1/2}(z) \qquad (35)$$

and

$$\hat{K}_\nu(z) = \left(\frac{2z}{\pi}\right)^{\frac{1}{2}} K_{\nu+1/2}(z) \qquad (36)$$

where I and K have their conventional meaning.

In the present problem, the waves associated with $\hat{I}_\nu(\gamma r)$ are regarded as being "incident," whereas the waves associated with $\hat{K}_\nu(\gamma r)$ are regarded as "reflected."

The surface impedance at $r=a_1$ is defined by

$$Z_\nu = \left[-\frac{E_\theta^\nu}{H_\phi^\nu}\right]_{r=a_1} \qquad (37)$$

which, in general, depends on ν. The medium between the limits $r=a_1$ and a_2 is now regarded as a nonuniform transmission line of length $l=a_1-a_2$. The transverse voltage V_T and the current I on the line are then analogous to the electric field $-E_\theta^\nu$ and the magnetic field H_ϕ^ν, respectively, where the superscript ν is to indicate the possible dependence on ν.

The characteristic impedance of the line looking inward is then

$$K^+(\gamma_1 r) = \eta_1 \frac{\hat{I}_\nu'(\gamma_1 r)}{\hat{I}_\nu(\gamma_1 r)} \qquad (38)$$

and the impedance looking outward is

$$K^-(\gamma_1 r) = -\eta_1 \frac{\hat{K}_\nu'(\gamma_1 r)}{\hat{K}_\nu(\gamma_1 r)}. \qquad (39)$$

In the above

$$\gamma_1 = [i\mu_1\omega(\sigma_1+i\epsilon_1\omega)]^{\frac{1}{2}}$$

and

$$\eta_1 = [i\mu_1\omega/(\sigma_1+i\epsilon_1\omega)]^{\frac{1}{2}}.$$

The line is now considered to be terminated by an impedance Z_t where

$$Z_t = \left[\eta_2 \frac{\hat{I}_\nu'(\gamma_2 r)}{\hat{I}_\nu(\gamma_2 r)}\right]_{r=a_2} \qquad (40)$$

From the analogy with transmission line theory, one readily finds that

$$Z_\nu = K^+(\gamma_1 a_1) \left[\frac{1+q_e \chi_e(a_2,a_1)\chi_e(a_1,a_2)}{1+q_h \chi_h(a_2,a_1)\chi_h(a_1,a_2)}\right] \qquad (41)$$

where

$$-q_e = \frac{(1/Z_t)-1/K^+(\gamma_1 a_1)}{(1/Z_t)+1/K^-(\gamma_1 a_1)}, \qquad (42)$$

$$-q_h = \frac{Z_t - K^+(\gamma_1 a_1)}{Z_t + K^-(\gamma_1 a_1)}, \qquad (43)$$

Surface Impedance of A Spherically Stratified Conductor

$$\overset{+}{\chi_e}(a_\mathbf{1}, a_\mathbf{2}) = \frac{a_1 \hat{I}'_\nu(\gamma_1 a_2)}{a_2 \hat{I}'_\nu(\gamma_1 a_1)}, \qquad (44)$$

$$\overset{-}{\chi_e}(a_\mathbf{2}, a_\mathbf{1}) = \frac{a_2 \hat{K}'_\nu(\gamma_1 a_1)}{a_1 \hat{K}'_\nu(\gamma_1 a_2)}, \qquad (45)$$

$$\overset{+}{\chi_h}(a_\mathbf{1}, a_\mathbf{2}) = \frac{a_1 \hat{I}_\nu(\gamma_1 a_2)}{a_2 \hat{I}_\nu(\gamma_1 a_1)}, \qquad (46)$$

$$\overset{-}{\chi_h}(a_\mathbf{2}, a_\mathbf{1}) = \frac{a_2 \hat{K}_\nu(\gamma_1 a_1)}{a_1 \hat{K}_\nu(\gamma_1 a_2)}. \qquad (47)$$

The preceding results can be greatly simplified under the assumption that $\gamma_1 a_1$ and $\gamma_2 a_2$ are large compared with unity. For example, noting that $\hat{I}_\nu(z)$ satisfies the equation

$$\frac{d^2 \hat{I}_\nu(z)}{dz^2} = \left[1 + \frac{\nu(\nu+1)}{z^2}\right] \hat{I}_\nu(z), \qquad (48)$$

it readily follows that $K^+(\gamma r)$ satisfies

$$K^2 + \eta_1 \frac{dK}{dz} = \left[1 + \frac{\nu(\nu+1)}{z^2}\right] \eta_1^2, \qquad (49)$$

where $K = K^+(\gamma_1 r)$ and $z = \gamma_1 r$. For a first approximation, the derivative term may be neglected. Thus

$$K^+(\gamma_1 r) \cong \eta_1 \left[1 + \frac{\nu(\nu+1)}{(\gamma_1 r)^2}\right]^{\frac{1}{2}}. \qquad (50)$$

This approximate result, obtained in a very simple fashion, corresponds to the use of the Debye or the second order result. The theory of Bessel functions indicates that its validity, in this form, depends on the conditions $\Re \gamma_1 r \gg 1$ and $|\gamma_1 r|^2$ somewhat greater than $\nu(\nu+1)$ [Sommerfeld, 1949]. Similarly,

$$K^-(\gamma_1 r) \cong \eta_1 \left[1 + \frac{\nu(\nu+1)}{(\gamma_1 r)^2}\right]^{\frac{1}{2}} \qquad (51)$$

and

$$K^+(\gamma_2 r) \cong \eta_2 \left[1 + \frac{\nu(\nu+1)}{(\gamma_2 r)^2}\right]^{\frac{1}{2}}. \qquad (52)$$

To within the same approximation, the reflection coefficients at $r = a_1$ may then be written

$$-q_e = \frac{(1/\eta_2)\left[1 - \left(\frac{\gamma_0 a_1}{\gamma_2 a_2} S\right)^2\right]^{-\frac{1}{2}} - (1/\eta_1)\left[1 - \left(\frac{\gamma_0 a_1}{\gamma_1 a_2} S\right)^2\right]^{-\frac{1}{2}}}{(1/\eta_2)\left[1 - \left(\frac{\gamma_0 a_1}{\gamma_2 a_2} S\right)^2\right]^{-\frac{1}{2}} + (1/\eta_1)\left[1 - \left(\frac{\gamma_0 a_1}{\gamma_1 a_2} S\right)^2\right]^{-\frac{1}{2}}} \qquad (53)$$

and

$$-q_h = \frac{\eta_2 \left[1 - \left(\frac{\gamma_0 a_1}{\gamma_2 a_2} S\right)^2\right]^{\frac{1}{2}} - \eta_1 \left[1 - \left(\frac{\gamma_0 a_1}{\gamma_1 a_2} S\right)^2\right]^{\frac{1}{2}}}{\eta_2 \left[1 - \left(\frac{\gamma_0 a_1}{\gamma_2 a_2} S\right)^2\right]^{\frac{1}{2}} + \eta_1 \left[1 - \left(\frac{\gamma_0 a_1}{\gamma_1 a_2} S\right)^2\right]^{\frac{1}{2}}} \qquad (54)$$

where $\nu(\nu+1) \cong \nu^2 \cong -\gamma_0^2 S^2 = k^2 S^2$. The corresponding (approximate) form for the product of the transmission coefficients is easily shown to be

$$\chi_e(a_2,a_1)\chi_e(a_1,a_2) \cong \chi_h(a_2,a_1)\chi_h(a_1,a_2)$$

$$\cong \exp\left[-2\gamma_1 \int_0^l \left[1 - \frac{\gamma_0^2 a_1^2 S^2}{\gamma_1^2 r^2}\right]^{\frac{1}{2}} dr\right] \quad (55)$$

where $l = a_1 - a_2$.

Using eqs (50), (53), (54), and (55), eq (41) for Z_r can be expressed in a fairly convenient form despite its complicated appearance. A great simplification can be made when the thickness l of the shell is small compared with a_1. Thus, the ratios a_1/a_2 and a_1/r in the preceding expressions may be replaced by unity. The surface impedance may then be written

$$Z_r = \eta_1 \left[1 - \frac{\gamma_0^2}{\gamma_1^2} S^2\right]^{\frac{1}{2}} \frac{\frac{\gamma_1^2}{\gamma_2^2}\left[\frac{\gamma_2^2-\gamma_0^2 S^2}{\gamma_1^2-\gamma_0^2 S^2}\right]^{\frac{1}{2}} + \tanh\left[(\gamma_1^2-\gamma_0^2 S^2)^{\frac{1}{2}} l\right]}{1 + \frac{\gamma_1^2}{\gamma_2^2}\left[\frac{\gamma_2^2-\gamma_0^2 S^2}{\gamma_1^2-\gamma_0^2 S^2}\right]^{\frac{1}{2}} \tanh\left[(\gamma_1^2-\gamma_0^2 S^2)^{\frac{1}{2}} l\right]} \quad (56)$$

where it has been assumed that $\mu_1 = \mu_2 = \mu_0$. This result is *identical* to what one would obtain on a planar two-layer earth for a plane wave incident at an angle of incidence arc sin S [Wait, 1958]. Consequently, in this limiting case the influence of earth curvature vanishes.

The specific results developed above for the surface impedance are restricted to TM waves. The corresponding results for TE waves are obtained by exactly the same method. In fact the results are completely analogous if admittances are used in place of impedances. The surface admittance defined by

$$Y_r = \frac{H_\phi}{E_\phi}\bigg]_{r=a_1}$$

has precisely the same form as eq (41) for Z_r if $K^+(\gamma_1 a)$ is now replaced by the admittance $M^+(\gamma_1 a)$ and q_e and q_h have the same form as eqs (53) and (54) except that η_1 and η_2 are replaced by their reciprocals everywhere. The bracketed terms are unchanged.

Under the assumptions that Re $\gamma_1 a_1$ and Re $\gamma_1 a_2 \gg 1$, $\mu_1 = \mu_2 = \mu_0$, and that $a_1 - a_2 \ll a_1$, one easily finds that

$$Y_r = \frac{1}{\eta_1}\left[1 - \frac{\gamma_0^2 S^2}{\gamma_1^2}\right]^{\frac{1}{2}} \frac{\left[\frac{\gamma_2^2-\gamma_0^2 S^2}{\gamma_1^2-\gamma_0^2 S^2}\right]^{\frac{1}{2}} + \tanh\left[(\gamma_1^2-\gamma_0^2 S^2)^{\frac{1}{2}} l\right]}{1 + \left[\frac{\gamma_2^2-\gamma_0^2 S^2}{\gamma_1^2-\gamma_0^2 S^2}\right]^{\frac{1}{2}} \tanh\left[(\gamma_1^2-\gamma_0^2 S^2)^{\frac{1}{2}} l\right]}. \quad (57)$$

This is equivalent to the result given by eq (26) in Note No. 3.

REFERENCES

SCHELKUNOFF, S.A., (1943), *Electromagnetic Waves*, Van Nostrand Publ. Co., New York

SOMMERFELD, A., (1949), *Partial Differential Equations*, Academic Press, New York.

WAIT, J.R., (1958), Transmission and reflection of electromagnetic waves in the presence of stratified media, J. Res. NBS, **61**, 205-232.

9
Excitation of the H.F. Surface Wave by Vertical and Horizontal Apertures

Introduction

Ever since the pioneering work by Zenneck [1907] and Sommerfeld [1909, 1926], the propagation of radio waves along the earth's surface has captured the fancy of theoreticians. The subject reached a state of some maturity in the 1930–1940 period when Van der Pol and Bremmer [1937, 1938, 1939] exposed a general theory for the diffraction of a dipole field by a finitely conducting spherical earth. Their approach also reached a pinnacle of complexity because they wished to retain mathematical generality. But they did clearly identify the nature of a number of approximations that led to simplified working formulas (i.e., the residue series) that have been used by many for numerical studies. This subject is very clearly and exhaustively described by Bremmer [1949] who also made important individual contributions.

Benefiting by hindsight, a series of papers by Fock [1945, 1965] and his Soviet colleagues appeared about this same time that essentially rederived the Van der Pol-Bremmer theory by introducing various physical approximations at the outset rather than at the end of the analysis. In a parallel development based on wartime research in the early 1940s, Booker and Walkinshaw [1946] showed that the residue series representations were really nothing more than sums of normal modes. When the problem was formulated in this fashion for a perfectly conducting earth, the modal spectrum was discrete and no need arose to introduce a continuous spectrum of modes that was an inherent feature of the earlier studies. Sommerfeld was apparently very conversant with this approach as can be evidenced by his most eloquent discussion of a related problem in his famous Munich lectures on theoretical physics [e.g., see Sommerfeld, 1949].

Here we present an exposition of the normal mode approach to ground wave propagation; this formulation is well adapted to the solution of the surface wave excitation problem. The principal end result is the residue-series representation for the vertical electric field over a spherical earth when excited by either a vertical or a horizontal antenna. The derivation incorporates the essential features of prior analyses without becoming embroiled in the mathematical niceties of branch cuts and intricate and tricky deformations of integration contours in the complex wave number planes. Previous papers that deal generally with the topic are as follows: Spies and Wait [1966], Wait [1967], Zucker [1969], Bahar [1970], Wait [1970], Wait [1971], Cho and King [1972], King, et al, [1974], Wait [1974], King and Wait [1976]. Many of these papers also contain additional references.

We will deal exclusively with an airless earth since the physics of the diffraction process is not changed when a smooth gravitationally stratified atmosphere is allowed for [Fock, 1965, Wait, 1972]. Furthermore, the earth's lower boundary is taken to be a homogeneous dissipative medium characterized by a specified conductivity σ and permittivity ε. Any irregularities of the surface profile of the earth's lower boundary are also ignored. We neglect any ferromagnetic effects and assume that the whole space is characterized by the free-space permeability μ_0.

Formulation

Our specific propagation model is illustrated in Fig. 1. In terms of spherical coordinates (r, θ, ϕ) the earth's surface is at $r=a$ where a is the actual earth's radius. A conically shaped source region is defined by $r>a$ and $\theta<\theta_0$; later we let θ_0 become vanishingly small.

An essential point of the present derivation is that complete azimuthal symmetry prevails in the sense that $\partial/\partial\phi=0$. Also the source itself is assumed to excite on T.M. (transverse magnetic) waves in the sense that the magnetic field has only an azimuthal or ϕ component H_ϕ. Simple physics tells us that the electric field has

*Reprinted in condensed form from: J.R. Wait, and D.A. Hill, *Radio Science*, 14, No. 5, 1979.

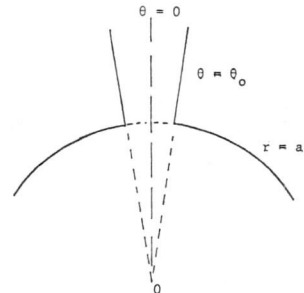

Fig. 1. Conical source region and a portion of earth's surface

only components E_r and E_θ. It is not surprising to learn that all three field components can actually be derived from a single scalar function U. For example, in the region $r > a$ and $\theta > \theta_0$, if we write

$$H_\phi = -i\varepsilon_0 \omega \partial U / \partial \theta \tag{1}$$

then U is what is known as a Debye potential. Here ε_0 is the permittivity of free space and ω is the angular frequency. The assumed time factor is $\exp(i\omega t)$. Then, from Maxwell,

$$E_r = (k^2 + \partial^2/\partial r^2)(rU) \tag{2}$$
$$E_\theta = (1/r)(\partial^2/\partial r \partial \theta)(rU) \tag{3}$$

As a further consequence, for $r > a$ and $\theta > \theta_0$,

$$(\nabla^2 + k^2)U = 0 \tag{4}$$

Solutions can be obtained by straight-forward separation of variables; however, we do not need to insist on singlevaluedness in the infinitely extended θ domain since we restrict attention to $\pi > \theta > \theta_0$. The radial functions are spherical Hankel functions $h_\nu(kr)$ of order ν that behave as $\exp(-ikr)$ as the argument $kr \to \infty$. The corresponding angular functions are Legendre functions $P_\nu(-\cos 0)$ also of order ν; the argument $-\cos\theta$ or $\cos(\pi - \theta)$ is chosen such that the fields remain finite at the antipode $\theta = \pi$. To be explicit we write

$$U = \sum_\nu A_\nu \frac{h_\nu(kr)}{h_\nu(ka)} P_\nu(-\cos\theta) \tag{5}$$

where the summation encompasses all required values of ν (to be specified below). Here A_ν is a coefficient yet to be determined.

The Surface Impedance Condition

We now invoke the surface impedance boundary condition that greatly simplifies the derivation. Simply stated, this requires that

$$E_\theta = -ZH_\phi (\text{at } r=a) \tag{6}$$

When Z is the surface impedance. A good value to choose for this parameter is

$$Z = (ik/\gamma)\left[1 + (k^2/\gamma^2)\right]^{1/2} \eta_0 \tag{7}$$

when $\gamma = [i\mu_0\omega(\sigma + i\varepsilon\omega)]^{1/2}$ is the propagation constant for the earth and $\eta_0 = (\mu_0/\varepsilon_0)^{1/2} = 120$ is the characteristic impedance of free space. Actually this form for Z would be exact for a vertically polarized wave

Excitation of the H.F. Surface Wave

at grazing incidence on a planar earth model. The so-called Leontovich boundary condition [e.g., see Fock 1965] is nothing more than the simpler version $Z=(ik/\gamma)\eta_0$ that has the additional constraint $|k^2/\gamma^2|\ll 1$. Actually the surface impedance formula can be made "exact" for the present problem by using

$$E_{\nu,*} = -Z_\nu H_{\nu,\phi} \tag{8}$$

where

$$Z_\nu = \frac{1}{\sigma+i\varepsilon\omega}\left[\frac{(\partial/\partial r)[rj_\nu(-i\gamma r)]}{rj_\nu(-i\gamma r)}\right]_{r=a} \tag{9}$$

is the "impedance" for spherical waves of order ν. Here j_ν is the spherical Bessel function that remains finite at $r=0$, the center of the earth! The limiting case (7) is obtained from (9) when the Debye approximation is made and noting that the important values of ν are near k (i.e., grazing waves). The reader can delve into this question by reading the author's textbook [Wait, 1972] and several review papers on the subject.[†] Here we will be content with the form given by (7) which has the great merit that Z the effective surface impedance of the earth does not depend on the characteristic value ν. As a consequence our boundary condition becomes

$$\left\{\frac{d}{dx}[\chi h_\nu(\chi)] - i\frac{Z}{\eta_0}\chi h_\nu(\chi)\right\}_{\chi=ka} = 0 \tag{10}$$

Another revealing form of this equation is obtained by rewriting it as follows

$$Z + Z_{e,\nu} = 0 \tag{11}$$

where

$$Z_{e,\nu} = -\frac{1}{i\varepsilon_0\omega}\left[\frac{(\partial/\partial r)[rh_\nu(kr)]}{rh_\nu(kr)}\right]_{r=a} \tag{12}$$

is an "external" wave impedance. Equation (11) can be termed the transverse resonance relation.

Airy Function Approximation

In spite of the compactness of (10) the determination of roots is a formidable task. In the early work on the subject such as by Vvedensky [1935, 1936, 1937] and Millington [1939] in the early 30s, the so-called tangent approximation was made which in effect replaced the spherical wave functions $h_\nu(kr)$ in (10) by their second order or Debye representation. Unfortunately this approximation breaks down just where it is needed most, i.e., for the lowest order roots where the attenuation is the least. It was primarily Van der Pol and Bremmer who surmounted this difficulty by utilizing a third order representation for the spherical wave functions in terms of Hankel functions of order 1/3. This process is described in a very comprehensive fashion by Bremmer [1949] in his book. A similar and, in many respects, a parallel development was carried out by Fock [1961] who used a very compact Airy function representation that was equivalent to the more cumbersome Hankel functions of order 1/3. We will adopt the latter form here; it amounts to making the substitution $\nu+1/2=(ka/2)^{1/3}t+ka$ and to writing

$$\chi h_\nu(\chi) \approx i(ka/2)^{1/6} w(t-y) \tag{13}$$

where

$$y=(2/ka)^{1/3}(\chi-ka), \chi=kr$$

[†]e.g., see list of references, notably King and Wait [1976].

and where the Airy function $w(t)$ satisfies the Stokes' differential equation

$$(d^2/dt^2)w(t) - tw(t) = 0 \tag{14}$$

In terms of contemporary Airy functions

$$w(t) = \pi^{1/2}[Bi(t) - iAi(t)]$$

Using (13) and the corresponding form for the derivative, we easily deduce that (10) or (11) is approximated by

$$[dw(t)/dt] - qw(t) = 0 \tag{15}$$

where $q = -i(ka/2)^{1/3} Z/\eta_0$.

Roots of this equation are denoted t_s where the subscript s assumes the values $1, 2, 3, \ldots$. The corresponding values of the roots ν_s are then obtained from

$$\nu_s + (1/2) \simeq ka + (ka/2)^{1/3} t_s$$

which is a highly accurate approximation for the lower order roots where $|\nu_s - ka| \ll (ka/2)^{2/3}$.

The Excitation Problem

A further key step in the analysis is to enforce the source condition which we can state as follows

$$I(r) \simeq 2\pi r \theta_0 H_\phi(r, \theta_0) \tag{16}$$

for $\theta_0 \ll 1$ and for $a < r < \infty$. Here $I(r)$ can be interpreted as the total axial current. Now, following Sommerfeld [1949], we note that for $\theta \ll 1$

$$P_\nu(-\cos\theta) \simeq \pi^{-1} \sin(\nu\pi) \ln\theta^2 \tag{17}$$

and

$$\partial P_\nu(-\cos\theta)/\partial\theta \simeq \pi^{-1} \sin(\nu\pi)(2/\theta) \tag{18}$$

Then using (1) and (5), the condition (16) becomes

$$I(r) = -4i\varepsilon_0 \omega \sum_\nu A_\nu \frac{rh_\nu(kr)}{h_\nu(ka)} \sin\nu\pi \tag{19}$$

where the summation is to include all roots ν or ν_s. An important aspect of the present development is that the radial wave functions for these discrete modes (i.e., $\nu = \nu_s$) are orthogonal in the sense that

$$\int_a^\infty h_\nu(kr) h_{\nu'}(kr) \, dr = 0 \tag{20}$$

if $\nu \neq \nu'$. This relationship holds exactly under the assumption that the surface impedance boundary condition given by (6) is valid for all discrete modes. Further discussion of this point appears elsewhere [Wait, 1968] where it is also shown that the normalizing integral can be approximated by

$$\int_{ka}^\infty \left[\frac{h_\nu(kr)}{h_\nu(ka)} \right]^2 d(kr) \simeq \left(\frac{ka}{2} \right)^{1/3} (t_s - q^2) \tag{21}$$

Excitation of the H.F. Surface Wave

for the roots $\nu=\nu_s$. It is now a simple matter to deduce that

$$A_\nu \simeq \frac{i\eta_0}{4\sin\nu\pi}\left(\frac{2}{ka}\right)^{1/3}\frac{1}{t_s-q^2}\int_a^\infty I(r)\frac{h_\nu(kr)}{h_\nu(ka)}\frac{1}{r}dr \qquad (22)$$

If we deal with a vertical (i.e., radially oriented) electric dipole source at $r=r_0$, this amounts to saying that

$$I(r)=I\,ds\,\delta(r-r_0) \qquad (23)$$

where $\delta(r-r_0)$ is the unit impulse function of $r=r_0$. Then (22) reduces to the approximate form

$$A_\nu \simeq \frac{i\eta_0}{4\sin\nu\pi}\left(\frac{2}{ka}\right)^{1/3}\frac{1}{t_s-q^2}\frac{I\,ds}{r_0}\frac{w(t_s-y_0)}{w(t_s)} \qquad (24)$$

where $y_0=k(2/ka)^{1/3}(r_0-a)$ and where the Airy function representations for the spherical Hankel functions is valid when $|kr_0-\nu|\ll(ka/2)^{2/3}$.

Expressions for the Field

The field quantity of most physical interest is the radial or vertical electric field E_r. Clearly from (2) this is given by

$$E_r = \frac{1}{r}\sum_{\nu=\nu_s} A_\nu \nu(\nu+1)\frac{h_\nu(kr)}{h_\nu(ka)}P_\nu(-\cos\theta) \qquad (25)$$

Again here the height-gain function can be represented in terms of Airy functions by

$$\frac{h_\nu(kr)}{h_\nu(ka)} \simeq \frac{w(t_s-y)}{w(t_s)} \qquad (26)$$

where $y=(2/ka)^{1/3}k(r-a)$ provided

$$|kr-\nu|\ll(ka/2)^{2/3}$$

for the applicable range of the heights $r-a$. Also in most cases we approximate the multiplicative factor

$$\nu(\nu+1)\simeq(\nu+1/2)^2\simeq[ka+(ka/2)^{1/3}t_s]^2\simeq(ka)^2 \qquad (27)$$

for the most important terms in (25). Also in this same vein

$$P_\nu(-\cos\theta)\simeq\left(\frac{2}{\pi\nu\sin\theta}\right)^{1/2}\cos\left[(\nu+1/2)(\pi-\theta)-\frac{\pi}{4}\right] \qquad (28)$$

provided $|\nu|\gg1$ and θ is not near π. In fact, the further simplification is possible in most cases:

$$P_\nu(-\cos\theta)\simeq\left(\frac{2}{\pi ka\theta}\right)^{1/2}\frac{1}{2}\exp\left[+i(\nu+1/2)(\pi-\theta)-i\frac{\pi}{4}\right] \qquad (29)$$

since $[-\text{Im}.\nu(2\pi-\theta)]\gg1$. Here we can identify $a\theta=d$ as the great circle distance between source dipole and observer.

Having made all the above approximations it is possible to write (25) in the following form

$$E_r \simeq E_0 \left(\frac{\pi x}{i}\right)^{1/2} \sum_s \frac{e^{-ixt_s}}{t_s - q^2} \frac{w(t_s - y_0)}{w(t_s)} \frac{w(t_s - y)}{w(t_s)} \tag{30}$$

where $x = (ka/2)^{1/3}\theta = (ka/2)^{1/3} d/a$ and where

$$E_0 = -i\mu_0 \omega I\, ds / (2\pi d)$$

can be identified as the corresponding vertical field of the electric dipole located on the surface of a flat perfectly conducting plane. The ratio E_r/E_0 is sometimes called the attenuation function and denoted W.

Discussion of Height Gain Functions

We now backtrack a bit and consider that we have a ground based vertical antenna of physical height h_0 with a specified current distribution $I(h)$. Then it easily follows that

$$E_r = -\left[\frac{i\mu_0 \omega}{2\pi d} \int_0^{h_0} I(h)\, dh\right] W \tag{31}$$

where the corresponding attenuation function is written

$$W = \left(\frac{\pi x}{i}\right)^{1/2} \sum_s \frac{e^{-ixt_s}}{t_s - q^2} G_s(y) H_s \tag{32}$$

where

$$G_s(y) = w(t_s - y)/w(t_s) \tag{33}$$

is the "height-gain" function for the observer and H_s is the "source excitation" function defined by

$$H_s = \int_0^{h_0} I(h) G_s(y)\, dh \Big/ \int_0^{h_0} I(h)\, dh \tag{34}$$

where

$$y = (2/ka)^{1/3} kh$$

It is now useful to note that the function $G(y) = w(t-y)/w(t)$ in general satisfies

$$d^2 G/dy^2 = (t - y) G \tag{35}$$

Also, of course, $G(0) = 1$ and because of the boundary condition (15)

$$[dG_s/dy]_{y=0} = -q \tag{36}$$

Then it is a simple matter to show that

$$G_s(y) = 1 - qy + \frac{t_s y^2}{2} - \frac{1 + t_s q}{6} y^3 + \ldots \tag{37}$$

or, more generally that

$$G_s(y) = \sum_{n=0,1,2\ldots} A_{s,n} y^n \tag{38}$$

Excitation of the H.F. Surface Wave

where the coefficients $A_{s,n}$ satisfy the recurrence relation

$$A_{s,n+2}(n+2)(n+1) = A_{s,n}t_s - A_{s,n-1} \tag{39}$$

when, by definition

$$A_0 = 1 \quad \text{and} \quad A_{-1} = A_{-2} = A_{-3} = 0$$

of course we can also write

$$G_s(y) = \sum_{n=0} a_{s,n} h^n \tag{40}$$

where

$$a_{s,n} = (y/h)^n A_{s,n} = \left[(2/ka)^{1/3} k\right]^n A_{s,n} \tag{41}$$

The "source excitation" function is now written

$$H_s = \sum_{n=0,1,2} a_{s,n} \int_0^{h_0} I(h) h^n \, dh \Big/ \int_0^{h_0} I(h) \, dh \tag{42}$$

This immediately illustrates that the higher order moments of the current distribution may contribute significantly to the total field.

We now represent the source current distribution as a sum of a finite number of exponentials. That is, we let

$$I(h) = \sum_{m=1,2,\ldots} p_m \exp(-\alpha_m h) \tag{43}$$

for $h_0 > h > 0$ and specify that $I(h) = 0$ for $h > h_0$. This is an obvious form for the source if we are attempting to optimize the launching of some type of surface wave. Then using (42) and (43), it follows readily that

$$H_s = \sum_m \sum_n a_n p_m \int_0^{h_0} e^{-\alpha_m h} h^n \, dh \Big/ \left[\sum_m \int_0^{h_0} e^{-\alpha_m h} \, dh \right] \tag{44}$$

or equivalently

$$H_s = \sum_m \sum_n a_n p_m \Lambda_{m,n} \Big/ \sum_m \Lambda_{m,0} \tag{45}$$

where

$$\Lambda_{m,n} = \int_0^{h_0} e^{-\alpha_m h} h^n \, dh \tag{46}$$

$$= \frac{n!}{\alpha_m^{n+1}} - e^{-\alpha_m h_0} \left[\frac{h_0^n}{\alpha_m} + \frac{n h_0^{n-1}}{\alpha_m^2} + \frac{n(n-1) h_0^{n-2}}{\alpha_m^3} \cdots + \frac{n! h_0}{\alpha_m^n} + \frac{n!}{\alpha_m^{n+1}} \right] \tag{47}$$

and

$$\Lambda_{m,0} = \int_0^{h_0} e^{-\alpha_m h} \, dh = \frac{1 - e^{-\alpha_m h_0}}{\alpha_m} \tag{48}$$

Extension to Horizontal Antenna Excitation

In formulating the problem we had assumed an azimuthally symmetric source that was specialized to a radially oriented vertical antenna carrying a filamental current. This is the most obvious choice of a transmitting antenna that is to launch a vertically polarized (or T.M.) ground wave over the spherical earth. But another strong contender is a horizontal antenna that carries a specified filamental current throughout its length. The natural building block here is the horizontal or tangentially-oriented electric dipole. Thus we could begin with the formal exact solution of the horizontal electric dipole in the vicinity of a homogeneous sphere [Wait, 1956]. This formalism, however, is not needed for present purposes if we make use of the reciprocity theorem and the prior solution for the vertical electric dipole.

For the vertical electric dipole an individual mode or term in (5) can be written

$$U = a_\nu h_\nu(kr) P_\nu(-\cos\theta) \tag{49}$$

where the radial function, of argument $\chi = kr$, satisfies

$$(d^2/d\chi^2)(\chi h_\nu) + [\chi^2 - \nu(\nu+1)]h_\nu = 0 \tag{50}$$

Also for the ranges of interest,

$$P_\nu(-\cos\theta) \simeq \text{const.} \times (\sin\theta)^{-1/2} \exp[-i(\nu+1/2)\theta] \tag{51}$$

is an adequate approximation. Thus, on using (1), (2), (3), (50) and (51) it follows that the corresponding field components for $r > a$ are to be obtained from

$$E_{\nu,r} = \nu(\nu+1)r^{-1}U_\nu \tag{52}$$

$$H_{\nu,\phi} \simeq -\varepsilon_0 \omega(\nu+1/2)U_\nu \tag{53}$$

$$E_{\nu,\theta} \simeq -i(\nu+1/2)\partial(rU_\nu)/\partial r \tag{54}$$

Now the "wave tilt" parameter W_ν for a given mode is defined and given by

$$W_\nu = \frac{E_{\nu,\theta}}{E_{\nu,r}} \simeq -\frac{i(\nu+1/2)}{\nu(\nu+1)} \frac{1}{U} \frac{\partial}{\partial r}(rU) \tag{55}$$

Using (52) this can also be written

$$W_\nu \simeq -\frac{i(\nu+1/2)}{\nu(\nu+1)rE_{\nu,r}} \frac{\partial}{\partial r}(r^2 E_{\nu,r}) \tag{56}$$

The corresponding "wave impedance" parameter is

$$Z_\nu = -\frac{E_{\nu,r}}{H_{\nu,\phi}} \simeq \frac{\nu(\nu+1)}{\varepsilon_0 \omega(\nu+1/2)r} \tag{57}$$

Now using the Airy function approximation given by (13) it follows without difficulty that

$$W_\nu \simeq \frac{\Delta}{Sq} \frac{w'(t_s - y)}{w(t_s - y)} \tag{58}$$

where Δ and S are defined by $q = -i(ka/2)^{1/3}\Delta$ or $\Delta = Z/\eta_0$ and $(\nu+1/2)^2 \simeq \nu(\nu+1) \simeq (kaS)^2$. Here, as before, t_s for mode s is related to ν (or ν_s) by $(\nu+1/2) \simeq ka + (ka/2)^{1/3}t_s$. In most cases S can be replaced by 1. Thus we see that $W_\nu \to \Delta$ as $y \to 0$ since $w'(t_s) = qw(t_s)$.

Excitation of the H.F. Surface Wave

An equivalent representation for the wave-tilt parameter for a given mode follows from (33), thus

$$W_v \simeq -\frac{\Delta}{S_q G_s(y)} G_s'(y) \tag{59}$$

On using the series for the height function $G_s(y)$ given by (37), it follows that

$$W_v \simeq \frac{\Delta}{S} \frac{1-t_s y/q + (1/2)(1+t_s q)y^2/q + \ldots}{1-qy+t_s y/2+\ldots} \tag{60}$$

Now if we write $S=(1-C^2)^{1/2} \simeq 1-(C^2/2)$ then for a given mode $t_s \simeq -C^2(ka/2)^{2/3}$. Keeping just first order terms in y we see that

$$W_v \simeq \frac{1}{S} \frac{\Delta + ik(r-a)C}{1+ik(r-a)} \tag{61}$$

This result can be shown to be quite consistent with the corresponding wave tilt for a vertically polarized wave with an angle of incidence of arc cos C or arc sin S (i.e., for a grazing angle of approximately C radians).

We now may apply the reciprocity theorem to determine the vertically polarized fields produced by a horizontal electric antenna. First of all we rewrite (30) specifically for the vertical field of a height h_2 and a range d vertical electric dipole VED of moment $I(h_1)dh_1$ at height h_1:

$$E_r = -\frac{i\mu_0 \omega I(h_1)dh_1}{2\pi d}\left(\frac{\pi x}{i}\right)^{1/2} \sum_s \frac{e^{-ixt_s}}{t_s-q^2} G_s(h_1)G_s(h_2)e^{-ikd} \tag{62}$$

where the height-gain functions are defined by

$$G_s(h_i) = w(t_s-y_i)/w(t_s); \quad i=1,2 \tag{63}$$

and where $y_i = (2/ka)^{1/3}kh_i$. As before $x=(ka/2)^{1/3}d/a$ and t_s are the roots of $w'(t_s)-qw(t_s)=0$ where $q=-i(ka/2)^{1/3}\Delta$ where Δ is the normalized surface impedance. Using the basic definition for the "wave tilt parameter" given by (55) it is evident that the corresponding expression for the total horizontal field E_θ for the VED is

$$E_\theta = -\frac{i\mu_0 \omega I(h_1)dh_1}{2\pi d}\left(\frac{\pi x}{i}\right)^{1/2} \sum_s \frac{e^{-ixt_s}}{t_s-q^2} W_s G_s(h_1)G_s(h_2)e^{-ikd} \tag{64}$$

where

$$W_s \simeq \Delta w'(t_s-y)/[qw(t_s-y)] \tag{65}$$

It is useful here to note that

$$W_s G_s(h_2) = \Delta g_s(h_2) \tag{66}$$

where

$$g_s(h_2) = \frac{w'(t_s-y_2)}{qw(t_s)} = \frac{w'(t_s-y_2)}{w'(t_s)} \tag{67}$$

We now can immediately invoke the reciprocity theorem to write down the expression for the vertical electric field at height h_2 of a horizontal electric dipole HED of moment $I dl_1$ at height h_1 for the same great circle range d. The result [after interchanging the indices 1 and 2 in (64)] is seen to be

$$E_r = -\frac{i\mu_0\omega I(h_1)\,dl_1}{2\pi d}\left(\frac{\pi x}{i}\right)^{1/2}\Delta\sum_s\frac{e^{-ixt_s}}{t_s-q^2}G_s(h_2)g_s(h_1)\cos\phi\, e^{-ikd} \tag{68}$$

where

$$g_s(h_1)=w'(t_s-y_1)/w'(t_s) \tag{69}$$

is the height-gain function for the HED. Here, of course, ϕ is the azimuthal angle at the receiver for the source HED oriented in the $\phi=0$ direction. In addition to the E_r field and the E_θ and H_ϕ derived therefrom there will be an H_r field and H_θ and E_ϕ derived therefrom. Here we will only be concerned with the TM (transverse magnetic) or vertically polarized fields that are fully characterized by (62).

We now consider a horizontal antenna at height h_1 of total length L with a current distribution $I(l)$. Then, clearly, the moment $I(h_1)\,dl_1$ in (1) is to be replaced by

$$\int_{-L/2}^{L/2} I(l)e^{ik_s l\cos\phi}\,dl \tag{70}$$

and moved inside the summation. Here k_s is the wave number ν/a for the mode in question; that is

$$k_s a = ka + (ka/2)^{1/3}t_s$$

In an analogy to the case of the vertical antenna we can define an antenna gain function in the following fashion

$$F_s = \int_{-L/2}^{L/2} I(l)e^{ik_s l\cos\phi}\,dl \bigg/ \int_{-L/2}^{L/2} I(l)\,dl \tag{71}$$

The working expression for the vertical electric field is thus

$$E_r = \left[-\frac{i\mu_0\omega}{2\pi d}\int_{-L/2}^{L/2} I(l)\,dl\right]\Delta\hat{W}\cos\phi \tag{72}$$

where

$$\hat{W} = \left(\frac{\pi x}{i}\right)^{1/2}\sum_s\frac{e^{-ixt_s}}{t_s-q^2}G_s(y)F_s g_s(h_1) \tag{73}$$

where

$$y(=y_2)=(2/ka)^{1/3}k(r-a)$$

An immediate example would be to assume a single travelling wave of current on the structure i.e., $I(l)=I_0\exp(-\Gamma l)$ for $-L/2<l<+L/2$. Then

$$\int_{-L/2}^{+L/2} I(l)\,dl = \frac{2I_0}{\Gamma}\sinh\frac{\Gamma L}{2} \tag{74}$$

and

$$F_s = \frac{\sinh[(\Gamma-ik_s\cos\phi)(L/2)]}{(\Gamma-ik_s\cos\phi)(L/2)}\cdot\frac{(\Gamma L/2)}{\sinh(\Gamma L/2)}$$

Excitation of the H.F. Surface Wave

Some Numerical Results and Final Remarks

The preceding formulation has been used to obtain extensive numerical results for various kinds of aperture distributions. We give a few examples here for a frequency of 10 MHz with propagation over land with a conductivity $\sigma_e = 10^{-2}$ mhos/m and a relative permittivity $\varepsilon_e/\varepsilon_0 = 6$. In each case the effective height of the receiving (vertical whip) antenna is zero and the great circle range is denoted d. Normal atmospheric refraction is accounted for by using an effective earth radius equal to four thirds times the actual earth's radius.

In Fig. 2 the attenuation function W as defined by (32), is plotted as a function of the range d for the Zenneck wave distribution [Hill and Wait, 1978] for the current on the ground-based vertical antenna of height h_0. In the present case this means that $I(h) = I_0 \exp(-\alpha h)$ where $\alpha = 0.02693 - i\, 0.03946 m^{-1}$. The values of h_0 shown range from 0 to 200m as indicated. The case $h_0 = 0$, of course, corresponds to the ground-based vertical electric dipole. These results show that the Zenneck wave distribution is not very good for launching ground waves over a spherical earth. The reduction of the field strength with increasing h_0 is due to phase cancellation of the contributions in the various current elements. This effect would not take place for the planar model (e.g., as in Hill and Wait, 1978).

In Fig. 3 results are shown for the corresponding attenuation function W_d when we have simply a vertical electric dipole source at height h. Here the results illustrate the expected great advantage of raising up the centroid of the current distribution so that full advantage can be taken of the height-gain function.

Finally in Fig. 4 we show the attenuation function $\hat{W}$ as defined by (73) for excitation by a horizontal antenna of length L. The effective height h_1 of this structure is taken to be zero and we also set $\phi = 0$ corresponding to the end-fire direction The lengths L vary from 0 to 195m. Here the current distribution is a travelling wave with a propagation constant $\Gamma = ik(1 - \Delta^2)^{1/2} = 0.0512 + i\, 0.2076 m^{-1}$ where k is the free space wave number. Again this corresponds to a Zenneck wave, a plane earth with the same electrical constants. The case $\Gamma = ik = 0.20958 m^{-1}$ of course corresponds to a travelling wave on the antenna structure with an assumed

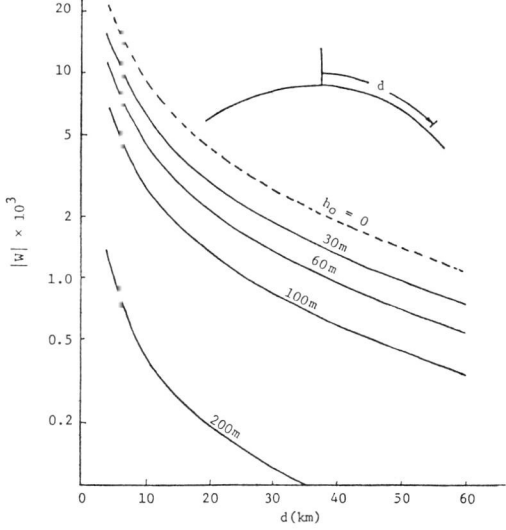

Fig. 2

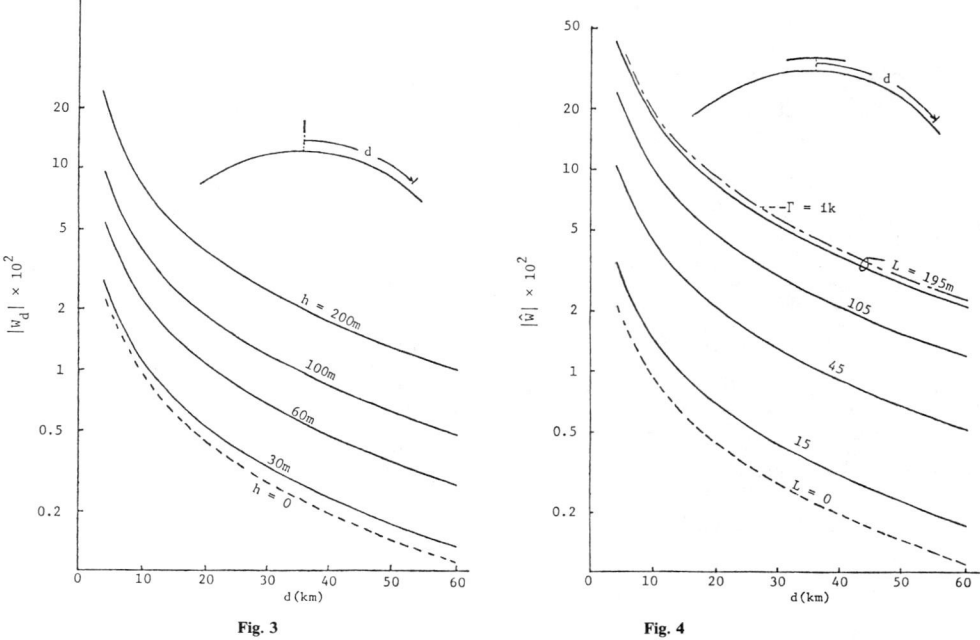

Fig. 3

Fig. 4

free space propagation constant. In the latter case the field strength at the receiver is actually higher than for the Zenneck wave distribution. Again this is not too surprising because the propagation constant of the dominant creeping waves over the spherical earth are nearer ik than to the corresponding wave value.

In general we may conclude that the Zenneck wave illumination is a poor choice for enhancing the ground wave field strength over a spherical earth.

References

BAHAR, E., 1970, Propagation of radio waves over a non-uniform layered medium, *Radio Science*, Vol. 5, No. 7, pp. 1069–1076.
BOOKER, H.G., and WALKINSHAW, W., 1946, The mode theory of tropospheric refraction and its relation to waveguides and diffraction, *Meteorological Factors in Radio-Wave Propagation*, Physical Society, London.
BREMMER, H., 1949, *Terrestrial Radio Waves*, Elsevier, New York.
CHO, S.H., and KING, R.J., 1972, EM ground-wave propagation over the curved lunar surface, *IEEE Trans. Geosci. Electronics*, Vol. OE-10, pp. 96–105.
FOCK, V.A., 1945, Diffraction of radio waves around the earth's surface, *Jour. Theo. and Exp. Physics*, Vol. 15, pp. 480–490.
FOCK, V.A., 1965, *Propagation and Diffraction of Electromagnetic Waves*, Pergamon Press, New York and Oxford.
HILL, D.A., and WAIT, J.R., 1978, Excitation of the Zenneck surface wave by a vertical aperture, *Radio Science*, Vol. 13, No. 6, pp. 967–977.

KING, R.J., CHO, S.H., JAGGARD, D.L., BRUCKNER, G.E., and HUSTIG, C.H., 1974, Experimental data for ground wave propagation over cylindrical surfaces, *IEEE Trans. on Antennas and Propagation*, Vol. AP-22, No. 4, 551–556.

KING, R.J., and WAIT, J.R., 1976, Electromagnetic groundwave propagation—theory and experiment, *Symposia Mathematica*, Vol. 18, pp. 107–208, Academic Press.

MILLINGTON, G., 1939, The diffraction of wireless waves around the earth, *Phil. Mag.*, Series 7, Vol. 27, No. 184, pp. 517–542.

SOMMERFELD, A.N., 1909, 1926, The propagation of waves in wireless telegraphy, *Ann. Phy.*, Series 4, Vol. 28, pp. 665–737 and Vol. 81, 1135–1153.

SOMMERFELD, A.N., 1949, *Partial Differential Equations*, Academic Press, New York.

SPIES, K.P., and WAIT, J.R., 1966, On the calculation of the ground wave attenuation factor at low frequencies, *IEEE Trans. on Antennas and Propagation*, Vol. AP-14, No. 4, pp. 515–517.

VAN DER POL, B., and BREMMER, H., 1937, 1938, 1939, The diffraction of electromagnetic waves from an electrical point source round a finitely conducting sphere, *Phil. Mag.*, Series 7, Vol. 24, pp. 141–176, pp. 825–864; Vol. 25, pp. 817–834; Vol. 26, pp. 261–275.

VVEDENSKY, B., 1935, 1936, 1937, The diffractive propagation of radio waves, *Tech. Physics* (USSR), Vol. 2, pp. 624–639; Vol. 3, pp. 913–325; Vol. 4, pp. 579–591.

WAIT, J.R., 1956, Low frequency radiation from a horizontal antenna over a spherical earth, *Can. Jour. Phys.*, Vol. 34, pp. 586–595.

WAIT, J.R., 1965, Theory of diffraction by a curved inhomogeneous body, *Jour. Math. Phys.*, Vol. 8, No. 4, pp. 920–925.

WAIT, J.R., 1968, Diffraction and scattering of the electromagnetic ground wave by terrain features, *Radio Science*, Vol. 3, No. 10, pp. 995–1003.

WAIT, J.R., 1970, Propagation of electromagnetic waves over a smooth multi-section curved earth—an exact theory, *Jour. Math. Phys.*, Vol. 11, No. 9, pp. 2850–2860.

WAIT, J.R., 1971, Theory of ground wave propagation, *Electromagnetic Probing in Geophysics*, Chap. 5, pp. 163–207, Golem Press, Boulder, Colo.

WAIT, J.R., 1972, *Electromagnetic Waves in Stratified Media*, 2nd Edition, Pergamon Press, New York and Oxford.

WAIT, J.R., 1974, Recent analytical investigations of electromagnetic ground wave propagation over inhomogeneous earth models, *Proc. IEEE*, Vol. 62, No. 8, pp. 1061–1072.

ZENNECK, J., 1907, Uber die Fortpflanzung ebener elektromagnetische Wellen einer ebener Leiterflache und ihre Beziehung zur drahtlosen Telegraphie. *Ann. Phys.*, Series 4, Vol. 23, pp. 846–866.

ZUCKER, F.J., 1969, Surface Wave Antennas, *Antenna Theory*, (ed. by R.E. Collin, and F.J. Zucker), Chap. 21, Part II, McGraw-Hill, New York.

10
Fields of a Dipole Over an Homogeneous Anisotropic Half-Space

INTRODUCTION

Since Sommerfeld's (1909) fundamental investigations over half a century ago, numerous papers have been written on various aspects of the boundary-value problem of a dipole in the vicinity of a plane boundary. Recently, an exhaustive text has been written by Baños (1966), which summarizes much of the past work in this field. Related problems, such as those involving stratified media, are discussed in a review article (Wait 1964) which also gives extensive references. No attempt will be made here to give priorities nor shall we describe the various controversies that have surrounded this subject.

In the present paper, we wish to discuss the influence of anisotropy in the conductivity and dielectric constant of the half-space. It seems surprising that little attention has been given to this aspect of the subject. A notable exception, of course, is the basic work of Arbel and Felsen (1963), who were concerned with dipole radiation in anisotropic plasma. However, their formulation is not readily applicable to the problems treated below. Also, we should mention some recent investigations of Tikhonov (1959) and Chetaev (1962) who were primarily interested in the quasi-static regime, wherein all relevant distances were small compared with the free-space wavelength. Their work has been extended recently by Sinha and Bhattacharya (1967), who treated the quasi-static problem of a dipole over a two-layer earth where the upper layer was anisotropic. In the solution given below, we shall include displacement currents in the analysis. This is necessary if the solutions are to be applied to radio propagation problems, because the horizontal range may be comparable or greater than a free-space wavelength. Then, at the same time, we are able to clarify where the quasi-static theory is applicable.

In what follows, we will be motivated by the hope that the derived results will have some application to electromagnetic wave propagation over the surface of the earth. While the models are highly idealized, it is believed that the influence of anisotropy in the electrical characteristics is demonstrated satisfactorily.

GENERAL CONSIDERATIONS

First of all, we consider an arbitrarily oriented electric dipole over a homogeneous anisotropic half-space. With respect to a cylindrical coordinate system (ρ, ϕ, z), the half-space $z > 0$ is assumed to be free space with dielectric constant ϵ_0 and permeability μ_0. The source is an electric dipole located at $\rho = 0$ and $z = h$. The dipole is of length ds and carries a current $I \exp(i\omega t)$ where ω is the angular frequency and t is the time. The conductivity (g) and dielectric constant (ϵ) of the lower half-space (i.e., $z < 0$) are tensors which have the form

$$(g) = \begin{bmatrix} g_h & 0 & 0 \\ 0 & g_h & 0 \\ 0 & 0 & g_v \end{bmatrix} \quad \text{and} \quad (\epsilon) = \begin{bmatrix} \epsilon_h & 0 & 0 \\ 0 & \epsilon_h & 0 \\ 0 & 0 & \epsilon_v \end{bmatrix}.$$

The magnetic permeability of the lower half-space is also taken to be μ_0. We shall refer to the combination $\sigma_h = g_h + i\epsilon_h\omega$ as the horizontal complex conductivity, while $\sigma_v = g_v + i\epsilon_v\omega$ is the vertical complex conductivity. Thus (σ) is also a diagonal tensor, but its elements are complex.

(Reprinted, in part, from J.R. Wait, Can. Jour. Phys., 44, 2387-2401, 1966).

Fields of a Dipole Over an Homogeneous Anisotropic Half-Space

Maxwell's equations in the lower half-space are written as

(1) $$\operatorname{curl} \boldsymbol{H} = (\sigma)\boldsymbol{E}$$

and

(2) $$\operatorname{curl} \boldsymbol{E} = -i\mu_0\omega\boldsymbol{H},$$

where $\boldsymbol{E}$ and $\boldsymbol{H}$ are the electric and magnetic fields. Introducing a vector potential $\boldsymbol{A}$ and scalar potential ψ in the usual manner, we have

(3) $$\boldsymbol{H} = \operatorname{curl} \boldsymbol{A}$$

and

(4) $$\boldsymbol{E} = -i\mu_0\omega\boldsymbol{A} - \operatorname{grad} \psi.$$

If we choose $\boldsymbol{A}$ and ψ such that

(5) $$\operatorname{div} \boldsymbol{A} + \sigma_h\psi = 0,$$

it is a simple matter to show that

(6) $$\nabla^2 A_x - \gamma^2 A_x = 0$$

and

(7) $$\nabla^2 A_y - \gamma^2 A_y = 0,$$

where

(8) $$\gamma^2 = i\mu_0\omega\sigma_h = i\mu_0\omega(g_h + i\epsilon_h\omega).$$

As a consequence,

(9) $$\nabla^2 A_z - (\sigma_h - \sigma_v)\frac{1}{\sigma_h}\frac{\partial}{\partial z}\operatorname{div} \boldsymbol{A} - \frac{\sigma_v}{\sigma_h}\gamma^2 A_z = 0,$$

which, of course, only reduces to the Helmholtz form if $\sigma_h = \sigma_v$.

In the upper half-space (i.e., $z > 0$), the Cartesian components A_{0j} of $\boldsymbol{A}$ all satisfy

(10) $$(\nabla^2 - \gamma_0^2)A_{0j} = 0, \qquad (j = x, y, z, \gamma_0^2 = -\epsilon_0\mu_0\omega^2),$$

except, of course, at the source itself. In what follows, we shall use a subscript 0 when reference is being made to the upper half-space $z > 0$.

VERTICAL ELECTRIC DIPOLE EXCITATION

When the source dipole is a vertical electric dipole, we may assume that $A_{0x} = A_{0y} = A_x = A_y = 0$. (If this were not the case, it would not be possible to satisfy the boundary conditions.) Proceeding in the usual manner, we then express A_{0z} as an integral of the form (Wait 1962)

(11) $$A_{0z} = \frac{Ids}{4\pi}\int_0^\infty \frac{\lambda}{u_0}[\exp[-u_0|z-h|] + R_{\|}(\lambda)\exp[-u_0(z+h)]]J_0(\lambda\rho)d\lambda,$$

where $u_0 = (\lambda^2 + \gamma_0^2)^{\frac{1}{2}}$, $J_0(\lambda\rho)$ is a zero-order Bessel function, and $R_{\|}(\lambda)$ is an unknown coefficient. For the lower half-space, we first note that

(12) $$\frac{K}{\rho}\frac{\partial}{\partial \rho}\rho\frac{\partial}{\partial \rho}A_z + \frac{\partial^2 A_z}{\partial z^2} - \gamma^2 A_z = 0,$$

where $K = \sigma_h/\sigma_v$. Thus, a suitable integral form for A_z is

(13) $$A_z = \frac{Ids}{4\pi}\int_0^\infty T(\lambda)\exp(vz)J_0(\lambda\rho)d\lambda,$$

where $v = (\lambda^2 K + \gamma^2)^{\frac{1}{2}}$.
The tangential field components are given by

$$E_{0\rho} = \frac{1}{i\epsilon_0 \omega} \frac{\partial^2 A_{0z}}{\partial \rho \partial z}, \qquad E_\rho = \frac{1}{\sigma_h} \frac{\partial^2 A_z}{\partial \rho \partial z},$$

(14)

$$H_{0\phi} = -\frac{\partial A_{0z}}{\partial \rho}, \qquad H_\phi = -\frac{\partial A_z}{\partial \rho}.$$

Using the fact that $E_{0\rho} = E_\rho$ and $H_{0\phi} = H_\phi$ at the interface $z = 0$, it follows immediately that

(15) $$R_\parallel(\lambda) = \frac{\gamma^2 u_0 - \gamma_0^2 v}{\gamma^2 u_0 + \gamma_0^2 v}$$

and

(16) $$T(\lambda) = \frac{2\gamma^2 \lambda}{\gamma^2 u_0 + \gamma_0^2 v} \exp(-u_0 h).$$

The factor $R_\parallel(\lambda)$ may be identified as a reflection coefficient for a plane wave incident onto a homogeneous anisotropic half-space with tensor conductivity (σ). In this case, the electric vector is *parallel* to the plane of incidence and $\lambda = -i\gamma_0 \sin \theta$, where θ is the (complex) angle of incidence.

When the lower half-space is isotropic, such that $\sigma_h = \sigma_v$, the results above reduce to the well-known Sommerfeld problem. In fact, some of the methods used to evaluate the integrals in the isotropic problem may be taken over with only small modification.

QUASI-STATIC LIMIT FOR VERTICAL DIPOLE

A particularly interesting limiting case is the quasi-static limit, where all distances are small compared with a free-space wavelength. In this case, it is permissible to neglect displacement currents in the air or upper half-space. Thus, setting $\gamma_0^2 = 0$, it is seen from (16) that for $z < 0$ and $h = 0$,

(17) $$A_z = \frac{Ids}{2\pi} \int_0^\infty e^{vz} J_0(\lambda \rho) d\lambda.$$

To evaluate this result, we now write Foster's integral (1931) in the following form:

(18) $$\int_0^\infty \frac{J_0(\lambda \rho)}{(\lambda^2 + \gamma^2/K)^{\frac{1}{2}}} \exp[(\lambda^2 + \gamma^2/K)^{\frac{1}{2}} K^{\frac{1}{2}} z] d\lambda$$
$$= I_0[(\gamma/2)K^{-\frac{1}{2}}(r + zK^{\frac{1}{2}})] K_0[(\gamma/2)K^{-\frac{1}{2}}(r - zK^{\frac{1}{2}})],$$

where $r = (\rho^2 + z^2 K)^{\frac{1}{2}}$, and where I_0 and K_0 are modified Bessel functions of order zero. Thus, it immediately follows that (17) may be written

(19) $$A_z = \frac{Ids}{2\pi} \frac{\partial}{\partial z_e} I_0 K_0 \qquad \text{where} \qquad z_e = zK^{\frac{1}{2}}$$

and where the arguments of I_0 and K_0 are as indicated in (18).

Using (3), (4), and (5), and the above formula for A_z, the field components may be expressed as follows:

(20) $$E_\rho = \frac{Ids}{2\pi (\sigma_h \sigma_v)^{\frac{1}{2}} r^3} V,$$

(21) $$\text{with} \quad V = r^3 \frac{\partial^3}{\partial \rho \partial z_e^2} I_0 K_0,$$

Fields of a Dipole Over an Homogeneous Anisotropic Half-Space   99

(22) $$E_z = \frac{Ids}{2\pi\sigma_v r^3} U,$$

(23) with $$U = r^3 \left(-\frac{\gamma^2}{K} + \frac{\partial^2}{\partial z_e^2}\right) \frac{\partial}{\partial z_e} I_0 K_0,$$

(24) $$H_\phi = -\frac{Ids}{2\pi r^2} W,$$

with

(25) $$W = r^2 \frac{\partial^2}{\partial \rho \partial z_e} I_0 K_0.$$

The dimensionless functions U, V, and W may be expressed in terms of the functions I_0, K_0, I_1, and K_1 without derivatives. The explicit form of these is identical with those appearing in the isotropic problem (Wait 1953). The numerical values for U, V, and W, as given in the earlier work (Wait 1953), are also applicable to this anisotropic problem. However, we should now note that the parameters θ and α are defined by

$$\theta = -\tan^{-1}(\rho/z_e) = -\tan^{-1}(K^{-\frac{1}{2}}\rho/z)$$

and

$$\alpha = (\gamma/K^{\frac{1}{2}})r = \gamma(\rho^2 K^{-1} + z^2)^{\frac{1}{2}}.$$

In the present problem, the normalization has been chosen such that at $\theta = 90°$, U and W approach unity in the static limit when $|\alpha|$ tends to zero.

Of special interest is the radial field E_ρ in the lower half-space. Using (20) and (21) and noting that A_z satisfies (12), we obtain

(26) $$E_\rho = \frac{Ids}{2\pi(\sigma_h\sigma_v)^{1/2}} \frac{\partial}{\partial \rho} \left[\frac{\gamma^2}{K} - \frac{1}{\rho}\frac{\partial}{\partial \rho}\rho\frac{\partial}{\partial \rho}\right] I_0 K_0.$$

Now, if $|\gamma\rho| \gg 1$ and $(-z) \ll \rho$, the Bessel functions I_0 and K_0 may be approximated by

(27) $$I_0 K_0 \cong (K^{\frac{1}{2}}/\gamma\rho) \exp(-\gamma|z|).$$

Thus,

$$E_\rho \cong -\frac{i\mu_0\omega Ids}{2\pi\gamma\rho^2} e^{-\gamma|z|},$$

which depends only on the horizontal conductivity σ_h. This result, which gives the coupling between a short vertical antenna on the surface and a short horizontal antenna within the conductor, is not influenced by the vertical conductivity σ_h. At least, this is the case when $|\gamma\rho| \gg 1$ and $\rho \gg |z|$.

Applying the quasi-static limit in the upper half-space leads immediately to

(28) $$A_{0z} = \frac{Ids}{4\pi} \int_0^\infty [e^{-\lambda|z-h|} + e^{-\lambda(z+h)}] J_0(\lambda\rho) d\lambda$$

$$= \frac{Ids}{4\pi} \left[\frac{1}{[\rho^2 + (z-h)^2]^{1/2}} + \frac{1}{[\rho^2 + (z+h)^2]^{1/2}}\right].$$

For the dipole on the surface (i.e., $h = 0$), we immediately obtain

(29) $$E_{0z} = \frac{-Ids}{2\pi i\epsilon_0\omega\rho^3} \quad \text{and} \quad H_{0\phi} = -\frac{Ids}{2\pi\rho^2},$$

whereas $E_{0\rho}$ is identical with E_ρ given by (26).

RADIATION FIELDS FOR VERTICAL ELECTRIC DIPOLE

To obtain asymptotic expressions of the fields valid for large horizontal ranges (i.e., $|\gamma_0 \rho| \gg 1$), we return to the exact integral formula for A_{0z} given by (11). The latter may be written in the equivalent form:

$$(30) \quad A_{0z} = \frac{I\,ds}{4\pi}\left[\frac{e^{-\gamma_0 R_0}}{R_0} + \frac{e^{-\gamma_0 R_1}}{R_1} - 2P\right],$$

where $R_0 = [(z-h)^2 + \rho^2]^{\frac{1}{2}}$, $R_1 = [(z+h)^2 + \rho^2]^{\frac{1}{2}}$, and

$$(31) \quad P = \int_0^\infty \frac{1 - R_\parallel(\lambda)}{2}\,\frac{\lambda}{u_0}\,e^{-u_0(z+h)}\,J_0(\lambda\rho)\,d\lambda.$$

It may be noted that

$$(32) \quad \frac{1 - R_\parallel(\lambda)}{2} = \frac{\gamma_0 \Delta(\lambda)}{u_0 + \gamma_0 \Delta(\lambda)},$$

where

$$\Delta(\lambda) = \frac{K^{\frac{1}{2}}}{\eta_0 \sigma_h}\left(\lambda^2 + \frac{\gamma^2}{K}\right)^{\frac{1}{2}} \quad \text{and} \quad \eta_0 = \frac{\gamma_0}{i\epsilon_0\omega} \cong 120\pi.$$

The integral P may be evaluated by the modified saddle-point method in an identical fashion with that employed in the corresponding isotropic problem. Thus, under the conditions that $|\gamma_0\rho| \gg 1$ and $|\Delta|^2 \ll 1$, we readily find that (Wait 1962)

$$(33) \quad P \cong i(\pi p)^{\frac{1}{2}} e^{-w}\,\mathrm{erfc}(iw^{\frac{1}{2}})\,\frac{e^{-\gamma_0 R_1}}{R_1},$$

where

$$(34) \quad w = p\left(1 + \frac{h+z}{\Delta R_1}\right)^2,$$

and

$$(35) \quad p = -(\gamma_0\rho/2)\Delta^2.$$

The factor Δ is the normalized surface impedance equal to $\Delta(\lambda)$ for $\lambda = -i\gamma_0 \sin\theta_0$, where $\theta_0 = \tan^{-1}[\rho/(h+z)]$. Thus,

$$(36) \quad \Delta = \frac{\gamma_0}{\gamma}\left(1 - \frac{\gamma_0^2}{\gamma^2} K \sin^2\theta_0\right)^{\frac{1}{2}}.$$

When the "numerical distance" p is large (i.e., $|p| \gg 1$), only the first term in the asymptotic expansion of the error function complement need be retained. In this case, it follows that

$$(37) \quad A_{0z} \cong (I\,ds/4\pi)[\psi_a + \psi_b],$$

where

$$(38) \quad \psi_a \cong \frac{e^{-\gamma_0 R_0}}{R_0} + \frac{C - \Delta}{C + \Delta}\,\frac{e^{-\gamma_0 R_1}}{R_1},$$

and

$$(39) \quad \psi_b \cong -\left\{\frac{1}{p(1 + (C/\Delta))^3} + \frac{1 \times 3}{2p^2(1 + (C/\Delta))^5}\right.$$
$$\left. + \frac{1 \times 3 \times 5}{4p^3(1 + (C/\Delta))^7} + \cdots\right\}\frac{e^{-\gamma_0 R_1}}{R_1},$$

where $C = \cos\theta_0 = (h+z)/R_1$. The contribution ψ_a corresponds to geometrical optics, since $(C-\Delta)/(C+\Delta)$ is the appropriate reflection coefficient for plane waves incident onto the lower half-space at an angle of incidence equal to θ_0. The contribution ψ_b is a Norton surface wave which is identical with the result expected for an isotropic lower half-space, if the appropriate value of Δ is employed.

Along the interface (i.e., $h = z = 0$), the geometrical optics contribution ψ_a vanishes and then, provided p is sufficiently large, we have

(40) $\qquad E_{0z} \cong -i\mu_0\omega A_{0z}$

(41) $\qquad \cong \dfrac{i\mu_0\omega I ds}{2\pi\rho} e^{-ik_0\rho} \left[-\dfrac{1}{2p} - \dfrac{1\times 3}{(2p)^2} - \dfrac{3\times 5}{(2p)^3} - \cdots \right],$

where $k_0 = -i\gamma_0$ and

(42) $\qquad \dfrac{1}{2p} = \dfrac{1}{(ik_0^3\rho/\gamma^2)(1+k_0^2 K/\gamma^2)}.$

The influence of the anisotropy of the lower half-space may be seen by expressing E_{0z} as a ratio to the field $E_{z0}^{(\text{iso})}$ for the corresponding isotropic half-space. Thus,

(43) $\qquad \dfrac{E_{0z}}{E_{0z}^{\text{iso}}} \cong \dfrac{1+k_0^2/\gamma^2}{1+k_0^2 K/\gamma^2},$

provided $|p|$ is sufficiently large that only the leading term in (41) is required. For sufficiently large values of $K(=\sigma_h/\sigma_v)$, this field ratio may be significantly less than unity. However, for propagation over the earth's surface, K itself would never be much greater than unity and as $|k_0^2/\gamma^2|$ is small, the field ratio would be near unity.

EXCITATION BY HORIZONTAL ELECTRIC DIPOLE

We now consider a horizontal electric dipole of current moment $I\,ds$ located at $z = h$ and oriented in the x direction. We now find, in order to satisfy the boundary conditions, that $\mathbf{A}$ has nonvanishing x and z components. To this end, we write, for the lower half-space, that

(44) $\qquad A_x = \dfrac{I ds}{4\pi} \displaystyle\int_0^\infty F(\lambda, z) J_0(\lambda\rho)\,d\lambda$

and

(45) $\qquad A_z = \dfrac{I ds}{4\pi} \dfrac{\partial}{\partial x} \displaystyle\int_0^\infty G(\lambda, z) J_0(\lambda\rho)\,d\lambda,$

where F and G are yet to be determined. Noting that A_x and A_z satisfy (6) and (9), respectively, it is a simple matter to show that F and G satisfy

(46) $\qquad \partial^2 F/\partial z^2 - (\lambda^2 + \gamma^2) F = 0$

and

(47) $$\partial^2 G/\partial z^2 - (\lambda^2 K + \gamma^2)G = (K-1)\partial F/\partial z,$$

where, as usual, $K = \sigma_h/\sigma_v$ and $\gamma^2 = i\mu_0\omega\sigma_h$.

Using rather elementary methods, we find that the solutions of the coupled system (46) and (47) are

(48) $$F = A(\lambda)\,e^{uz} \quad \text{where} \quad u = (\lambda^2 + \gamma^2)^{1/2}$$

and

(49) $$G = B(\lambda)\,e^{vz} - (u/\lambda^2)A(\lambda)\,e^{uz} \quad \text{where} \quad v = (\lambda^2 K + \gamma^2)^{1/2}$$

where A and B are to be determined from the boundary conditions.

Suitable integral representations for the potentials in the upper and the lower half-space are now written:

(50) $$A_{0x} = \frac{Ids}{4\pi}\int_0^\infty \frac{\lambda}{u_0}[e^{-u_0|z-h|} + R_\perp(\lambda)e^{-u_0(z+h)}]J_0(\lambda\rho)d\lambda,$$

(51) $$A_{0z} = \frac{Ids}{4\pi}\frac{\partial}{\partial x}\int_0^\infty S(\lambda)e^{-u_0(z+h)}J_0(\lambda\rho)d\lambda,$$

(52) $$A_x = \frac{Ids}{4\pi}\int_0^\infty A(\lambda)e^{uz}J_0(\lambda\rho)d\lambda,$$

(53) $$A_z = \frac{Ids}{4\pi}\frac{\partial}{\partial x}\int_0^\infty [B(\lambda)e^{vz} - (u/\lambda^2)A(\lambda)e^{uz}]J_0(\lambda\rho)d\lambda.$$

The field components, for $z < 0$, are to be obtained from

(54) $$E_x = \frac{1}{\sigma_h}\left[-\gamma^2 A_x + \frac{\partial}{\partial x}\left(\frac{\partial A_x}{\partial x} + \frac{\partial A_z}{\partial z}\right)\right],$$

(55) $$E_y = \frac{1}{\sigma_h}\frac{\partial}{\partial y}\left(\frac{\partial A_x}{\partial x} + \frac{\partial A_z}{\partial z}\right),$$

(56) $$E_z = \frac{1}{\sigma_h}\left[-\gamma^2 A_z + \frac{\partial}{\partial z}\left(\frac{\partial A_x}{\partial x} + \frac{\partial A_z}{\partial z}\right)\right],$$

(57) $$H_x = \frac{\partial A_z}{\partial y},\quad H_y = \frac{\partial A_x}{\partial z} - \frac{\partial A_z}{\partial x},\quad H_z = -\frac{\partial A_x}{\partial y}.$$

These field expressions also apply to the upper half-space (i.e., $z > 0$), if σ_h is replaced by $i\epsilon_0\omega$ and γ by γ_0.

Continuity of the tangential field components at $z = 0$ immediately leads to the results that

(58) $$A(\lambda) = (\lambda/u_0)(1 + R_\perp(\lambda))\,e^{-u_0 h},$$

where

(59) $$R_\perp(\lambda) = (u_0 - u)/(u_0 + u),$$

and

Fields of a Dipole Over an Homogeneous Anisotropic Half-Space 103

(60) $$S(\lambda) = [B(\lambda) - (u/\lambda^2) A(\lambda)] e^{u_0 h},$$

where

(61) $$B(\lambda) = \frac{1}{\lambda^2} \frac{\gamma_0^2 + u_0 u + \lambda^2}{u_0 + \gamma_0^2 v/\gamma^2} A(\lambda).$$

Within the lower half-space we now have the following exact integral formulas:

(62) $$A_x = \frac{Ids}{2\pi} \int_0^\infty \frac{\lambda}{u_0 + u} e^{uz} e^{-u_0 h} J_0(\lambda\rho) d\lambda$$

and

(63) $$A_z = \frac{Ids}{2\pi} \frac{\partial}{\partial x} \int_0^\infty \left[\frac{\gamma^2 u_0}{\gamma^2 u_0 + \gamma_0^2 v} e^{vz} - \frac{u}{u_0 + u} e^{uz} \right] \frac{e^{-u_0 h}}{\lambda} J_0(\lambda\rho) d\lambda.$$

QUASI-STATIC LIMIT FOR HORIZONTAL DIPOLE

In the quasi-static limit (i.e., $\gamma_0^2 = 0$), and assuming that $h = 0$, (62) and (63) reduce to

(64) $$A_x = \frac{Ids}{2\pi} \int_0^\infty \frac{\lambda}{\lambda + u} e^{uz} J_0(\lambda\rho) d\lambda$$

and

(65) $$A_z = \frac{Ids}{2\pi} \frac{\partial}{\partial x} \int_0^\infty \left[e^{vz} - \frac{u}{\lambda + u} e^{uz} \right] J_0(\lambda\rho) \lambda^{-1} d\lambda.$$

The horizontal electric fields are then obtained from

(66) $$E_x = -i\mu_0 \omega A_x - \partial\psi/\partial x$$

and

(67) $$E_y = -\partial\psi/\partial y,$$

where

(68) $$\psi = -\frac{1}{\sigma_h} \left(\frac{\partial A_x}{\partial x} + \frac{\partial A_z}{\partial z} \right).$$

The potential ψ is now expressed as the sum of two parts ψ_1 and ψ_2 such that ψ_2 vanishes in the isotropic limit. Thus,

(69) $$\psi_1 = -\frac{Ids}{2\pi\sigma_h} \frac{\partial}{\partial x} \int_0^\infty e^{uz} J_0(\lambda\rho) d\lambda$$

and

(70) $$\psi_2 = \frac{Ids}{2\pi\sigma_h} \frac{\partial}{\partial x} \int_0^\infty \left(\frac{ue^{uz} - ve^{vz}}{\lambda} \right) J_0(\lambda\rho) d\lambda.$$

First of all, it is interesting to note that in the static limit (i.e., $\omega = 0$), $\mathbf{E} = -\text{grad } \psi$, where

(71) $$\psi = -\frac{Ids}{2\pi\sigma_h} K^{\frac{1}{2}} \frac{\partial}{\partial x} \int_0^\infty \exp(\lambda K^{\frac{1}{2}} z) J_0(\lambda\rho) d\lambda$$

(72) $$= -\frac{Ids}{2\pi(\sigma_h\sigma_v)^{1/2}} \frac{\partial}{\partial x} \left[\frac{1}{\rho^2 + Kz^2}\right]^{\frac{1}{2}}$$

(73) $$= \frac{Ids}{2\pi(\sigma_h\sigma_v)^{1/2}} \frac{x}{(\rho^2 + Kz^2)^{3/2}}, \quad K = \sigma_h/\sigma_v.$$

This result is consistent with potential theory for static fields in anisotropic media (e.g., Smythe 1968).

For the quasi-static case where $\omega \neq 0$, for $z < 0$, it is convenient to write the electric field in the form

(74) $$\mathbf{E} = \mathbf{E}_1 + \mathbf{E}_2$$

where, again, $\mathbf{E}_2$ vanishes in the isotropic limit. For the isotropic component, we see that

(75) $$E_{1x} = -i\mu_0\omega A_x - \partial\psi_1/\partial x$$

and

(76) $$E_{1y} = -\partial\psi_1/\partial y.$$

The resulting integrals may be expressed in closed form as demonstrated previously. For example,

(77) $$E_{1x} = -\frac{Ids}{2\pi\sigma_h} \left(\frac{\partial^3 N}{\partial y^2 \partial z} + \frac{\partial^2 P}{\partial z^2}\right)$$

and

(78) $$E_{1y} = \frac{Ids}{2\pi\sigma_h} \frac{\partial^3 N}{\partial x \partial y \partial z},$$

where

(79) $$P = \exp(-\gamma r_1)/r_1,$$

(80) $$N = I_0[(\gamma/2)(r_1 + z)] K_0[(\gamma/2)(r_1 - z)],$$

and

$$r_1 = (\rho^2 + z^2)^{\frac{1}{2}} = (x^2 + y^2 + z^2)^{\frac{1}{2}}.$$

The anisotropic component may be evaluated by using the basic integral (Ryzhik and Gradshtein 1951):

(81) $$\int_0^\infty \frac{e^{uz}}{u} J_1(\lambda\rho) d\lambda = \frac{1}{\gamma\rho} [e^{\gamma z} - \exp(-\gamma(\rho^2 + z^2)^{\frac{1}{2}})],$$

which is valid for $z < 0$ and $\text{Re}\,\gamma > 0$.

By writing the potential ψ_2 in the form

(82) $$\psi_2 = -\frac{Ids}{2\pi\sigma_h} \cos\phi \frac{\partial}{\partial z} \int_0^\infty (e^{uz} - e^{vz}) J_1(\lambda\rho) d\lambda,$$

where $\cos \phi = x/\rho$, it follows immediately that, for $z < 0$,

$$(83) \quad \psi_2 = -\frac{Ids \cos \phi}{2\pi\sigma_h\rho} \frac{\partial}{\partial z} \left[\frac{\exp[-\gamma(\rho^2 + z^2)^{\frac{1}{2}}]z}{(\rho^2 + z^2)^{1/2}} - \frac{\exp[-\gamma(\rho^2 K^{-1} + z^2)^{\frac{1}{2}}]z}{(\rho^2 K^{-1} + z^2)^{1/2}} \right].$$

Then, of course,

$$(84) \quad E_{2x} = -\partial \psi_2/\partial x \quad \text{and} \quad E_{2y} = -\partial \psi_2/\partial y.$$

QUASI-STATIC FIELDS IN THE INTERFACE FOR THE HORIZONTAL DIPOLE

An important limiting case is that when $h = z = 0$, which corresponds to the situation where both source and observer are in the interface. In this case, we may write in cylindrical coordinates

$$(85) \quad E_\rho = -i\mu_0\omega A_x \cos \phi - (\partial/\partial\rho)\psi$$

and

$$(86) \quad E_\phi = i\mu_0\omega A_x \sin \phi - (1/\rho)(\partial/\partial\phi)\psi,$$

where

$$(87) \quad A_x = \frac{Ids}{2\pi\gamma^2\rho^3}[1 - (1 + \gamma\rho)e^{-\gamma\rho}],$$

and

$$(88) \quad \psi = \frac{Ids}{2\pi\sigma_h\rho^2} \cos \phi [1 - \exp(-\gamma\rho) + K^{\frac{1}{2}}\exp(-\gamma\rho K^{-\frac{1}{2}})].$$

The corresponding expressions for the tangential electric fields in the interface may be written

$$(89) \quad E_\rho = \frac{Ids}{2\pi\sigma_h\rho^3} \cos \phi [p_1 + p_2],$$

where

$$(90) \quad p_1 = 1 + (1 + \gamma\rho) e^{-\gamma\rho},$$

$$(91) \quad p_2 = 2[K^{\frac{1}{2}} \exp(-\gamma\rho K^{-\frac{1}{2}}) - \exp(-\gamma\rho)] + \gamma\rho[\exp(-\gamma\rho K^{-\frac{1}{2}}) - \exp(-\gamma\rho)],$$

and

$$(92) \quad E_\phi = \frac{Ids}{2\pi\sigma_h\rho^3} \sin \phi [q_1 + q_2],$$

where

$$(93) \quad q_1 = 2 - (1 + \gamma\rho) e^{-\gamma\rho},$$

and

$$(94) \quad q_2 = K^{\frac{1}{2}} \exp(-\gamma\rho K^{-\frac{1}{2}}) - \exp(-\gamma\rho).$$

For an isotropic half-space, both p_2 and q_2 vanish.

A rather simple expression is also obtained for the vertical component of the magnetic field in the interface. Using (87), we find that, in the quasi-static approximation,

$$H_z = -\frac{\partial A_x}{\partial y}$$ (95)

$$= \frac{Ids}{2\pi\gamma^2\rho^4}[3 - (3 + 3\gamma\rho + \gamma^2\rho^2)e^{-\gamma\rho}]\sin\phi,$$

which does not depend on K. It has the same form as an isotropic medium with conductivity σ_h.

RADIATION FIELDS OF THE HORIZONTAL DIPOLE

Our final topic is a short discussion of the far fields of the horizontal dipole. To simplify the treatment, we choose $z = h = 0$. As it turns out for this case, A_x is expressible in closed form without approximation. Unfortunately, however, this is not the case for A_z.

From (62) we see that, for $z = h = 0$,

$$A_{0x} = A_x = \frac{Ids}{2\pi}\int_0^\infty \frac{\lambda}{u_0 + u} J_0(\lambda\rho) d\lambda$$ (96)

$$= \frac{Ids}{2\pi(\gamma_0^2 - \gamma^2)}\int_0^\infty \lambda(u_0 - u) J_0(\lambda\rho) d\lambda.$$ (97)

Noting that

$$\int_0^\infty \frac{\lambda}{u_0} J_0(\lambda\rho) d\lambda = \frac{e^{-\gamma_0\rho}}{\rho}$$ (98)

and using the property

$$\left(\frac{1}{\rho}\frac{\partial}{\partial\rho}\rho\frac{\partial}{\partial\rho} + \lambda^2\right) J_0(\lambda\rho) = 0,$$ (99)

it readily follows that

$$A_x = \frac{Ids}{2\pi(\gamma^2 - \gamma_0^2)\rho^3}[(1 + \gamma_0\rho)e^{-\gamma_0\rho} - (1 + \gamma\rho)e^{-\gamma\rho}],$$ (100)

which is exact. From this, it follows that

$$H_z = \frac{Ids\sin\phi}{2\pi(\gamma^2 - \gamma_0^2)\rho^4}[(3 + 3\gamma_0\rho + \gamma_0^2\rho^2)e^{-\gamma_0\rho} - (3 + 3\gamma\rho + \gamma^2\rho^2)e^{-\gamma\rho}],$$ (101)

which, not surprisingly, has the same form as a horizontal dipole on the surface of an isotropic half-space (Wait 1961). In the far zone, $|\gamma_0\rho| \gg 1$, it is seen that

$$H_z \cong E_\phi/\eta_0 \cong \frac{Ids\gamma_0^2}{2\pi(\gamma^2 - \gamma_0^2)\rho^2} e^{-\gamma_0\rho} \sin\phi,$$ (102)

which characterizes the "horizontally polarized" ground wave from the horizontal dipole on the surface.

Of more practical interest is the vertically polarized ground wave radiated from the horizontal dipole. This is related directly to A_z. Again choosing $z = h = 0$, it is evident from (61) that

$$(103) \quad A_z = A_{0z} = \frac{Ids}{2\pi} \frac{\partial}{\partial x} \int_0^\infty \frac{u_0^2\gamma^2 - \gamma_0^2 uv}{(u_0\gamma^2 + \gamma_0^2 v)(u_0 + u)} \frac{J_0(\lambda\rho)}{\lambda} d\lambda.$$

For $|\gamma_0\rho| \gg 1$, this may be approximated by (Wait 1961)

$$(104) \quad A_{0z} \cong -\frac{Ids}{2\pi\rho} \cos\phi \Delta_0 [1 - i(\pi p_0)^{\frac{1}{2}} e^{-p_0} \operatorname{erfc}(ip_0^{\frac{1}{2}})],$$

where $p_0 = -(\gamma_0\rho/2)\Delta_0^2$ and $\Delta_0 = (\gamma_0/\gamma)(1 - \gamma_0^2 K/\gamma^2)^{\frac{1}{2}}$. This result is valid when $|\Delta_0^2| \ll 1$. The corresponding field components are then to be obtained from

$$(105) \quad E_{0z} \cong -\eta_0 H_{0\phi} \cong -i\mu_0\omega A_{0z}.$$

The factor in square brackets in (104) is the usual Sommerfeld attenuation function which is appropriate for vertically polarized ground waves over a flat earth of surface impedance Δ_0.

Within the limits of the approximations stated above, the vertical electric field component in the air at the surface may be expressed simply by

$$(106) \quad E_{0z} \cong \Delta_0 \cos\phi E_{0z}^{(v)},$$

where $E_{0z}^{(v)}$ is the vertical electric field of a vertical electric dipole of current moment $I\,ds$. This result is what one would expect on the basis of previous analyses for an isotropic conducting half-space. In the anisotropic case, the "wave-tilt" factor Δ_0 depends on the conductivity ratio K. However, the dependence is rather slight as $|(\gamma_0/\gamma)^2|$ is usually small and consequently $\Delta_0 \cong (\gamma_0/\gamma)$, which does not involve K.

CONCLUDING REMARKS

The results given in this paper should shed some light on the influence of anisotropy of ground conductivity in radio propagation. While in most cases the effect would seem to be small, it is interesting to note that, in certain configurations, the effect of the anisotropy may be displayed in an explicit fashion. This is particularly the case for the quasi-static fields of the x-directed horizontal dipole in the interface $z = 0$. For example, from (87) and (88), it is a simple matter to show that, for $z = 0$,

$$(107) \quad E_y = \frac{Ids\,xy}{2\pi\sigma_h\rho^5} 3[1 + \Lambda],$$

where

$$(108) \quad 3\Lambda = (3K^{\frac{1}{2}} + \gamma\rho)\exp(-\gamma\rho K^{-\frac{1}{2}}) - (3 + \gamma\rho)\exp(-\gamma\rho).$$

In the isotropic limit, $K = 1$ and Λ vanishes. It is also evident that Λ vanishes even for the anisotropic case, if $|\gamma\rho| \gg 1$. On the other hand, the low frequency limit yields

$$E_y = \frac{3Ids xy}{2\pi(\sigma_h\sigma_v)^{1/2}\rho^5}[1-\gamma\rho(K^{\frac{1}{2}}-K^{-\frac{1}{2}})+0(\gamma^2\rho^2)].$$

Then, if displacement currents are negligible, σ_h, σ_v, and K are real, whereas $\gamma\rho = |\gamma\rho|\exp(i\pi/4)$, with $|\gamma\rho| \cong (\sigma_h\mu_0\omega)^{\frac{1}{2}}\rho$. Consequently, the presence of anisotropy manifests itself in a frequency dependence of the transfer ratio E_y/I. This immediately suggests that a measurement of the E_y component as a function of frequency should yield diagnostic information of the anisotropy of a conducting half-space.

To illustrate the frequency dependence of the transfer ratio E_y/I, the amplitude and the phase of $1 + \Lambda$ is plotted in Fig. 1(a), (b), respectively. A small inset in Fig. 1(b) shows the geometry. The value of the anisotropy parameter K is indicated on the curves. As expected, when K is equal to or near unity, the frequency dependence of the factor $1 + \Lambda$ is very small. On the other hand, values of K in excess of or less than unity, lead to a significant frequency dependence. It should be stressed that the curves in Fig. 1(a), (b) are valid only when all displacement currents are negligible. A planned sequel to this paper

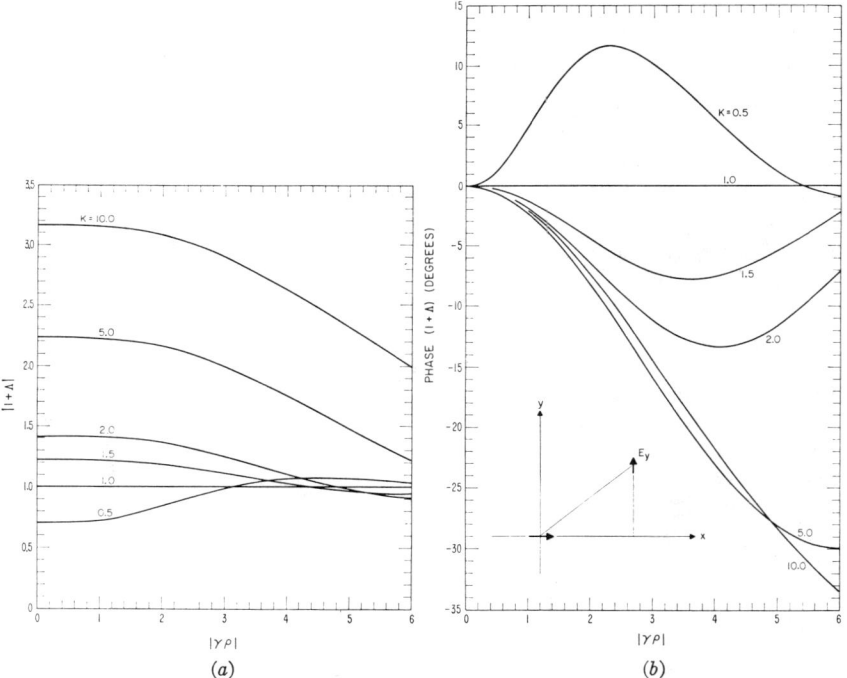

FIG. 1(a, b). The normalized transfer impedance for two perpendicularly oriented dipoles on the surface of an anisotropic conducting half-space $[K = \sigma_h/\sigma_v, |\gamma\rho| \cong (\sigma_h\mu_0\omega)^{\frac{1}{2}}\rho]$.

will contain an analysis for a stratified anisotropic half-space with dipolar excitation. Unfortunately, however, the resulting integrals cannot be evaluated in closed form.

REFERENCES

ARBEL, E., and FELSEN, L.B., (1963), *Electromagnetic Wave Theory*, edited by E.C. Jordan, Pergamon Press, Oxford, pp. 391-420.

BANOS, A., (1966), *Dipole Radiation in the Presence of a Conducting Half-Space*, Pergamon Press, Oxford.

CHETAEV, D.C., (1962), On the field of a low frequency electric dipole situated on the surface of a anisotropic conducting half-space, *Zhur. Tekh. Fiz.*, 32, No. 11, 1342-1352.

FOSTER, R.M., (1931), Mutual impedance of grounded wires lying on the surface of the earth, *Bell System Tech. J.*, 10, 408-419.

RYZHIK, I.M. and GRADSHTEIN, I.S., (1951), *Tables of Integrals, Sums and Products*, Gostekhizdat, Moscow-Leningrad.

SINHA, A.K., and BHATTACHARYA, P.K., (1967), Electric dipole over an anisotropic and inhomogeneous earth, *Geophysics*, 32, No. 4, 652-662.

SMYTHE, W.R., (1968), *Static and Dynamic Electricity*, 3rd ed., McGraw-Hill, N.Y..

SOMMERFELD, A., (1909), Propagation of radio waves, *Ann. Physik*, 28, 665-695.

TIKHONOV, A.N., (1959), The propagation of continuous electromagnetic wave in laminary anisotropic medium, *Doklady Akad. Nauk S.S.S.R.*, 4, 566-578.

WAIT, J.R., (1953), The fields of an electric dipole in a semi-infinite conducting medium, *J. Geophys. Res.*, Vol. 58, 20-28.

WAIT, J.R., (1961), The electromagnetic fields of a horizontal dipole in the presence of a conducting half-space, *Can. J. Phys.*, Vol. 39, 1017-1028.

WAIT, J.R., (1964), Electromagnetic surface waves, in *Advances in Radio Research*, Vol. 1, edited by J.A. Saxton, Academic Press, London, pp. 157-217.

WAIT, J.R., (1966), Electromagnetic fields of a dipole over an anisotropic half-space, *Can. J. Phys.*, 44, 2387-2401.

11
Fields of an Horizontal Dipole Over a Stratified Anisotropic Half-Space

Here we outline a general theory for the radiation of a horizontal dipole over a horizontally stratified anisotropic medium. The model is a lower half-space which consists of N homogeneous layers and an upper half-space with properties identical to those of free space. The fields vary everywhere according to $\exp(i\omega t)$. The anisotropy in the lower half-space is of a type such that, for the nth layer, the complex conductivity tensor (σ) is of the form

$$(\sigma_n) = \begin{bmatrix} \sigma_{hn} & 0 & 0 \\ 0 & \sigma_{hn} & 0 \\ 0 & 0 & \sigma_{vn} \end{bmatrix} \qquad (1)$$

where σ_{hn} and σ_{vn} are the (complex) conductivity in the horizontal, and the vertical directions, respectively. We see that $\sigma_{hn} = g_{hn} + i\varepsilon_{hn}\omega$ and $\sigma_{vn} = g_{vn} + i\varepsilon_{vn}\omega$ where g_{hn}, ε_{hn}, g_{vn}, and ε_{vn} are all real. To simplify the problem, we assume that the magnetic permeability μ_o is constant everywhere and equal to that of free space.

The situation is illustrated in Fig. 1 where the y axis is perpendicular to the paper. A horizontal dipole of length ds with current I is located on the z axis at $z = h$ and it is oriented in the x direction. The problem is to obtain an expression for the fields in the upper half-space in terms of the properties of the layers.

The method outlined here is an extension of that used for analyzing radiation from dipoles in the presence of a homogeneous, isotropic half-space. An excellent review for problems of this type has been published recently by Baños [1965]. Related studies involving stratified media have also been discussed extensively [Wait, 1970]. Then, of course, numerous investigations of

Fields of an Horizontal Dipole

plane wave propagation in stratified anisotropic media have appeared. An excellent treatise on this subject was written by Budden [1961], who has made notable contributions to the subject.

Our starting point is Maxwell's equations which, for the nth layer, are written

$$\operatorname{curl} H_n = (\sigma_n) E_n \tag{2}$$

and

$$\operatorname{curl} E_n = -i\mu_0 \omega H_n \tag{3}$$

where E_n and H_n are the electric and magnetic fields. Introducing a vector potential A_n and a scalar potential ψ in the usual manner, we have

$$H_n = \operatorname{curl} A_n \tag{4}$$

and

$$E_n = -i\mu_0 \omega A_n - \operatorname{grad} \psi_n. \tag{5}$$

In a straightforward manner, it is found that

$$(\nabla^2 - \gamma_n^2) \begin{matrix} A_{zn} \\ A_{yn} \end{matrix} = 0 \tag{6}$$

where $\gamma_n^2 = i\sigma_{hn}\mu_0\omega$ and provided $\operatorname{div} A_n + \sigma_{hn}\psi_n = 0$. As a consequence,

$$\nabla^2 A_{zn} - (\sigma_{hn} - \sigma_{vn}) \frac{1}{\sigma_{hn}} \frac{\partial}{\partial z} \operatorname{div} A_n$$
$$- \gamma_n^2 \frac{\sigma_{vn}}{\sigma_{hn}} A_{zn} = 0. \tag{7}$$

In the upper half-space, all three components of the vector potential satisfy

$$(\nabla^2 - \gamma_0^2) A_{j0} = 0 \tag{8}$$

where $j = x, y, z,$ and $\gamma_0^2 = -\varepsilon_0\mu_0\omega^2 = -\omega^2/c^2$, where ε_0 is the dielectric constant of free space and c is the velocity of light in vacuum.

For the present problem, it turns out that the boundary conditions may all be satisfied if the vector potential A_n has only x and z components. We write the latter in the integral form

$$A_{zn} = \int_0^\infty F_n(\lambda, z) J_0(\lambda \rho) d\lambda \tag{9}$$

and

$$A_{zn} = \frac{\partial}{\partial x} \int_0^\infty G_n(\lambda, z) J_0(\lambda \rho) d\lambda \tag{10}$$

where J_0 is a Bessel function of order zero, where $\rho = (x^2 + y^2)^{1/2}$, and where F_n and G_n are yet to be determined. Noting that A_{xn} and A_{zn} satisfy (6) and (7), respectively, it may be shown without difficulty that

$$\frac{\partial^2 F_n}{\partial z^2} - (\lambda^2 + \gamma_n^2)F_n = 0 \qquad (11)$$

and

$$\frac{\partial^2 G_n}{\partial z^2} - (\lambda^2 \kappa_n + \gamma_n^2)G_n = (\kappa_n - 1)\frac{\partial F_n}{\partial z} \qquad (12)$$

where $\kappa_n = \sigma_{hn}/\sigma_{vn}$.

The general solutions of (12) may be written in the form

$$F_n = A_n(\lambda)\exp(u_n z) + a_n(\lambda)\exp(-u_n z) \qquad (13)$$

where $u_n = (\lambda^2 + \gamma_n^2)^{1/2}$. This may be inserted into the right-hand side of (12) which is then an elementary inhomogeneous wave equation. Its solution is readily found to be

$$G_n = B_n(\lambda)\exp(v_n z) + b_n(\lambda)\exp(-v_n z)$$
$$- (u_n/\lambda^2)A_n(\lambda)\exp(u_n z)$$
$$+ (u_n/\lambda^2)a_n(\lambda)\exp(-u_n z) \qquad (14)$$

where $v_n = (\lambda^2 \kappa_n + \gamma_n^2)^{1/2}$. This form holds for all N layers except that $b_N(\lambda)$ and $a_N(\lambda)$ are zero, because the fields must vanish as $z \to -\infty$. Also, to be definite, we choose the real parts of u_n and v_n to be positive.

In anticipation of the form of the final results we write, for the upper half-space

$$A_{x0} = \frac{Ids}{4\pi}\int_0^\infty [\exp(-u_0|z-h|)$$
$$+ R_\perp(\lambda)\exp(-u_0(z+h))]\frac{\lambda}{u_0}J_0(\lambda\rho)d\lambda \qquad (15)$$

and

$$A_{z0} = \frac{Ids}{4\pi}\frac{\partial}{\partial x}\int_0^\infty C(\lambda)$$
$$\cdot \exp(-u_0(z+h))J_0(\lambda\rho)d\lambda, \qquad (16)$$

where $u_0 = (\lambda^2 + \gamma_0^2)^{1/2}$, while $R_\perp(\lambda)$ and $C(\lambda)$ are coefficients yet to be determined. It may be remarked that, in the absence of the lower half-space, both $R_\perp(\lambda)$ and $C(\lambda)$ would be zero and the remaining result for A_{xo} is merely equal to $(Ids)/(4\pi R_0)\exp(-\gamma_0 R_0)$ where $R_0 = [\rho^2 + (z-h)^2]^{1/2}$.

Fields of an Horizontal Dipole

The requirement that the tangential fields are continuous at the nth interfaces in turn requires that, for $z = z_n$,

$$A_{x,n-1} = A_{x,n}, \quad \frac{\partial A_{z,n-1}}{\partial z} = \frac{\partial A_{z,n}}{\partial z},$$

$$A_{z,n-1} = A_{z,n}.$$

and

$$\frac{1}{\gamma_{n-1}^2}\left(\frac{\partial A_{x,n-1}}{\partial x} + \frac{\partial A_{z,n-1}}{\partial z}\right)$$
$$= \frac{1}{\gamma_n^2}\left(\frac{\partial A_{x,n}}{\partial x} + \frac{\partial A_{z,n}}{\partial z}\right). \tag{17}$$

By applying these four boundary equations to the bottom interface (i.e., at $z = z_{N-1}$) and working to the top (i.e., at $z = 0$), it is possible to successively eliminate the coefficients A_n, B_n, a_n, b_n as n progresses from N to 1. The final result of this process may be written succinctly as follows:

$$R_\perp(\lambda) = (N_0 - Y_1)/(N_0 + Y_1), \tag{18}$$

where $N_0 = u_0/(i\mu_0\omega)$ and where

$$Y_n = N_n \frac{Y_{n+1} + N_n \tanh u_n h_n}{N_n + Y_{n+1} \tanh u_n h_n},$$
$$Y_n = u_n/(i\mu\omega) \text{ for } n = 1, 2, 3, \cdots, N-1$$

and $Y_N = N_N = u_N/i\mu_0\omega)$. In addition, it is found that

$$C(\lambda) = [R_\perp(\lambda) + R_\parallel(\lambda)]\lambda^{-1} \quad \text{where}$$
$$R_\parallel(\lambda) = (K_0 - Z_1)/(K_0 + Z_1), \tag{19}$$

where $K_0 = u_0/(i\varepsilon_0\omega)$ and where

$$Z_n = K_n \frac{Z_{n+1} + K_n \tanh v_n h_n}{K_n + Z_{n+1} \tanh v_n h_n},$$

$K_n = v_n/\sigma_{hn}$ for $n = 1, 2, 3, \ldots, N-1$ and $Z_N = K_N = v_N/\sigma_{hN}$.

A special case of the foregoing general formulation is when the lower half-space is homogeneous, corresponding to an infinite thickness of the upper layer (i.e., $z_1 \to \infty$). Then $R_\perp(\lambda) = (u_0 - u_1)/(u_0 + u_1)$ and $R_{||}(\lambda) = (\gamma_1^2 u_0 - \gamma_0^2 v_1)/(\gamma_1^2 u_0 + \gamma_0^2 v_1)$ and, therefore, $\lambda C(\lambda) = 2(\gamma_1^2 u_0)/(\gamma_1^2 u_0 + \gamma_0^2 v_1) - 2u_1(u_0 + u_1)$.

This is in agreement with a result derived previously [Wait, 1966]. In this special case, $R_\perp(\lambda)$ and $R_{||}(\lambda)$ may be identified as the Fresnel-type reflection coefficients for polarization perpendicular, and parallel, respectively, to the plane of incidence for a plane wave incident at an angle $\sin^{-1}(\lambda c/\omega)$ onto a homogeneous anisotropic half-space. In the general case, $R_\perp(\lambda)$ and $R_{||}(\lambda)$, as defined by (18) and (19), may be interpreted as plane-wave reflection coefficients for stratified anisotropic media.

An explicit evaluation of the integrals in (9) and (10), for the general case, is a complicated business. However, if the individual layers are sufficiently highly conducting, the waveguide modes within these layers are highly attenuated. Then the asymptotic procedure for evaluating the far-zone field is identical to that used for isotropic stratified media [Wait, 1970]. Choosing both the source and the observer to be on the surface of the half-space (i.e., $z = h = 0$), we have, in cylindrical coordinates,

$$H_z \cong E_\phi/\eta_0 \cong \frac{Ids\gamma_0{}^3}{2\pi(\gamma_1{}^2 - \gamma_0{}^2)\rho^2} \cdot \exp(-\gamma_0\rho)Q_\perp{}^2 \sin\phi, \qquad (20)$$

where $Q_\perp \simeq N_1/Y_1\big|_{\lambda = \omega/c = -i\gamma_0}$ and

$$E_z \cong -\eta_0 H_\phi \cong \frac{i\mu_0\omega Ids}{2\pi\rho}\Delta_0 Q_{||} \cdot [1 - i(\pi p)^{1/2}e^{-p}\text{erfc}(ip^{1/2})]\cos\phi, \qquad (21)$$

where $\Delta_0 = (\gamma_0/\gamma_1)(1 - \gamma_0^2\kappa_1/\gamma_1^2)^{1/2}$, $p = -(\gamma_0\rho/2)\Delta_0^2 Q_{||}{}^2$, $Q_{||} = Z_1/K_1\big|_{\lambda = \omega/c = -i\gamma_0}$, and erfc is the complement of the error function.

The quantities $Q_\perp$ and $Q_{||}$ are identical in form to the "stratification factors" encountered in isotropic stratified media [Wait, 1962]. The normalization is such that they approach unity in the case of a homogeneous half-space. The general definition of $Q_\perp$ and $Q_{||}$, as given above, indicates that they are to be evaluated as the ratio of admittances and impedances, respectively, for $\lambda = \omega/c$ where $\omega/c = 2\pi/(\text{free-space wavelength})$.

When the "numerical distance" $|p|$ is small compared with unity, the attenuation function given by the square bracket in (21) is approximately unity. In this important case, E_z varies as ρ^{-1} whereas H_z varies as ρ^{-2}. Thus, we recover the well-known fact that a low-frequency horizontal dipole radiates predominantly a vertically polarized ground wave.

REFERENCES

BAÑOS, A., (1965), *Dipole Radiation in the Presence of a Conducting Half-Space*, New York:Pergamon.

BUDDEN, K.G., (1961), *Radio Waves in the Ionosphere*, Cambridge, England: Cambridge University Press.

SINHA, A.K., (1969), Vertical electric dipole over and inhomogeneous and anisotropic earth, *Pure and Appl. Geophy.*, 72, 123-147.

WAIT, J.R., (1966), Fields of a horizontal dipole over a stratified anisotropic half-space, *IEEE Trans.*, AP-14, 790-792.

WAIT, J.R., (1966), Electromagnetic fields of a dipole over an anisotropic half-space, *Can. J. Phys*, 44, 2387-2401.

WAIT, J.R., (1970, 2nd Edition) *Electromagnetic Waves in Stratified Media*, New York:Pergamon.

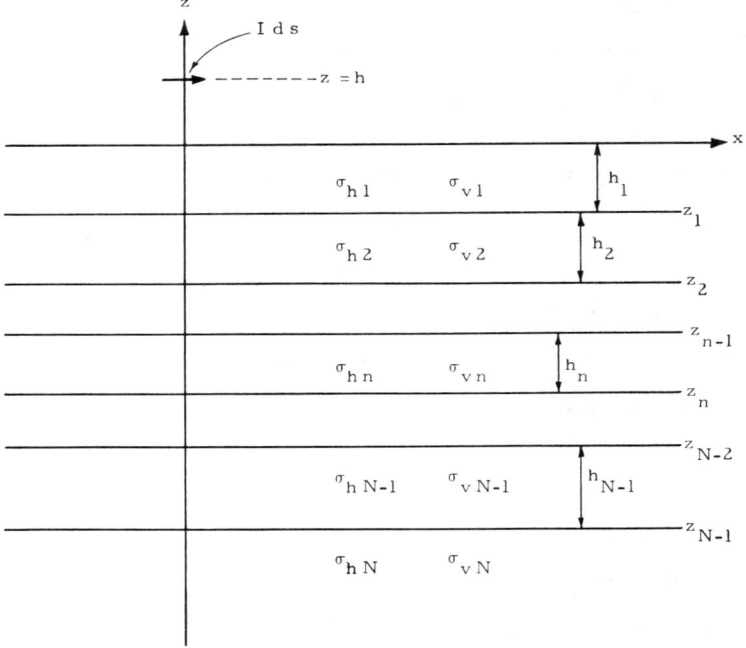

Fig. 1. Horizontal electric dipole over an N layered stratified half-space. (The general layer is indicated by a subscript n.)

12
Asymptotic Evaluation of the Field of a Vertical Dipole Over an Impedance Plane Surface

STATEMENT OF PROBLEM

We consider the idealized model of a vertical electric dipole of current moment Ids located at a height h over a plane surface that can be characterized by surface impedance $Z_s = \eta_o \Delta$ ohms. With respect to a cylindrical coordinate system (ρ,ϕ,z) with z axis downward, the surface is the plane $z = 0$ and the dipole is located at $z = -h$. The fields for the space $z < 0$ for the usual time factor $\exp(i\omega t)$, can be derived from

$$E_\rho = \partial^2 \Pi / \partial \rho \partial z \quad (1), \qquad E_z = (k_o^2 + \partial^2/\partial z^2)\Pi \quad (2)$$

and
$$H_\phi = -i\varepsilon_o \omega \partial \Pi / \partial \rho \quad (3)$$

where $k_o = (\varepsilon_o \mu_o)^{1/2} \omega$. As we showed before, the Hertz potential can be written

$$\Pi = \frac{Ids}{4\pi i \varepsilon_o \omega} \int_0^\infty \left[e^{-u_o|z+h|} + R(\lambda) e^{u_o(z-h)} \right] \frac{\lambda}{u_o} J_o(\lambda \rho) d\lambda \quad (4)$$

where
$$u_o = (\lambda^2 - k_o^2)^{1/2} \quad \text{and} \quad R(\lambda) = \frac{u_o - ik_o \Delta}{u_o + ik_o \Delta} \quad (5)$$

in the appropriate reflection coefficient. It can be readily confirmed that the boundary condition

$$[E_\rho = \eta_o \Delta H_\phi]_{z=0} \quad (6)$$

is satisfied.

An alternative form for (4) is clearly

$$\Pi = \frac{Ids}{4\pi i \varepsilon_o \omega} \left[\frac{e^{-ik_o R_o}}{R_o} + \frac{e^{-ik_o R}}{R} - 2P \right] \quad (7)$$

where
$$P = \int_0^\infty \frac{ik_c \Delta \lambda e^{-u_o(h-z)}}{(u_o + ik_o)u_o} J_o(\lambda \rho) d\lambda \quad (8)$$

(Reprinted, in part in revised form, from J.R. Wait, Electromagnetic Waves in Stratified Media, New York, 1970).

$$R_o = [\rho^2 + (z+h)^2]^{1/2} \quad \text{and} \quad R = [\rho^2 + (z-h)^2]^{1/2}$$

Here we have used the identity

$$\frac{e^{-ik_oR}}{R} = \int_0^\infty \frac{\lambda}{u_o} e^{-u_o(h-z)} J_0(\rho\lambda)\, d\lambda \tag{9}$$

THE MODIFIED SADDLE POINT METHOD

We now wish to discuss the asymptotic evaluation of P. Before proceeding, certain conditions and restrictions must be clearly stated. At a later stage of the analysis these may be relaxed somewhat. For the moment

$$|\Delta| < 1$$
$$0 < \arg \Delta < \frac{\pi}{4}$$

and

$$k_0\rho \gg 1$$

Since

$$J_0(x) = \tfrac{1}{2}[H_0^{(1)}(x) + H_0^{(2)}(x)] \quad \text{and} \tag{10}$$

$$H_0^{(2)}(x) = -H_0^{(1)}(-x) \tag{11}$$

where $H_0^{(1)}$ and $H_0^{(2)}$ are Hankel functions of the first and second kind, respectively, it follows that

$$P = \frac{ik_0\Delta}{2}\int_{-\infty}^{\infty} \frac{\lambda\, e^{-u_0(h-z)}}{(u_0 + ik_0\Delta)u_0} H_0^{(1)}(\lambda\rho)\, d\lambda \tag{12}$$

Introducing the substitutions

$$\lambda = k_0 \cos\alpha \quad \text{and} \quad \Delta = \sin\alpha_0$$

it is seen that

$$P = \frac{ik_0\Delta}{4}\int_{-i\infty}^{\pi+i\infty} \frac{\cos\alpha\, e^{-ik_0(h-z)\sin\alpha} H_0^{(1)}(k_0\rho\cos\alpha)}{\sin\frac{\alpha+\alpha_0}{2}\cos\frac{\alpha-\alpha_0}{2}}\, d\alpha \tag{13}$$

Essentially, P is now a spectrum of plane waves travelling away from the ground plane such that the vertical component of the propagation constant is $k_0 \sin\alpha$. Again, complex angles are included in the spectrum. When $|k_0\rho \cos\alpha| \gg 1$, the first term of the asymptotic expansion of the Hankel function can be employed. That is,

$$H_0^{(1)}(k_0\rho\cos\alpha) \cong \left[\frac{2}{\pi k_0\rho\cos\alpha}\right]^{1/2} e^{ik_0\rho\cos\alpha} e^{-i\pi/4} \tag{14}$$

When this is inserted into Eq. (13), the equation of P becomes

$$P = e^{i\pi/4} \frac{\Delta}{2} \left(\frac{k_0}{2\pi\rho}\right)^{1/2} \int_{-i\infty}^{\pi+i\infty} \frac{(\cos \alpha)^{1/2} e^{-ik_0 R \cos(\theta-\alpha)}}{\sin \frac{\alpha + \alpha_0}{2} \cos \frac{\alpha - \alpha_0}{2}} d\alpha \tag{15}$$

where R and θ are defined by

$$h - z = R \sin \theta \quad \text{and} \quad \rho = -R \cos \theta$$

In the usual case, the separation ρ between transmitter and receiver is large compared to their respective heights, h and $-z$, and consequently, θ is slightly less than π. It can be seen that the exponential factor in the integrand of P is rapidly varying except for a region near $\alpha = \theta \cong \pi$. This is, of course, the saddle point of the integrand or the point of stationary phase. The important part of the integrand is in the region near the saddle point. In fact, this is the justification for employing the asymptotic expansion for the Hankel functions $H_0^{(1)}$; the argument $k_0\rho \cos \alpha$ is always large in the region near the saddle point if $k_0\rho$ itself is large and $h - z$ is less than ρ.

The integral for P is now in a form where the saddle point method of integration can be applied. The usual technique [Sommerfeld, 1949] is to deform the contour (which in the present case is along the negative imaginary axis, the real axis from 0 to π, and then along a line parallel to the positive imaginary axis) to a path of steepest descent. For example, if we let

$$\cos(\alpha - \theta) = 1 - i\tau^2$$

and let τ range from $-\infty$ to $+\infty$ through real values, an integral of the type

$$\int_{-i\infty}^{\pi+i\infty} G(\cos \alpha) e^{-ik_0 R \cos(\alpha-\theta)} d\alpha \tag{16}$$

where $G(\cos \alpha)$ is slowly varying at $\alpha = \theta$, is transformed to

$$(2i)^{1/2} e^{-ik_0 R} \int_{-\infty}^{+\infty} \frac{G(\cos \alpha)}{\sqrt{(1 - i\tau^2/2)}} e^{-k_0 R \tau^2} d\tau \tag{17}$$

In this deformation of the contour, account must be taken of the singularities of the integrand that are crossed. In the case of poles, $2\pi i$ times the sum of the residues is added to the new integral. The final step in this classical procedure is to expand $G(\cos \alpha)/\sqrt{(1 - i\tau^2/2)}$ in a power series in τ enabling the integration to be carried out term by term. The leading term of this resulting asymptotic expansion is

$$(2\pi i)^{1/2} G(\cos \theta) e^{-ik_0 R}(k_0 R)^{-1/2} \tag{18}$$

and succeeding terms contain $(k_0 R)^{-3/2}$, $(k_0 R)^{-5/2}$ and so on.

As is often the case, the contribution from the saddle point (which is the branch point in the original λ plane) cannot be separated from the pole(s) of the integrand. For example, in the present problem,

$$G(\cos \alpha) = \frac{(\cos \alpha)^{1/2} e^{-ik_0 R \cos(\theta - \alpha)}}{\sin \dfrac{\alpha + \alpha_0}{2} \cos \dfrac{\alpha - \alpha_0}{2}} \tag{19}$$

has a pole at $\alpha = \pi + \alpha_0$ which is near the saddle point $\alpha = 0$ since α_0 is small and θ is near π. In other words, the integrand is not slowly varying near the saddle point on account of the factor $\cos(\alpha - \alpha_0)/2$. The integral we have to contend with is then of the form

$$I = \int_{-i\infty}^{\pi + i\infty} \frac{e^{-ik_0 R \cos(\theta - \alpha)}}{\cos \dfrac{\alpha - \alpha_0}{2}} d\alpha \tag{20}$$

after the slowly varying factors have been separated out. The necessary modification of the saddle point method to treat integrals of this type was devised by Van der Waerden [1950] and Clemmow [1950]. On making the usual deformation of the contour via the subsitution $\cos(\theta - \alpha) = 1 - i\tau^2$, the integral now becomes

$$I \cong -2 e^{-i\pi/4} e^{-ik_0 R} \sqrt{2} \cdot \cos\left(\frac{\theta - \alpha_0}{2}\right) \int_{-\infty}^{+\infty} \frac{e^{-k_0 R \tau^2} d\tau}{\tau^2 + i2 \left[\cos\left(\dfrac{\theta - \alpha_0}{2}\right)\right]^2} \tag{21}$$

It can readily be verified that no poles of the integrand are crossed in this deformation for the argument of α_0(or Δ) less than 45 degrees. The integral to consider is now of the type

$$A = \int_{-\infty}^{+\infty} \frac{e^{-x\tau^2}}{\tau^2 + c} d\tau \tag{22}$$

where, for real τ, $x = k_0 R$ and

$$c = i2 \left[\cos\left(\frac{\theta - \alpha_0}{2}\right)\right]^2$$

Since $\displaystyle\int_{-\infty}^{+\infty} e^{-x\tau^2} d\tau = \left(\frac{\pi}{x}\right)^{1/2}$ it follows that

$$A = \frac{1}{c}\left(\frac{\pi}{x}\right)^{1/2} - \frac{2}{c}\int_0^\infty \frac{\tau^2}{\tau^2 + c} e^{-x\tau^2} d\tau \tag{23}$$

Now, it can be seen readily that

$$\frac{d}{dx} e^{-xc} A = -\left(\frac{\pi}{x}\right)^{1/2} e^{-xc} \tag{24}$$

and an integration with respect to x from x to ∞ leads to

$$e^{-xc}A = \int_x^\infty \left(\frac{\pi}{x}\right)^{1/2} e^{-xc}\,dx \tag{25}$$

or, after a change of variable,

$$A = 2(\pi/c)^{1/2} e^{xc} \int_{\sqrt{(xc)}}^\infty e^{-z^2}\,dz \tag{26}$$

which holds for *all* c in the c plane with a cut along the negative real axis. This can be written

$$A = \frac{\pi}{\sqrt{c}} e^{xc}\,\text{erfc}\,\sqrt{(xc)} \tag{27}$$

where erfc is the complement of the error integral given by

$$\text{erfc}\,z_0 = \frac{2}{\sqrt{\pi}} \int_{z_0}^\infty e^{-z^2}\,dz \tag{28}$$

where the integration in the z plane is directed from z_0 towards the right to $+\infty$. Therefore

$$I = -i2\pi\,e^{-ik_0R}\,e^{-w}\,\text{erfc}(iw^{1/2}) \tag{29}$$

where

$$w = -k_0Rc = -2ik_0R\left[\cos\left(\frac{\theta - \alpha_0}{2}\right)\right]^2$$

and finally,

$$P = \int_0^\infty \frac{(ik_c\lambda\Delta)}{(u_0 + ik_0\Delta)u_0}\,e^{-u_0(h-z)}J_0(\lambda\rho)\,d\lambda \tag{30}$$

$$\cong i(\pi p)^{1/2}\,e^{-w}\,\text{erfc}(iw^{1/2})\,e^{-ik_0R}/R$$

where

$$w \cong p\left(1 + \frac{(h-z)^2}{\Delta R}\right)^2 \tag{31}$$

and

$$p = -\frac{ik_0R}{2}\,\Delta^2 = |p|\,e^{ib}$$

While the derivation was carried out for the argument of Δ lying in the range 0 to $\pi/4$ or b in the range 0 to $-\pi/2$, the formula is actually valid for all values of b if the above definition of the error function complement is used. The justification for this step is based on the principle of analytical continuation. If, in the initial derivation, the argument of Δ was allowed to exceed $\pi/4$, we would find that a pole would be crossed in the deformation of the contour in the α plane. There would be a corresponding change in the integral A but, after a change of the sign of the variable τ, the expression yielded for I would be unchanged in form if we are consistent in the definition of the complex error function.

ALTERNATIVE METHOD

As mentioned in the preceding section, the formula for P given by Eq. (30) is valid only if $|\Delta|$ is small compared with unity. When $|\Delta|$ is no longer small it is necessary to employ an alternative method. Actually, the approach is simpler since the branch point at $\lambda = k_0$ is no longer close to the pole of the integrand. Therefore, the conventional saddle-point method may be used. To apply this method it is convenient to express P in the equivalent form

$$P = \frac{e^{-ik_0 R}}{R} + ik_0 Q(\alpha, \Delta) \tag{33}$$

where

$$Q(\alpha, \Delta) = \int_0^\infty e^{-ik_0 C\alpha}(C + \Delta)^{-1} J_0(k_0 \rho S) S \, dS \tag{34}$$

where

$$\alpha = h - z \quad \text{and} \quad C = (1 - S^2)^{1/2}$$

On deforming the integration contour into the steepest descent path, attention must be paid to the location of the pole at $C = -\Delta$. The resulting asymptotic approximation is readily shown to be

$$Q(\alpha, \Delta) \cong \frac{\cos \theta}{\cos \theta + \Delta} \frac{e^{-ik_0 R}}{(-ik_0 R)} + q Q_s(\alpha, \Delta) \tag{35}$$

where

$$\cos \theta = \frac{\alpha}{(\alpha^2 + \rho^2)^{1/2}} = \frac{(h-z)}{R}$$

and

$$Q_s(\alpha, \Delta) = -i\pi \Delta e^{ik_0 R \Delta \cos\theta} H_0^{(2)}[k_0 R \sin \theta \sqrt{(1 - \Delta^2)}] \tag{36}$$

$$q = 1 \quad \text{for} \quad \arg\left[\sin\frac{1}{2}\left(\frac{\pi}{2} - \theta + \arcsin \Delta\right)\right] > \frac{\pi}{4} \tag{37}$$

$$= 0 \quad \text{for} \quad \arg\left[\sin\frac{1}{2}\left(\frac{\pi}{2} - \theta + \arcsin \Delta\right)\right] < \frac{\pi}{4}$$

The first term in Eq. (24) is the contribution from the saddle point, and the second term is the residue of the pole which may or may not be captured in the resulting deformation of the contour. The above expression for $Q(\alpha, \Delta)$ and the corresponding one for P are valid in the far field such that terms which vary as $1/R^2$, $1/R^3$, etc. may be neglected.

Rigorous expansions for the integral P have been given by Furutsu [1959]. His results would indicate that our asymptotic forms are adequate for most cases of practical interest.

FURTHER DISCUSSION

We shall confine our attention to the case where $|\Delta|$ is small. This covers the situations of most practical interest.

The expression for P given by (20) can also be written

$$P = \left(\frac{p}{w}\right)^{1/2} [1 - F(w)] \frac{e^{-ik_0 R}}{R} \tag{38}$$

where

$$F(w) = 1 - i(\pi w)^{1/2} e^{-w} \operatorname{erfc}(iw^{1/2}) \tag{39}$$

It is noted that when $z = h = 0$

$$\psi_0 = \frac{I\,ds}{2\pi i\omega\varepsilon_0} \frac{e^{-ik_0\rho}}{\rho} F(p) \tag{40}$$

and the vertical electric field is given by

$$E_{0z} \cong \frac{i\mu\omega I\,ds}{2\pi\rho} e^{-ik_0\rho} F(p) \quad \text{for } k\rho \gg 1 \tag{41}$$

The function $F(p)$ can be regarded as the correction to the field of a dipole on the surface of a perfectly conducting plane. For $|p| \ll 1$ it approaches unity. $F(p)$ has the same functional form as the ground wave attenuation of Norton [1935–1941] who presented numerical values for the case where (in the present notation), b is in the quadrant 0 to -180 degrees. In the case of a homogeneous ground Δ has a phase angle in the range 0 degrees (for a perfect dielectric) to 45 degrees (for a good conductor). The corresponding values of b are -90 degrees and 0 degrees respectively. In the case of a stratified ground, however, the phase angle of Δ may be outside this range. In the particular case of a two-layer conducting half-space the parameter p can be written

$$p = p_e = p_1 Q^2 \quad \text{where} \quad p_1 = \frac{-ik_0\rho}{2} (\eta_1/\eta_0)^2 \tag{42}$$

where

$$Q = |Q| e^{iq} \cong \frac{(\gamma_1/\gamma_2) + \tanh\gamma_1 h_1}{1 + (\gamma_1/\gamma_2)\tanh\gamma_1 h_1} \tag{43}$$

in terms of the propagation constants γ_1 and γ_2 of the upper and lower layers and the thickness h_1 of the upper layer. This function Q was discussed in Note. No. 2. There it was seen that for a highly conducting substratum q can become positive. Consequently, the phase angle b can exceed 0 and would approach 90 degrees for

and
$$|\gamma_2/\gamma_1| \to \infty$$
$$|\gamma_1 h_1| \ll 1$$

In analogy to the terminology introduced by Sommerfeld for the homogeneous half-space, p_e is here described as an effective numerical distance. Numerical values for the attenuation function $F(p_e)$ can be obtained from the power series expansion

$$F(p_e) = 1 - i(\pi p_e)^{1/2} - 2p_e + i\pi^{1/2} p_e^{3/2} + \ldots$$

The curves for $b < 0$ correspond to those computed by Norton and it is interesting to note that for this region $|F_e|$ never exceeds unity. On the other hand when $b > 0$ $|F_e|$ may exceed unity and when $b = 90$ degrees this effect is most pronounced. Apparently, in the case where b is positive, the energy is being guided to some extent along the surface. This effect can be seen in the asymptotic development of $F(p_e)$ which for $|p_e| \gg 1$ reads

$$F(p_e) \cong -\frac{1}{2p_e} - \frac{1 \times 3}{(2p_e)^2} - \frac{1 \times 3 \times 5}{(2p_e)^3} - \frac{1 \times 3 \times 5 \times 7}{(2p_e)^4} \cdots \quad (45)$$

when
$$-2\pi < b < 0$$

and

$$F(p_e) \cong -2i\sqrt{(\pi p_e)}\, e^{-p_e} - \frac{1}{2p_e} - \frac{1 \times 3}{(2p_e)_2} - \frac{1 \times 3 \times 5}{(2p_e)^3} \cdots \quad (46)$$

when
$$2\pi > b > 0$$

The term $-2i\sqrt{(\pi p_e)}\, e^{-p_e}$ has all the characteristics of a surface wave. It is not present in the asymptotic development when b is negative. At $b = 0$, 2π or -2π, p_e is real and this term vanishes asymptotically. The trapped surface wave is most predominant when $b = 90$ degrees which corresponds to Δ or Z_1 being positive imaginary. For example, if the surface is coated by a thin dielectric film of thickness h_1 with a dielectric constant ε_1,

$$Z_1 \cong i\mu\omega h_1[1 - \varepsilon_0/\varepsilon_1] \cong iX \quad (47)$$

which is purely imaginary. Then since

$$p_e = \frac{-ik_0\rho}{2}(Z_1/\eta_0)^2 \quad (48)$$

$p_e = |p_e|\, e^{i\pi/2}$ or $b = \pi/2$ rad $= 90$ degrees. The attenuation factor now has the asymptotic development

$$F(p_e) \cong -2\, e^{i3\pi/4}\sqrt{(\pi|p_e|)}\, e^{-i|p_e|} + \frac{i}{2|p_e|} + \frac{3}{4|p_e|^2} - \frac{15i}{8|p_e|^3} \cdots \quad (49)$$

where
$$|p_e| = \frac{k_0\rho}{2}\frac{X^2}{\eta_0^2}$$

and the vertical electric field is

$$E_{0z} \cong \frac{i\mu\omega I\,ds}{2\pi\rho} e^{-ik_0\rho} F(p_e) \tag{50}$$

which shows that the (trapped) surface wave component varies as

$$\frac{1}{\sqrt{(k_0\rho)}} \exp\left[-ik_0\left(1 + \frac{X^2}{2\eta_0^2}\right)\rho\right]$$

which has the characteristics of a cylindrical wave travelling in the positive ρ with a phase velocity

$$\left(1 + \frac{X^2}{2\eta_0^2}\right)^{-1}$$

times that of light.

The (trapped) surface wave is not excited when b is negative and then the distant electric field varies asymptotically as

$$\frac{1}{(k_0\rho)^2} e^{-ik_0\rho}$$

which has a phase velocity equal to that of light.

When the source dipole and the observer are both raised above the surface (i.e. $h > 0$ and $-z > 0$), the field in the upper half space can be expressed conveniently in the asymptotic sense as the sum of three partial fields in the manner

$$\psi_0 = [\psi_a + \psi_b + \varepsilon\psi_s]\frac{I\,ds}{4\pi i\omega\varepsilon_0} \tag{51}$$

where
$$\varepsilon = 0 \text{ for } \arg w < 0$$
and
$$\varepsilon = 1 \text{ for } \arg w > 0$$

An asymptotic development shows that

$$\psi_a \cong \frac{e^{-ikR_0}}{R_0} + \left(\frac{C - \Delta}{C + \Delta}\right)\frac{e^{-ikR}}{R} \tag{52}$$

$$\psi_b = -\left\{\frac{1}{p(1 + C/\Delta)^3} + \frac{1\times 3}{2p^2(1 + C/\Delta)^5} + \frac{1\times 3\times 5}{4p^3(1 + C/\Delta)^7} + \ldots\right\}\frac{e^{-ikR}}{R} \tag{53}$$

and

$$\psi_s \cong -\frac{2\Delta}{\Delta + C}[2i\sqrt{(\pi w)}\,e^{-w}]\frac{e^{-ikR}}{R} \tag{54}$$

where
$$C = \frac{h - z}{R} = \frac{h + |z|}{R}, \quad k = 2\pi/\text{wavelength} = k_0$$

and
$$\Delta = Z_1/\eta_0$$
In terms of ψ_0, the vertical electric field components for $k_0\rho \gg 1$ are given by
$$E_{0z} \cong k_0^2(1 - C^2)\psi_0 \tag{55}$$

In the above ψ_a can be identified as a geometrical optical term being just the primary field e^{-ikR_0}/R_0 and a specularly reflected component e^{-ikR} R modified by a reflection coefficient $(C - \Delta)/(C + \Delta)$. ψ_b is an asymptotic expansion containing terms varying as R^{-2}, R^{-3}, etc. ψ_s which is only present for positive b values, represents a (trapped) surface wave and has a phase velocity less than that of light. When b is negative there is no (trapped) surface wave present.

Sometimes ψ_a itself is called a space wave and ψ_b a surface wave. This usage would correspond to that of Norton [1935–1941]. This would be an obvious designation when the half space is homogeneous since ψ_s is zero and there is no trapped surface wave excited.

Equations (52), (53) and (54) enable one to compute the field in the upper half space ($z < 0$) for an electric dipole located at $z = -h$ in terms of the ratio Δ of the surface impedance Z_1 to that of free space η_0. As a first approximation Z_1 could be replaced by $Z_1(\lambda)$ for $\lambda = 0$ which would be the normal surface impedance. Clearly, a better value would be $Z_1(\lambda_s)$ where λ_s is the saddle point of the integral in Eqs. (32) or (33). This would mean that

$$\Delta = \frac{Z_1(\lambda_s)}{\eta_0} \tag{56}$$

where
$$\lambda_s = k_0 S$$
where
$$S = \rho/R = \sqrt{(1 - C^2)}$$
In the case of a homogeneous half space, with $\mu_1 = \mu_0$
$$\Delta = \frac{K_1(\lambda_s)}{\eta_0} = \frac{u_1(\lambda_s)}{(\sigma_1 + i\omega\varepsilon_1)\eta_0} \tag{57}$$
$$= \frac{n_1}{n_0}\left(1 - \frac{n_1^2}{n_0^2} S^2\right)^{\frac{1}{2}} \tag{58}$$

The coefficient $(C - \Delta)/(C + \Delta)$ in Eq. (52) now can be identified as the Fresnel reflection coefficient for a plane wave with parallel polarization incident at angle θ. The equations are now in complete agreement with the results of Norton [1935–1941] for a vertical electric dipole over a homogeneous conducting ground.

The corresponding treatment for a vertical magnetic dipole located at $z = -h$ over the M layered half space is almost identical to the above. The quantity ψ_0 is then to be identified with the z component of the magnetic Hertz vector and then Δ is to be defined by

$$\Delta = \eta_0 Y_1 \tag{59}$$

where Y_1 is the surface admittance at the interface $z = 0$.

Further (numerical) discussion of inductive surface effects are given later when we deal with the spherical earth model.

REFERENCES

CLEMMOW, P.C., (1950), Some extensions to the method of integration by steepest descents, *Quart. J. Mech.*, 3, 241-256.

FURUTSU, K., (1959), On the excitation of the waves of proper solutions, *Trans. I.R.E.*, Special Supplement, AP-7, 209-218.

NORTON, K.A., (1935), Propagation of radio waves over a planar earth, *Nature*, 135, 954-955.

NORTON, K.A., (1936), The propagation of radio waves over the surface of the earth and in the upper atmosphere, *Proc. I.R.E.*, 24, Pt. I, 1367-1387.

NORTON, K.A., (1937), The propagation of radio waves over the surface of the earth and in the upper atmosphere, *Proc. I.R.E.*, 25, Pt. II, 1203-1236.

NORTON, K.A., (1941), The calculation of ground-wave field intensity over a finitely conducting spherical earth, *Proc. I.R.E.*, 29, 623-639.

SOMMERFELD, A.N., (1949), *Partial Differential Equations*, Academic Press, New York (includes good summary of the author's pioneering investigations of the dipole-over-the-half-space problem).

VAN DER WAERDEN, B.L., (1950), On the method of saddle points, *Appl. Sci. Res.*, B-2, No. 7, 33-45.

ADDITIONAL RELATED REFERENCES

ATWOOD, S., (1951), Surface wave propagation over a coated plane conductor, *J. Appl. Phys.*, 22, 504-509.

BARLOW, H.E.M., and CULLEN, A.L., (1953), Surface waves, *Proc. I.E.E.*, 100, Pt. III, 321-331.

BARLOW, H.E.M., and FERNANDO, W.M.G., (1956), An investigation of the properties of radial surface waves launched over flat reactive surfaces, *Proc. I.E.E.*, 103, 307-318.

BOUWKAMP, C.J., (1950), On Sommerfeld's surface wave, *Phys. Rev.*, 80, 294.

BRICK, D.B., (1954), The radiation of a Hertzian dipole over a coated conductor, Monograph No. 113, Institution of Electrical Engineers (London).

CULLEN, A.L., (1945), The excitation of plane surface waves, *Proc. I.E.E.*, 101, Pt. IV, 225-235.

ELIASSEN, K.E., (1957), A survey of ground conductivity and dielectric constant in Norway within the frequency range 012-10 Mc/s, *Geophys. Publ.*, 19, No. 11, 1-20 (Norske Meteorologiske Institutt, Oslo). (experimental confirmation of the predicted layered earth effects promulgated by Wait).

GOUBAU, G., (1950), Surface waves and their application to transmission lines, *J. Appl. Phys.*, 21, 1119-1128.

GOUBAU, G., (1952), On the excitation of surface waves, *Proc. I.R.E.*, 40, 865-868.

GROSSKOPF, J., and VOGT, K., (1940), On the measurement of earth conductivity, *Telegr.-u. Fernspr. Techn.*, 29, 164-172.

GROSSKOPF, J., and VOGT, K., (1941), The measurement of electrical conductivity for a stratified ground, *Hochfrequenztech. u. Electroakust.*, 58, 52-57. (As in the previous paper, the formula for the wave-tilt is based on the Zenneck wave which can be in error for poorly conducting ground.)

HACK, F., (1908), The propagation of electromagnetic waves over a plane conductor, *Ann. Phy.*, 27, 43.

ROLF, B., (1930), Graphs to Prof. Sommerfeld's attenuation formula for radio waves, *Proc. I.R.E.*, 18', 391-402. (These results are spurious since they are based on Sommerfeld's 1909 paper that has the famous error in sign.)

ROTMAN, W., (1951), A study of single surface corrugated guides, *Proc. I.R.E.*, 39, 952-959.

SOMMERFELD, A.N., (1899), Uber die Fortpfanzung electrodynamischer Wellen langs eines Drahtes, *Ann. Phys., u. Chem.*, 67, 233.

SOMMERFELD, A.N., (1909 and 1926), The propagation of waves in wireless telegraphy, *Ann. Phys.*, Series 4, 28, 665; 81, 1135.

STANLEY, G.M., (1960), Layered earth propagation in the vicinity of Point Barrow, Alaska, *J. Res. Nat. Bur. Stand.*, 64D, 95-99.

TAI, C.T., (1951), The effect of a grounded slab on radiation from a line source, *J. Appl. Phys.*, 22, 405.

VOGLER, L.E., (1964), Note on the attenuation function for propagation over a flat layered ground, *I.E.E.E. Trans.*, AP-12, 240-242 (first correct calculations of the attenuation function $F(\rho_e)$ for the case where b is positive).

WAIT, J.R., (1953), Radiation from a vertical electric dipole over a stratified ground, *Trans. I.R.E.*, Vol. AP-1, 9-12, Part I.

WAIT, J.R., RASER, W.C.G., (1956), Radiation from a vertical dipole over a stratified ground, Part II, *Trans. I.R.E.*, Vol. AP-3, No. 4.

WAIT, J.R., (1954), Note on the theory of radio propagation over an ice-covered sea, Def. Res. Tele. Est., Radio Physics Lab., Project Report 18-0-7.

WAIT, J.R., (1957), Excitation of surface waves on conducting, stratified dielectric-clad and corrugated surfaces, *J. Res. NBS*, 59, No. 6, 365-377.

WAIT, J.R., (1959), Guiding of electromagnetic waves by uniformly rough surfaces, *I.R.E. Trans.*, Vol. AP-7, Pts. I, II, S154-S168.

WAIT, J.R., and SCHLAK, G.A., (1967), New asymptotic solution for the electromagnetic fields of a dipole over a stratified medium, *Elec. Letrs.*, 3, No. 9, 421-422.

13
Fields of a Circular Loop of Current Buried in a Two-Layer Earth

INTRODUCTION

In communicating with and/or locating trapped miners, we had earlier suggested that a feasible source would be a wire loop that could be excited by a portable transmitter. A great deal of effort has gone into the problem by various groups in the U.S. and elsewhere. An accessible and very well written account can be found in the prize-winning paper by Large, Ball and Farstad [1973].

In a previous analysis, we considered the problem of a small horizontal loop or vertical magnetic dipole located in the bottom region of a two-layer earth. The results were quasi-static in the sense that all significant distances in the problem were small compared with a free space wavelength. Using numerical integration, the magnitude of the ratio of the horizontal to the vertical magnetic field was examined for an observer on the earth's surface. This was shown to have diagnostic features that could be used as the basis of a source location technique, in spite of the fact that the curves were modified to some extent by the layer structure. [Wait, 1971; Wait and Spies, 1971a].

Here we extend the earlier analysis to allow for the finite extent of the source loop. Also, we derive explicit expressions for the fields that are valid everywhere. In order to render the problem some generality, displacement currents in the air and in the ground are retained at least in the initial formulation.

First of all we deal with a circular loop of radius a. When a tends to zero we then recover the field expressions given earlier for a magnetic dipole. We then consider a loop of finite size with any specified shape. In particular,

Fields of a Circular Loop of Current

we deal with one of rectangular form.

BASIC FORMULATION

The model for the circular loop is illustrated in Fig. 1. The earth is taken to be a two layer structure where the upper layer or slab of thickness b is characterized by electrical properties σ_1, ε_1, and μ_1. The lower semi-infinite region has corresponding properties σ_2, ε_2, and μ_2. With respect to a polar coordinate system (ρ,ϕ,z), the earth's surface is $z = 0$ and the lower interface of the upper slab is $z = -b$. The loop of radius a, carrying a total uniform current I, is located in the plane $z = -h$ and is coaxial with the z axis. The region $z > 0$ is assumed to be free space with electrical constants (ε_o, μ_o).

The source current and all associated electromagnetic fields are taken to vary as $\exp(i\omega t)$.

The source condition for the surface current density $j_\phi(\rho)$ in the plane $z = -h$ is given as follows

$$j_\phi(\rho) = I\delta(\rho - a) \tag{1}$$

where $\delta(\rho - a)$ is the unit impulse function. Here we find it convenient to apply the Fourier-Bessel transform in order to obtain the desired spectral form for the source field. Thus we have the pair

$$j_\phi(\rho) = \int_0^\infty f(\lambda) J_n(\lambda\rho) \lambda d\lambda \tag{2}$$

and

$$f(\lambda) = \int_0^\infty j_\phi(\rho) J_n(\lambda\rho) \rho d\rho \tag{3}$$

where n is an integer to be selected and J_n is a Bessel function of order n. Clearly, for the filamental current model,

$$f(\lambda) = I J_n(\lambda a) a \tag{4}$$

By symmetry it is evident that the electromagnetic fields have only non-vanishing components E_ϕ, H_ρ and H_z. A subscript 1, 2, or 0 is added to these quantities when they refer specifically to the upper slab, lower region or free space respectively. In particular, we have the source conditions

$$H_{2\rho}(-h + 0) - H_{2\rho}(-h - 0) = j_\phi(\rho) \tag{5}$$

$$E_{2\phi}(-h + 0) - E_{2\phi}(-h - 0) = 0 \tag{6}$$

that must be met.

In view of the symmetry of the problem, it is apparent that the fields in any of the regions can be obtained from a magnetic Hertz vector that has only an axial component Π^* that we dub the "Hertz potential". Thus

$$E_\phi = i\mu\omega \partial \Pi^*/\partial \rho \tag{7}$$

$$H_\rho = \partial^2 \Pi^*/\partial \rho \partial z \tag{8}$$

$$H_z = (-\gamma^2 + \partial^2/\partial z^2)\Pi^* \tag{9}$$

where

$$\gamma^2 = i\mu\omega(\sigma + i\varepsilon\omega).$$

Now because Π^* satisfies

$$\left(\frac{1}{\rho}\frac{\partial}{\partial \rho}\rho\frac{\partial}{\partial \rho} + \frac{\partial^2}{\partial z^2} - \gamma^2\right)\Pi^* = 0 \tag{10}$$

we easily deduce that solutions, finite at $\rho = 0$, have the form

$$\Pi^* = J_0(\lambda \rho)\exp(\mp uz) \tag{11}$$

where

$$u = (\lambda^2 + \gamma^2)^{1/2}.$$

Here λ is the transverse wave number than can be identified with the variable λ in equ. (2). In particular, we note that H_ρ is made up of linear combinations of solutions of the form $J_1(\lambda\rho)\exp(\mp uz)$ [on noting that $dJ_0(x)/dx = -J_1(x)$]. Now it is clearly evident we should choose $n = 1$ in (2), (3), and (4). Thus, our source current density, in the plane $z = -h$, should be given by

$$j_\phi(\rho) = Ia \int_0^\infty J_1(\lambda a) J_1(\lambda \rho) \lambda d\lambda \tag{12}$$

We are now in the position to write down the desired forms for the Hertz potentials in the three regions:

Fields of a Circular Loop of Current

$$\Pi_1^* = p \int_0^\infty A[e^{-u_1 z} + R_o e^{u_1 z}] \frac{J_1(\lambda a)}{u_2} J_o(\lambda \rho) d\lambda \tag{13}$$

$$\Pi_2^* = p \int_0^\infty [e^{-u_2|z+h|} + R_\perp e^{2u_2 b} e^{u_2(z-h)}] \frac{J_1(\lambda a)}{u_2} J_o(\lambda \rho) d\lambda \tag{14}$$

$$\Pi_o^* = p \int_0^\infty T_o e^{-u_o z} \frac{J_1(\lambda a)}{u_2} J_o(\lambda \rho) d\lambda \tag{15}$$

where the five functional coefficients p, A, R_o, $R_\perp$ and T are not yet known. Other multiplicative factors have been inserted to simplify the interpretation of the final results.

In order that (5) and (6) be satisfied, we readily deduce that $p = Ia/2$. The other four coefficients can be deduced by insisting that the tangential fields E_ϕ and H_ρ be continuous at the two interfaces $z = 0$ and $z = -b$. This algebraic process yields, for the reflection coefficients,

$$R_o = (N_1 - N_o)/(N_1 + N_o) \tag{16}$$

and

$$R_\perp = (N_2 - Y_1)/(N_2 + Y_1) \tag{17}$$

where $N_1 = u_1/i\mu_1\omega$, $N_2 = u_2/i\mu_2\omega$, $N_o = u_o/i\mu_o\omega$ and Y_1 are appropriate admittances. Here

$$Y_1 = -\left[\frac{H_{1\rho}}{E_{1\phi}}\right]_{z=-b} = N_1 \frac{1 - R_o e^{-2u_1 b}}{1 + R_o e^{-2u_1 b}} \tag{18}$$

Actually these results are immediately obvious if one bears in mind the transmission line analogy for such problems.

To relate A and T_o we write down the matching equations for continuity of H_ρ at $z = 0$ and $z = -b$:

$$u_1(1 - R_o)A = u_o T_o \tag{19}$$

$$u_1[e^{u_1 b} - R_o e^{-u_1 b}]A = u_2[e^{u_2 b} - R_\perp e^{u_2 b}]e^{-u_2 h} \tag{20}$$

Thus

$$T_o = \frac{u_2}{u_o}[(1 - R_o)(1 - R_\perp)e^{u_2 b}/(e^{u_1 b} - R_o e^{-u_1 b})]e^{-u_2 h} \tag{21}$$

and

$$A = [(u_2/u_1)(1 - R_\perp)e^{u_2 b}/(e^{u_1 b} - R_o e^{-u_1 b})]e^{-u_2 h} \tag{22}$$

which completes our task of finding the unknown coefficients.

Now using the identity

$$1 - R_+ = 2Y_1/(N_2 + Y_1) \tag{23}$$

it is not difficult to show that

$$\frac{T_o}{u_2} = \frac{1}{u_o}\left(\frac{2N_o}{N_1 + N_o}\right)\left(\frac{2N_1}{N_2 + N_1}\right)[1 - R_oR_2e^{-2u_1b}]^{-1} e^{-u_2(h-b)} e^{-u_1b} \tag{24}$$

where

$$R_2 = (N_1 - N_2)/(N_1 + N_2) \tag{25}$$

The right hand side of (24) can clearly be interpreted as the product of two transmission coefficients (in large parentheses) and a propagation factor $\exp[-u_2(h-b) - u_1b]$ corresponding to the one way transmission loss from the source to the bottom of the upper slab and then through the slab to the surface. The bracketed factor, when expanded, can be interpreted as the effect of multiple reflections within the slab, e.g.,

$$[1 - R_oR_2e^{-2u_1b}]^{-1} = 1 + R_oR_2e^{-2u_1b} + (R_oR_2)^2e^{-4u_1b} + \ldots \tag{26}$$

The field quantities of special interest are the magnetic components $H_{o\rho}$ and H_{oz} in the free space region. These are obtained from

$$H_{o\rho} = \partial^2\Pi_o^*/\partial\rho\partial z \tag{27}$$

and

$$H_{oz} = (-\gamma_o^2 + \partial^2/\partial z^2)\Pi_o^* \tag{28}$$

where $\gamma_o^2 = -\varepsilon_o\mu_o\omega^2$.

A special case of particular interest is when all distances are small compared with the free space wavelength. Thus, we can say that $|\gamma_o\ell| \ll 1$ where ℓ is a typical distance. This amounts to setting $\gamma_o = 0$ in the general formulae derived above. Furthermore, to simplify the discussion we set $\mu_1 = \mu_2 = \mu_o$ then, for example, $N_1/N_2 = u_1/u_2$ and $N_o/N_1 = u_o/u_1 = \lambda/u_1$.

Using the preceding simplifications, we find without difficulty that

$$T_o = \frac{u_2}{u_o} T = \frac{u_2}{\lambda} T \tag{29}$$

where
$$T = \frac{\left(\dfrac{2\lambda}{u_1 + \lambda}\right)\left(\dfrac{2u_1}{u_1 - u_2}\right)\exp[-u_2(h-b) - u_1 b]}{1 - \left(\dfrac{u_- - \lambda}{u_- + \lambda}\right)\left(\dfrac{u_1 - u_2}{u_1 + u_2}\right)\exp[-2u_1 b]} \tag{30}$$

where $u_1 = (\lambda^2 + \gamma_1^2)^{1/2}$ and $u_2 = (\lambda^2 + \gamma_2^2)^{1/2}$. The desired field components, in the quasi-static approximation, can be written

$$H_{o\rho} = b_o P \quad \text{and} \quad H_{oz} = b_o Q \tag{31}$$

where
$$b_o = I(\pi a^2)/(2\pi h^3)$$

$$P = \frac{h^3}{2} \int_0^\infty \lambda^2 T(\lambda) e^{-\lambda z} J_1(\lambda \rho) \frac{2J_1(\lambda a)}{\lambda a} d\lambda \tag{32}$$

and
$$Q = \frac{h^3}{2} \int_0^\infty \lambda^2 T(\lambda) e^{-\lambda z} J_0(\lambda \rho) \frac{2J_1(\lambda a)}{\lambda a} d\lambda \tag{33}$$

The limiting case of a magnetic dipole is obtained by letting a become vanishingly small whence
$$2J_1(\lambda a)/(\lambda a) \to 1$$
in the above expressions.

HORIZONTAL LOOP OF ANY SHAPE

It is actually a very simple matter to generalize the above results to a horizontal loop whose profile is non-circular. To this end we write down the Hertz potential $d^2 \Pi_o^*$ in the free space region for a current loop of elemental area $dx'dy'$ located at $(x',y',-h)$ with reference to the cartesian system of coordinates (x,y,z). Thus,

$$d^2 \Pi_o^* = \frac{I dx' dy'}{4\pi} \int_0^\infty T(\lambda) e^{-\lambda z} J_0\left(\lambda[(x-x') + (y-y')^2]^{1/2}\right) d\lambda$$

for the usual quasi-static approximation wherein $\gamma_o = 0$ and $u_o = \lambda$. Then, of course,

$$\Pi_o^* = \frac{IA}{4\pi} \int_0^\infty T(\lambda) e^{-\lambda z} f(\lambda) d\lambda \tag{35}$$

where
$$f(\lambda) = \frac{1}{A} \iint_{\text{Area}} \left(J_0 \lambda[(x-x')^2 + (y-y')^2]^{1/2}\right) dx'dy' \quad (36)$$

and A is the area of the loop.

The desired magnetic field components expressed in cartesian coordinates are now obtained from

$$H_{ox} = \partial^2 \Pi_o^*/\partial x \partial z \quad (37)$$

$$H_{oy} = \partial^2 \Pi_o^*/\partial y \partial z \quad (38)$$

and
$$H_{oz} = \partial^2 \Pi_o^*/\partial z^2 \quad (39)$$

In carrying out these operations we note that

$$\partial J_0(\lambda\hat{\rho})/\partial x = -J_1(\lambda\hat{\rho})\lambda\partial\hat{\rho}/\partial x = -J_1(\lambda\hat{\rho})\lambda(x-x')/\hat{\rho}$$

where $\hat{\rho} = [(x-x')^2 + (y-y')^2]^{1/2}$. Then we may write

$$H_{ox} = b_o P_x \quad (40)$$

$$H_{oy} = b_o P_y \quad (41)$$

$$H_{oz} = b_o P_z \quad (42)$$

where

$$P_{\substack{x\\y}} = \frac{h^3}{2} \int_0^\infty \lambda^2 T(\lambda) e^{-\lambda z} f_{\substack{x\\y}}(\lambda) d\lambda \quad (43)$$

and

$$P_z = \frac{h^3}{2} \int_0^\infty \lambda^2 T(\lambda) e^{-\lambda z} f(\lambda) d\lambda \quad (44)$$

where $b_o = IA/(2\pi h^3)$

Here
$$f_x(\lambda) = \frac{1}{A} \iint_{\text{Area}} J_1(\lambda\hat{\rho}) \frac{x-x'}{\hat{\rho}} dx'dy' \quad (45)$$

and
$$f_y(\lambda) = \frac{1}{A} \iint_{\text{Area}} J_1(\lambda\hat{\rho}) \frac{y-y'}{\hat{\rho}} dx'dy' \quad (46)$$

and $f(\lambda)$ is given by (36). One may note that

Fields of a Circular Loop of Current

$$f_x(\lambda) = -\lambda^{-1}\partial f(\lambda)/\partial x \qquad (47)$$

and

$$f_y(\lambda) = -\lambda^{-1}\partial f(\lambda)/\partial y. \qquad (48)$$

An approach to evaluate (36) is to introduce new variables (r,θ) defined by

$$x' - x = r \cos\theta$$
$$y' - y = r \sin\theta$$

The element of area then becomes $r\,dr\,d\theta$ and

$$f(\lambda) = \frac{1}{A}\int_0^{2\pi}\int_0^{r_0(\theta)} J_0(\lambda r)\, r\, dr\, d\theta \qquad (49)$$

where $r = r_0(\theta)$ is the equation for the profile of the loop that is contained in the plane $z = -h$. Using the Bessel function identity

$$\int_0^\alpha J_0(\alpha)\alpha\, d\alpha = \alpha J_1(\alpha) \qquad (50)$$

it follows that

$$f(\lambda) = \frac{1}{A\lambda}\int_0^{2\pi} J_1\!\left(\lambda r_0(\theta)\right) r_0(\theta)\, d\theta \qquad (51)$$

Of course, if $r_0(\theta) = a$, we obtain

$$f(\lambda) = \frac{2}{\lambda a} J_1(\lambda a) \qquad (52)$$

which is the expected value.

Another case of some interest is the rectangular loop, since this shape might be quite feasible for underground workings. Thus, for a rectangle with dimensions $2s$ by 2ℓ, the vertices in the plane $z = -h$ would be located at $(s,\ell,)$, $(-s,\ell)$, $(-s,-\ell)$ and $(s,-\ell)$. In the (r,θ) system the vertices would be located at (r_1,θ_1), (r_2,θ_2), (r_3,θ_3) and (r_4,θ_4) where

$$r_1 = [(s-x)^2 + (\ell-y)^2]^{1/2}, \quad r_2 = [(s+x)^2 + (\ell-y)^2]^{1/2}$$

$$r_3 = [(s+x)^2 + (\ell+y)^2]^{1/2} \text{ and } r_4 = [(s-x)^2 + (\ell-y)^2]^{1/2}.$$

The corresponding angles are defined by

$$\cot\theta_1 = \frac{s-x}{\ell-y}, \quad \cot\theta_2 = \frac{-s-x}{\ell-y}, \quad \cot\theta_3 = \frac{s+x}{\ell+y} \quad \text{and} \quad \cot\theta_4 = \frac{s-x}{-\ell-y}$$

Also we note that

$$\begin{aligned}
r_o(\theta) &= \frac{\ell-y}{\sin\theta} \quad \text{for } \theta_1 \text{ to } \theta_2 \\
&= -\frac{s+x}{\cos\theta} \quad \text{for } \theta_2 \text{ to } \theta_3 \\
&= -\frac{\ell+y}{\sin\theta} \quad \text{for } \theta_3 \text{ to } \theta_4 \\
&= \frac{s-x}{\cos\theta} \quad \text{for } \theta_4 \text{ to } \theta_1
\end{aligned} \tag{53}$$

Thus (51) for the rectangular loop takes the form

$$f(\lambda) = \frac{1}{A\lambda} \left\{ \int_{\theta_1}^{\theta_2} J_1\left[\lambda \frac{(\ell-y)}{\sin\theta}\right] \frac{\ell-y}{\sin\theta} d\theta + \int_{\theta_2}^{\theta_3} J_1\left[-\lambda \frac{s+x}{\cos\theta}\right]\left[\frac{-(s+x)}{\cos\theta}\right] d\theta \right.$$
$$\left. + \int_{\theta_3}^{\theta_4} J_1\left[-\lambda \frac{(\ell+y)}{\sin\theta}\right]\left[\frac{-(\ell+y)}{\sin\theta}\right] d\theta + \int_{\theta_4}^{\theta_1} J_1\left[\lambda \frac{s-x}{\cos\theta}\right]\left[\frac{s-x}{\cos\theta}\right] d\theta \right\} \tag{54}$$

These integrals are of two types that can be defined as

$$\int_0^{\theta_o} \frac{1}{\cos\theta} J_1\left(\frac{\beta}{\cos\theta}\right) d\theta = C(\beta,\theta_o) \tag{55}$$

and

$$\int_{\theta_o}^{\pi/2} \frac{1}{\sin\theta} J_1\left(\frac{\beta}{\sin\theta}\right) d\theta = S(\beta,\theta_o) \tag{56}$$

These can be handled by using the series formulas for the Bessel function, i.e.,

$$J_1(\alpha) = \sum_{m=0}^{\infty} B_m \alpha^{2m+1} \tag{57}$$

where

$$B_m = (-1)^m \frac{1}{m!(m+1)!2^{2m+1}}$$

Then clearly

$$C(\beta,\theta_o) = \sum_{m=0}^{\infty} B_m \beta^{2m+1} C_m(\theta_o) \tag{58}$$

and

$$S(\beta,\theta_o) = \sum_{m=0}^{\infty} B_m \beta^{2m+1} S_m(\theta_o) \tag{59}$$

Fields of a Circular Loop of Current

where
$$C_m(\theta_o) = \int_0^{\theta_o} \frac{1}{(\cos \theta)^{2m+2}} d\theta \tag{60}$$

and
$$S_m(\theta_o) = \int_{\theta_o}^{\pi/2} \frac{1}{(\sin \theta)^{2m+2}} d\theta \tag{61}$$

These are integrals of a standard type and we may write

$$C_o(\theta_o) = \tan \theta_o \tag{62}$$

$$C_1(\theta_o) = \tan \theta_o + \frac{\tan^3 \theta_o}{3} \tag{63}$$

$$C_2(\theta_o) = \frac{\sin \theta_o}{5 \cos^5 \theta_o} + \frac{4}{15} \frac{\sin \theta_o}{\cos^3 \theta_o} + \frac{8}{15} \tan \theta_o \tag{64}$$

For higher order values we have the recurrence formula

$$C_m(\theta_o) = \frac{\sin \theta_o}{(2m+1)\cos^{2m+1} \theta_o} + \frac{2m}{2m+1} C_{m-1}(\theta_o) \tag{65}$$

Similarly
$$S_o(\theta_o) = \cot \theta_o \tag{66}$$

$$S_1(\theta_o) = \cot \theta_o + \frac{\cot^3 \theta_o}{3} \tag{67}$$

$$S_2(\theta_o) = \frac{\cos \theta_o}{5 \sin^5 \theta_o} + \frac{4}{15} \frac{\cos \theta_o}{\sin^3 \theta_o} + \frac{8}{15} \cot \theta_o \tag{68}$$

and, for higher values,

$$S_m(\theta_o) = \frac{\cos \theta_o}{(2m+1)\sin^{2m+1} \theta_o} + \frac{2m}{2m+1} S_{m-1}(\theta_o) \tag{69}$$

Note: Equ. (49) is not perfectly general. The limits need to be modified when the observer is outside the loop. However, in the case of the rectangular loop it appears (54) is valid. (This comment provided by Dr. D.A. Hill, 11 Oct. 1979).

NUMERICAL RESULTS FOR THE SMALL LOOP

While (32) and (33) can be evaluated analytically in certain limiting cases, the most direct approach is numerical integration. Convergence of the integrals is assured by the exponential factor in the numerator of the right-hand side of (30). Also, we note that P and Q are functions of the dimensionless parameters $B = (\sigma_2\mu_0\omega)^{1/2} b$, $H = (\sigma_2\mu_0\omega)^{1/2} h$, $D = \rho/h$, $Z = z/h$, and $K = (\sigma_1/\sigma_2)^{1/2}$. A digital computer program has been written for evaluating P and Q, and results have been obtained for a wide range of the parameters [Wait and Spies, 1971b].

A quantity of special interest is the magnitude of the ratio H_ρ/H_z at $z = 0$ on the earth's surface. The magnitude of this ratio is $|P/Q|$ and typical plots are indicated in Figs. 2-4 for selected values of the parameters. The abscissa in each case is the distance parameter D or ρ/h. The limiting case designated $H = 0$ corresponds to purely static conditions (i.e., $\omega \to 0$). In this case, it is not difficult to verify that, for $z = 0$, (32) and (33) reduce to

$$P = (3/2)DR^{-5}$$

$$Q = (1/2)[3R^{-5} - R^{-3}]$$

where $R = (D^2 + 1)^{1/2}$. The curves in Figs. 2-4 are consistent with these static field formulas for the small values of H. In particular, we see that the null for Q at $D = \sqrt{2}$ leads to an infinity for $|P/Q|$. This pronounced peak, however, is somewhat blurred when H is different from zero. In fact, the behavior of the ratio $|P/Q|$ becomes modified drastically for the larger values of H.

In each case, we see that, for sufficiently small values of H, the values of P and Q tend toward the limiting static forms. Another check was to note that for large values of H and B the results are consistent with the

Fields of a Circular Loop of Current

formulas

$$P \sim e^{i\pi/4} [3/(HK)] D^{-4} [2K/(K+1)]$$
$$\times \exp[-e^{i\pi/4}(H-B) - e^{i\pi/4} KB]$$

and

$$P/Q \simeq e^{i\pi/4} HDK/3$$

To consider a specific example, we choose $f = \omega/2\pi = 10^2$ Hz, $\sigma_2 = 10^{-3}$ mho/m, and $h = 200$ m; thus, $H = 0.7$. In this case, the upper layer of thickness $(B/H) \times 200$ m, has a conductivity $K^2 \times 10^{-3}$ mho/m. As indicated in Figs. 2 and 3, the presence of this masking layer deteriorates the ideal static-like behavior of the field ratio $|P/Q|$.

Using the analytical results from this study, it is possible to elucidate the effect of the stratification in the overburden and to provide improved criteria for direction finding techniques. The results given here are considered only as an example of the many possibilities that exist.

APPENDIX A

The Homogeneous Half-Space and Analytical Approximations (Quasi-Static Case)

Considering the loop, whose area $\times$ turns product is IA, as a vertical magnetic dipole, we can obtain the resultant magnetic fields in the air (i.e., $z>0$) from the following as a special case.

$$i\mu_0\omega H_\rho = \partial^2 F/(\partial\rho\partial z) \quad (1)$$

and

$$i\mu_0\omega H_z = \partial^2 F/\partial z^2 \quad (2)$$

where μ_0 is the permeability of free space, ω is the angular frequency, and F is a scalar function. †

$$F = \frac{i\mu_0\omega IA}{2\pi} \int_0^\infty \frac{\lambda}{u+\lambda} e^{-\lambda z} e^{-uh} J_0(\lambda\rho) d\lambda \quad (3)$$

where $u = (\lambda^2+\gamma^2)^{1/2}$, $\gamma^2 = i\sigma\mu_0\omega$, and J_0 is a Bessel function of order zero.

Using (1) and (2) along with (3), we find that the fields at the location (ρ, z) for $z>0$ are

$$H_\rho = b_0 P \quad (4)$$

and

$$H_z = b_0 Q \quad (5)$$

where

$$P = h^3 \int_0^\infty \frac{\lambda^3}{\lambda+u} e^{-\lambda z} e^{-uh} J_1(\lambda\rho) d\lambda \quad (6)$$

and

$$Q = h^3 \int_0^\infty \frac{\lambda^3}{\lambda+u} e^{-\lambda z} e^{-uh} J_0(\lambda\rho) d\lambda \quad (7)$$

and the normalizing factor is

$$b_0 = IA/(2\pi h^3). \quad (8)$$

By changing the variable of integration to $x = \lambda h$, the integrals P and Q are expressible in the form

$$P = \int_0^\infty \frac{x^3 e^{-xZ}}{x + (x^2 + iH^2)^{1/2}} e^{-(x^2+iH^2)^{1/2}} J_1(xD) dx \quad (9)$$

---

† Note F is related to the earlier Hertz potential Π^* by $F = i\mu_0\omega\Pi^*$

Fields of a Circular Loop of Current

and

$$Q = \int_0^\infty \frac{x^2 e^{-xZ}}{x + (x^2 + iH^2)^{1/2}} e^{-(x^2 + iH^2)^{1/2}} J_0(xD)dx \quad (10)$$

where $H = (\sigma\mu_0\omega)^{1/2} h$, $D = \rho/h$, and $Z = z/h$ are dimensionless parameters. Equations (9) and (10) are in a form suitable for numerical integration, and a computer program has been prepared to accomplish this task. Before discussing the results, some special limiting cases will be mentioned.

LIMITING CASES

When the conductivity σ and/or the frequency ω are sufficiently low, the parameter H is effectively zero. Then, without difficulty, the integrals P and Q are expressible in the closed forms

$$P = (3/2)D(Z+1)R^{-5} \quad (11)$$

and

$$Q = (1/2)[3(Z+1)^2 R^{-5} - R^{-3}] \quad (12)$$

where $R = [D^2 + (Z+1)^2]^{1/2}$. These simple results for the normalized magnetic fields P and Q correspond to static conditions.

The other limiting case of interest is when the conductivity σ and/or the frequency ω are sufficiently high that the integrals P and Q may be approximated in an asymptotic sense. Essentially, this amounts to arguing that when $|\gamma\rho| \gg 1$, the exponential factor $\exp(-uh)$ in (3), (6), and (7) may be adequately approximated by $\exp(-\gamma h)$. Then, treating both H and D as large parameters, we find that

$$P \sim - e^{i\pi/4}(3/H)D^{-4}\exp[-e^{i\pi/4}H] \quad (13)$$

and

$$Q \sim + e^{i\pi/2}(9/H^2)D^{-5}\exp[-e^{i\pi/4}H]. \quad (14)$$

A precise description of the validity of these asymptotic forms is subtle, but on comparison with numerical integration or the exact forms, it appears that they are useful when both H and HD are $\gg 1$.

Another interesting check is to let h be identically zero. Then for $z = 0$, we have

$$F = \frac{i\mu\omega I A}{2\pi} \int_0^\infty \frac{\lambda}{u + \lambda} J_0(\lambda\rho)d\lambda \quad (15)$$

$$= \frac{i\mu\omega I A}{2\pi} \frac{1}{\gamma^2 \rho^3} [1 - (1+\gamma\rho)e^{-\gamma\rho}] \quad (16)$$

where we have made use of a known identity.† Then, again for $z=0$, we have

$$E_\phi = \partial F/\partial \rho = \frac{i\mu_0\omega IA}{2\pi} \cdot \frac{3}{\gamma^2\rho^4}\alpha \qquad (17)$$

where

$$\alpha = 1 - [1 + \gamma\rho + (\gamma^2\rho^2/3)]e^{-\gamma\rho}. \qquad (18)$$

If we now assume $|\gamma\rho| \gg 1$ and utilize the approximate (Leontovich) boundary condition, we have

$$H_\rho \simeq \eta E_\phi, \qquad \eta = i\mu_0\omega/\gamma \qquad (19)$$

$$\simeq -\frac{IA}{2\pi} \cdot \frac{3}{\gamma\rho^4}. \qquad (20)$$

This is consistent with (13) given above if we note that the right-hand side of (20) is $\lim_{h\to 0} b_0 P$. Also, we can obtain an exact expression for H_z at $z=0$ by using $i\mu_0\omega H_z = -(1/\rho)(\partial/\partial\rho)\rho E_\phi$ and (17); thus $H_z = -IA\rho/(2\pi\gamma^2\rho^5)\beta$ where

$$\beta = 1 - [1 + \gamma\rho + (4/9)\gamma^2\rho^2 + (1/9)\gamma^3\rho^3]e^{-\gamma\rho}. \qquad (21)$$

Again, if $|\gamma\rho| \gg 1$, $\beta \simeq 1$, and (21) is consistent with (14).

As a matter of theoretical interest, we have one final special case which deserves mention. Specifically, we take $\rho=0$, which means the observer is directly above the source loop. Then, of course, $P=0$, but Q is finite. If, in fact $z=0$, we have

$$Q = Q_* = h^3 \int_0^\infty \lambda^3(u+\lambda)^{-1}e^{-uh}d\lambda. \qquad (22)$$

Now,

$$(u+\lambda)^{-1} = (u-\lambda)(u^2-\lambda^2)^{-1} = \gamma^{-2}(u-\lambda),$$

and therefore

$$Q_* = \frac{h^3}{\gamma^2}\left[\int_0^\infty u\lambda^3 e^{-uh}d\lambda - \int_0^\infty \lambda^4 e^{-uh}d\lambda\right]. \qquad (23)$$

We can now verify that

$$Q_* = \frac{h^3}{\gamma^2}\left[\frac{\partial^4 p}{\partial h^4} - \gamma^2\frac{\partial^2 p}{\partial h^2} + \frac{\partial^5 q}{\partial h^5} - 2\gamma^2\frac{\partial^3 q}{\partial h^3} + \gamma^4\frac{\partial q}{\partial h}\right] \qquad (24)$$

where

---

† Note that
$$\lim_{\alpha \to 0} \int_0^\infty (\lambda/u)J_0(\lambda\rho)e^{-u\alpha}\,d\lambda = e^{-\gamma\rho}/\rho$$

$$p = \int_0^\infty e^{-uh} \frac{\lambda}{u} d\lambda \qquad (25)$$

and

$$q = \int_0^\infty \frac{e^{-uh}}{u} d\lambda. \qquad (26)$$

These latter two integrals are known to be

$$p = e^{-\gamma h}/h \qquad (27)$$

and

$$q = K_0(\gamma h) \qquad (28)$$

where K_0 is a modified Bessel function of order zero. Using only the well-known identities $K_0'(Z) = -K_1(Z)$ and $K_1'(Z) = -[K_0(Z)+Z^{-1}K_1(Z)]$, it turns out that (24) can be reduced to the compact form

$$Q_s = 2e^{-x}x^{-2}[(12 + 12x + 5x^2 + x^3)] \\ - 3[(x + (8/x))K_1(x) + 4K_0(x)] \qquad (29)$$

where $x = \gamma h = \sqrt{i}H$. This provides an alternative method to compute the field directly above the source dipole.

Discussion of Results and Final Remarks

As indicated above, P and Q are the normalized components of the horizontal and the vertical magnetic field on the surface for the buried loop source. The magnitudes $|P|$ and $|Q|$ plotted as a function of the normalized distance D (or ρ/h) are illustrated in Figs. 2 and 3, respectively. The parameter on the curve is the normalized depth parameter H (or $(\sigma\mu_0\omega)^{1/2}h$). This can be calculated from the simple formula $H = 0.086(\sigma_{mmho}f)^{1/2}h_{km}$ where σ_{mmho} is the earth conductivity in millimhos per meter, f is the frequency in hertz, and h_{km} is the burial depth in kilometers. For example, if $\sigma_{mmho} = 1$, $f = 10^2$, and $h_{km} = 0.2$, we have $H = 0.17$. In this case, we can make the rather important observation that the character of the $|P|$ versus D curves in Fig. 2 are essentially the same as for the case H=0. Thus, the conductivities of the overburden are probably neglible for this configuration of the receiving loop provided the burial depths are not more than 0.5 km and the operating frequencies are in the low audio range.

APPENDIX B

The Three-Layer Case

With reference to Fig. 6, it not difficult to show that the function $T(\lambda)$ analogous to (30) is given by

$$T(\lambda) = D^{-1}\left(\frac{2u_2}{u_2 + u_3}\right)\left(\frac{2u_1}{u_1 + u_2}\right)\left(\frac{2\lambda}{u_1 + \lambda}\right) \exp[-u_3(h - c) - u_2(c - b) - u_1 b] \quad (1)$$

where

$$D = 1 - \left(\frac{u_2 - u_3}{u_2 + u_3}\right)\left(\frac{u_2 - u_1}{u_2 + u_1}\right) \exp[-2u_2(c - b)]$$

$$- \left(\frac{u_1 - \lambda}{u_1 + \lambda}\right)\left(\frac{u_1 - u_2}{u_1 + u_2}\right) \exp[-2u_1 b] \quad (2)$$

$$- \left(\frac{u_1 - \lambda}{u_1 + \lambda}\right)\left(\frac{u_2 - u_3}{u_2 + u_3}\right) \exp[-2u_1 b - 2u_2(c - b)]$$

where $u_m = (\lambda^2 + \gamma_m^2)^{1/2}$, $\gamma_m^2 = i\sigma_m \mu_0 \omega$ for $m = 1,2,3$. Here $\mu_m = \mu_0$ and all displacement currents are ignored.

The three quantities in large parentheses on the right-hand side of (1) are the transmission coefficients for the interfaces at $z = -c$, $-b$ and 0 respectively. The total complex phase is given by the exponential term in (1). The total effect of the internal reflections are included in the expression for D given by (2).

The expression for D was given incorrectly in the report by Wait and Spies [1971b]. I am grateful to Dr. Thurlow W.H. Caffey, who pointed this out.

Exercise: Give an explicit derivation of the preceding equations for $T(\lambda)$ and the denominator D.

REFERENCES

LARGE, D.G., BALL, L., and FARSTAD, A.J., (1973), Radio transmission to and from underground coal mines - theory and experiment, *IEEE Trans.*, COM-21, No. 3, 194-202.

WAIT, J.R., and SPIES, K.P. (1971a), Electromagnetic fields of a small loop buried in a stratified earth, *IEEE Trans.*, AP-19, No. 5, 717-718.

WAIT, J.R., and SPIES, K.P., (1971b), Evaluation of the surface EM fields for a buried magnetic dipole, *AFCRL Rep. 52*, (available from N.T.I.S., Springfield, Va. 22151).

WAIT, J.R., (1971), Electromagnetic induction technique for locating a buried source, *IEEE Trans.*, GE-9, No. 2, 95-98.

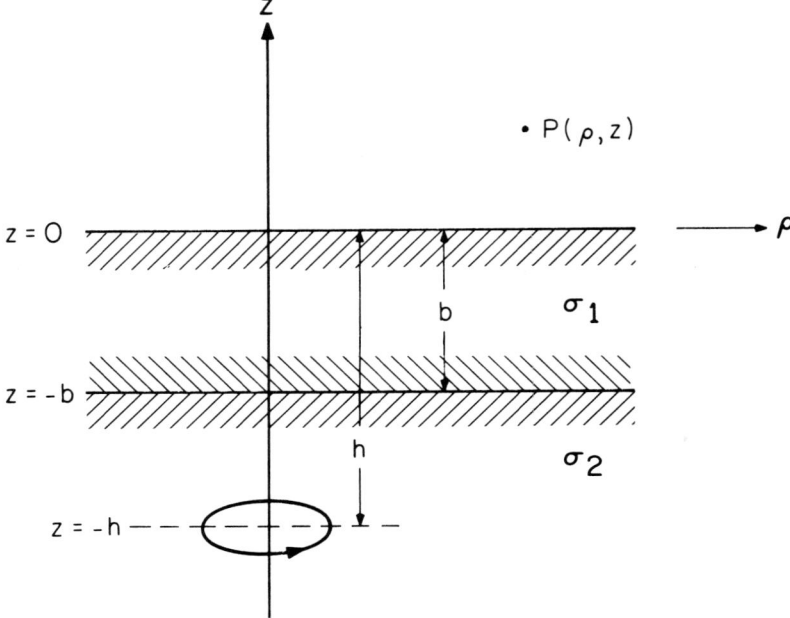

Fig. 1. Circular loop of radius a buried in a two layer earth.

Fields of a Circular Loop of Current

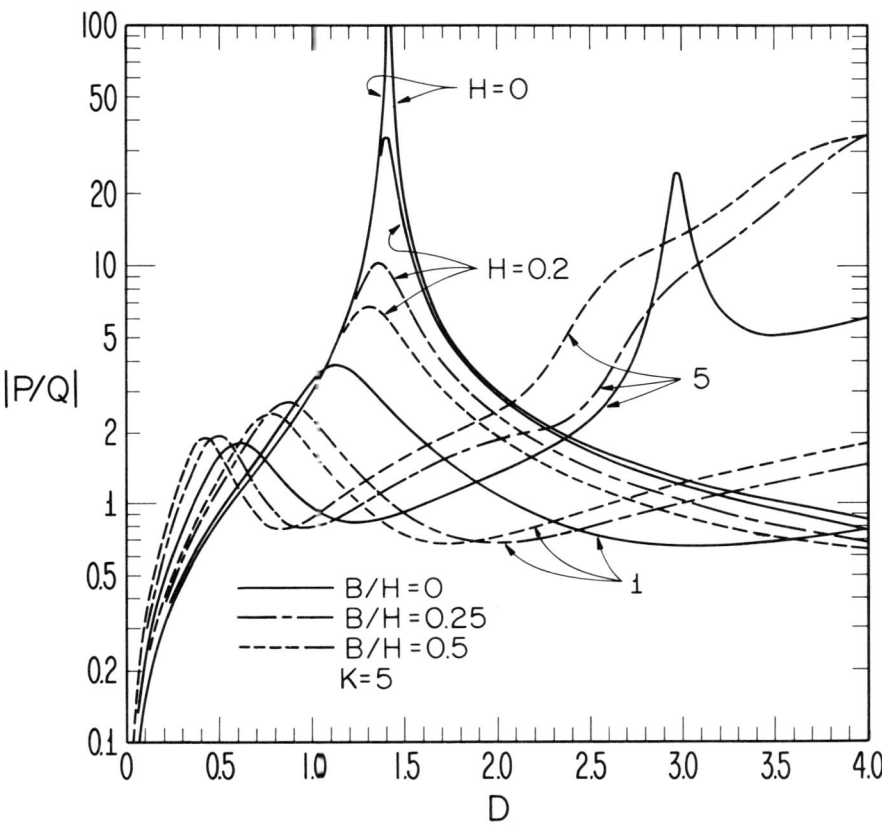

Fig. 2. Field ratio on surface for various values of $H = (\sigma_2 \mu_0 \omega)^{1/2} h$ and $B/H = b/h$ for fixed $K = (\sigma_1/\sigma_2)^{1/2} = 5$.

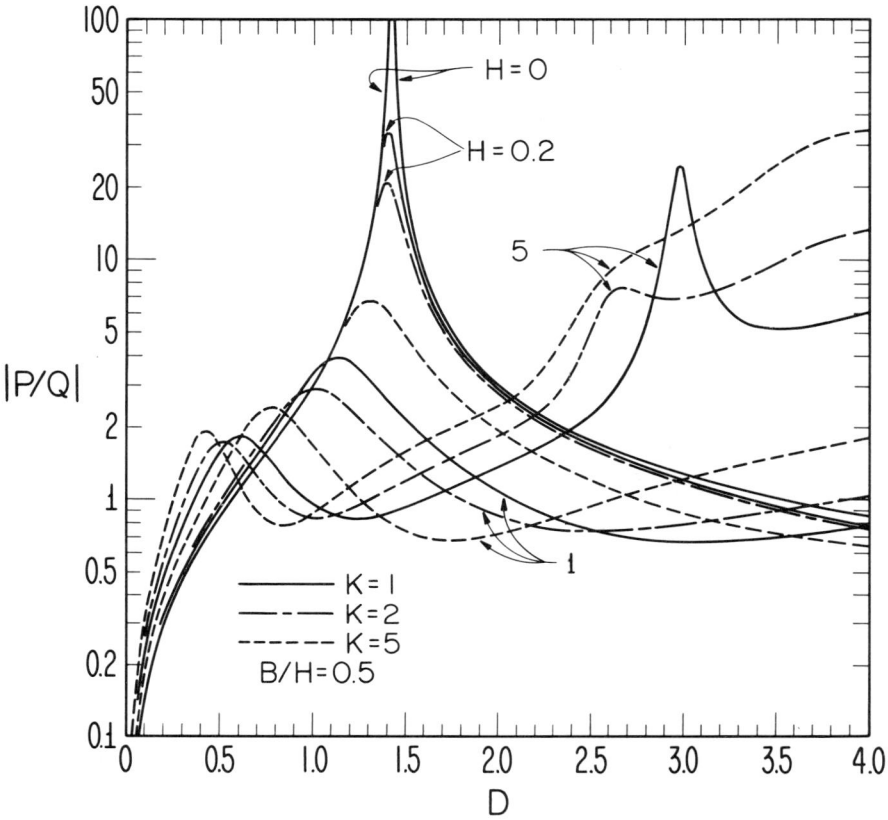

Fig. 3. Field ratio for various values of H and K for fixed B/H = b/h = 0.5.

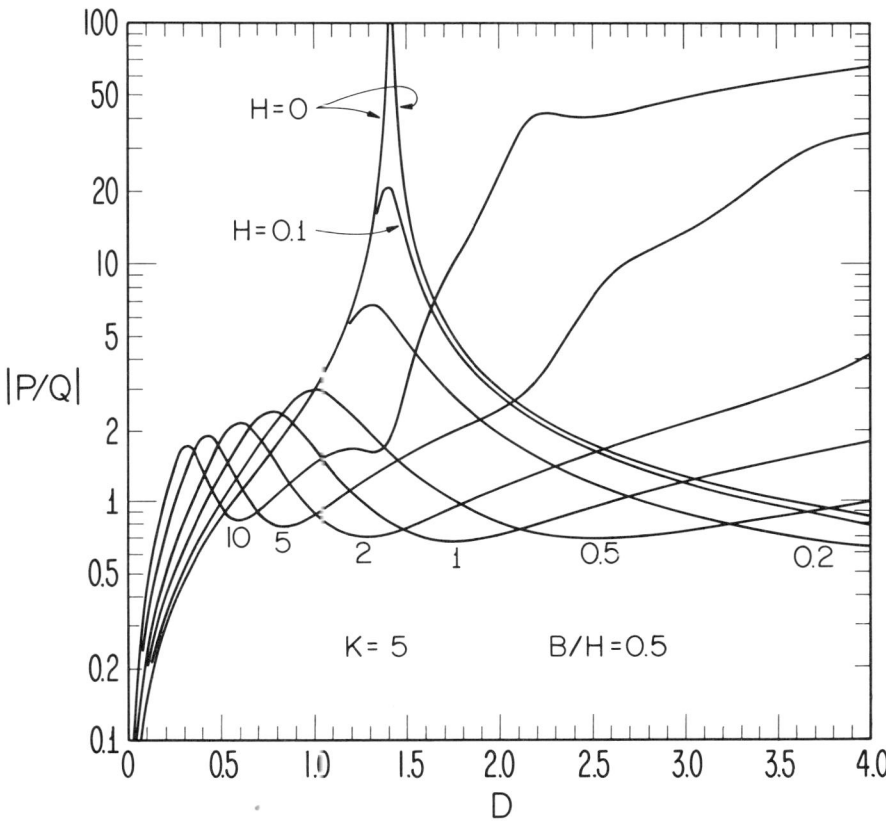

Fig. 4. Field ratio for various values of H for fixed K = $(\sigma_1/\sigma_2)^{1/2}$ = 5 and B/H = b/h = 0.5.

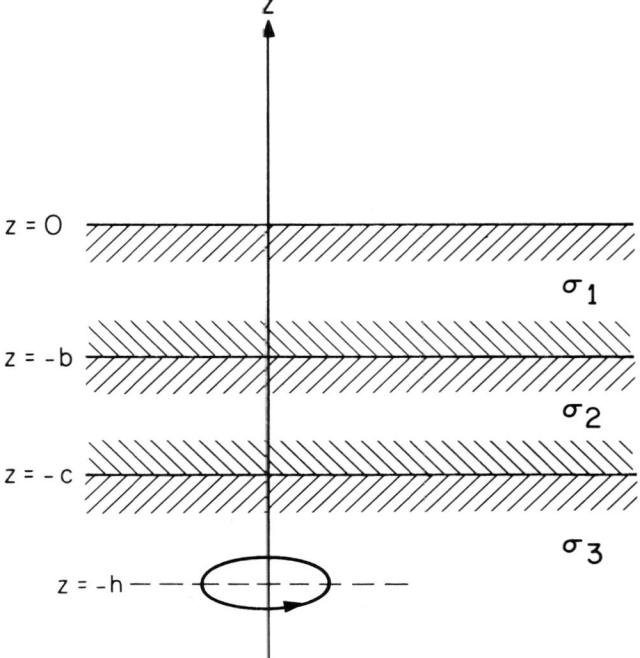

Fig. 5. Magnetic dipole buried in three-layer earth.

14
Transmissions in an Idealized Earth Crust Waveguide

Introduction

There is evidence that extended layers of low-conductivity material exist in the earth's crust. The possibility that these layers may guide electromagnetic waves with low loss is intriguing. The analogy with VLF electromagnetic waves in the earth-ionosphere waveguide immediately suggests itself. A major shortcoming, however, arises in that the excitation mechanism for subsurface waveguides is inherently poor unless, of course, the source itself is located within the semi-insulating stratum. Also, there is a fundamental difficulty in that nearly all geological materials *in situ* are electrically conductive by virtue of connate waters in the pore structure.

It is the purpose of this paper to present some theoretical solutions which should provide insight into the mechanisms of propagation in subsurface waveguides. To reduce the complexity, the model is highly idealized. Nevertheless, it is believed that the inherent features of this type of subsurface propagation are adequately displayed.

Formulation

We start with a simple model which is illustrated in figure 1. As indicated, the subsurface waveguide is bounded by two semi-infinite homogeneous media. The source is taken to be a vertical electric dipole which, for the moment, is in the upper medium. The conductivity, permittivity, and permeability of the media are σ_i, ϵ_i, and μ_i, respectively. The subscript i takes the values 1, 2, and 3 when referring to the upper, middle, and lower regions, respectively.

With reference to a cylindrical coordinate system (ρ, ϕ, z), the waveguide (i.e., the middle region) is bounded by the planes $z=0$, and $z=h_2$. The source dipole, which has a current moment Ids, is located at $z=-h$. The fields, in any of the regions, may be derived from a Hertz vector which has only a z component Π_{iz} ($i=1, 2,$ or 3). In what follows, the time factor is taken to be $\exp(i\omega t)$.

The formal solution for the problem, as stated, is known [Wait, 1962]. For the upper region,

$$\Pi_{1z} = \frac{Ids}{4\pi(\sigma_1 + i\epsilon_1\omega)} \left[\frac{\exp(-ik_1R)}{R} - \frac{\exp(-ik_1R')}{R'} + F \right], \tag{1}$$

where

$$R = [\rho^2 + (z+h)^2]^{1/2}, \quad R' = [\rho^2 + (z-h)^2]^{1/2},$$

$ik_1 = [(i\mu_1\omega)(\sigma_1 + i\epsilon_1\omega)]^{1/2}$, and F is an integral which is a spectrum of plane waves. If the upper surface of the waveguide were effectively a perfect magnetic conductor, F would be identically zero. However, in general, it is given exactly by the integral

$$F = \int_{-\infty}^{+\infty} \frac{K_1}{K_1 + Z_2} H_0^{(2)}(\lambda\rho) e^{-u_1(h-z)} \frac{\lambda}{u_1} d\lambda, \tag{2}$$

where

$$Z_2 = K_2 \frac{K_3 + K_2 \tanh(u_2 h_2)}{K_2 + K_3 \tanh(u_2 h_2)},$$

where

$$K_i = \frac{u_i}{\sigma_i + i\epsilon_i\omega}, \quad u_i = (\lambda^2 - k_i^2)^{1/2}, \quad k_i^2 = -(i\mu_i\omega)(\sigma_i + i\epsilon_i\omega),$$

(Reprinted, in part, from J.R. Wait, *Radio Science*, 1, 913-924, 1966)

are chosen so that the real parts are taken positive. On the other hand, the real part of k_i is positive and the imaginary part of k_+ is negative.

The integrand in (2) has a rather complicated analytical form, and any closed-form evaluation is out of the question. However, we may achieve some simplification if we deal with the case where the conductivity σ_2 of the middle region is small compared with the conductivities σ_1 and σ_3. Thus, the situation is similar to the earth-ionosphere waveguide and, therefore, we anticipate that the important contributions to the received field come from the waveguide modes propagating in region 2. However, some important differences exist. In the first place, the source may be *within* the upper highly conducting region. Also, we wish to investigate whether or not the (usually neglected) lateral waves will contribute in any significant way. Thus, it seems that a careful evaluation of the integral F is warranted.

The Residue Series

To recast the solution in a form for displaying the waveguide character of the problem, we write

$$\frac{2K_1}{K_1+Z_2}=f(\lambda)\,\frac{1}{1-R_u R_l \exp(-2u_2 h_2)}, \tag{3}$$

where R_u and R_l are reflection coefficients defined by

$$R_u=\frac{K_2-K_1}{K_2+K_1}, \quad R_l=\frac{K_2-K_3}{K_2+K_3}, \quad \text{and} \quad f(\lambda)=\frac{2K_1(K_2+K_3\tanh u_2 h_2)(1+e^{-2u_2 h_2})}{(K_2+K_1)(K_3+K_2)}.$$

From the identity given by (3), it is evident that the integrand of F has poles at $\lambda=\lambda_j$ where λ_j are solutions of

$$1-R_u R_l \exp(-2u_2 h_2)=0. \tag{4}$$

By making use of this modal condition, it is a simple matter to show that

$$f(\lambda_j)=\frac{4K_1 K_2}{K_2^2-K_1^2}\bigg]_{\lambda=\lambda_j}. \tag{5}$$

The locations of the pole singularities λ_j in the complex λ plane are indicated in figure 2 by small crosses. They are in two sets located (skew-symmetrically) in the second and fourth quadrants. The integrand also has branch points at $\lambda=\pm k_1$, $\pm k_2$, and $\pm k_3$. Following the usual prescription, the original contour is deformed in such a manner as to enclose the pole and branch point singularities. The process is illustrated in figure 2. For reasons which are evident later on, the branch lines are drawn from k_1, k_2, and k_3 down to infinity in the lower half plane. (This is appropriate when the distance ρ is somewhat greater than both h and z.) The contour is closed by a semicircle in the lower half plane. By an application of Jordan's lemma, it may be shown that the contribution to the integral from the latter vanishes as the radius of the semicircle approaches infinity. Thus, we are left with the contribution from the portion which encloses the poles λ_j and the contributions from the branch lines. It may be demonstrated that for the present problem, the branch line integration associated with k_2 is identically zero. Thus, it does not play any further role. (Note that Z_2 is an even function of u_2).

The pole contributions to F are designated F_p and, by the theorem of residues, it is given by

$$F_p = -i\pi \sum_j f(\lambda_j) \frac{H_0^{(2)}(\lambda_j\rho)\lambda_j(u_{1,j})^{-1} \exp[-u_{1,j}(h-z)]}{\left[\frac{\partial}{\partial \lambda}(1-R_uR_l\exp(-2u_2h_2))\right]_{\lambda=\lambda_j}}, \qquad (6)$$

where $u_{1,j}=(\lambda_j^2-k_1^2)^{1/2}$. In order to exploit the analogy with the earth-ionosphere waveguide, we write

$$\lambda_j = k_2 S_j = k_2(1-C_j^2)^{1/2},$$

where S_j and C_j are the sine and cosine of a complex angle associated with the waveguide modes of order j.

By carrying out the differentiation indicated in (6) and making use of (4), it follows that

$$F_p \cong \frac{-i2\pi}{h_2} \frac{k_2^2}{k_1^2} \sum_j \frac{C_j^2}{C_j^2-\Delta_j^2} \delta_j H_0^{(2)}(k_2 S_j\rho) \exp[-u_{1,j}(h-z)], \qquad (7)$$

where

$$\delta_j = \frac{1}{\left[1+i\frac{\partial R_u R_l/\partial C}{2k_2h_2R_uR_l}\right]_{C=C_j}}, \qquad (8) \qquad \Delta_j = \frac{\eta_1}{\eta_2}\left(1-\frac{k_2^2}{k_1^2}S_j^2\right)^{1/2}, \qquad (9)$$

$$\eta_1 = \frac{ik_1}{\sigma_1+i\epsilon_1\omega} \quad \text{and} \quad \eta_2 = \frac{ik_2}{\sigma_2+i\epsilon_2\omega}.$$

The vertical electric field in region 1 which corresponds to the sum of the modes is obtained from

$$E_{1z}^p = \left(k_1^2 + \frac{\partial^2}{\partial z^2}\right)\left[\frac{Ids}{4\pi(\sigma_1+i\epsilon_1\omega)}F_p\right]. \qquad (10)$$

Thus, under the condition that $\mu_1=\mu_2=\mu_3=\mu_0$, we readily find that

$$E_{1z}^p = E_0(k_2/k_1)^4 W_1, \qquad (11)$$

where

$$E_0 = \frac{-Ids i\mu_0\omega}{2\pi}\frac{e^{-ik_2\rho}}{\rho}, \qquad (12)$$

and

$$W_1 = \frac{-i\pi\rho}{h_2} e^{ik_2\rho} \sum_j \delta_j \frac{C_j^2 S_j^2}{C_j^2-\Delta_j^2} H_0^{(2)}(k_2 S_j\rho) \exp[-u_{1,j}(h-z)]. \qquad (13)$$

The corresponding expression for the field E_{2z}^p is obtainable from the general solution or, more directly, it is gotten by noting that

$$[k_2^2 E_{2z} = k_1^2 E_{1z}]_{z=0}$$

and by remembering that in the middle region Π_{2z} must satisfy $(\nabla^2+k_2^2)\Pi_{2z}=0$. Thus, we readily find that

$$E_{2z}^p = E_0(k_2/k_1)^2 W_2, \qquad (14)$$

where

$$W_2 = -\frac{i\pi\rho}{h_2} e^{ik_2\rho} \sum_j \delta_j \frac{C_j^2 S_j^2}{C_j^2 - \Delta_j^2} H_0^{(2)}(k_2 S_j \rho) \hat{f}_j(z) \exp[-u_{1,j}h], \qquad (15)$$

where

$$\hat{f}_j(z) = \frac{e^{ik_2 C_j z} + R_u^{(j)} e^{-ik_2 C_j z}}{1 + R_u^{(j)}}, \qquad (16)$$

and where

$$R_u^{(j)} = \frac{C_j - \Delta_j}{C_j + \Delta_j}. \qquad (17)$$

Another simple extension of the solution is that when the source itself is located *in* the waveguide. The corresponding solutions are then

$$E_{1z}^\rho = E_0(k_2/k_1)^2 W_1', \qquad (18)$$

$$E_{2z}^\rho = E_0 W_2', \qquad (19)$$

where the primed quantities W_1' and W_2' have the same form as W_1 and W_2 except that $\exp[-u_{1,j}h]$ is now replaced by $f_j(h')$ where $z = h'$ is the location of the source. The function W_2' is identical in the form to the corresponding result derived for the earth-ionosphere waveguide.

The Lateral Waves

We shall now discuss the (nonvanishing) branch line contributions. Specifically, we shall consider the portion of the integration which runs from $k_3 - i\infty$ to k_3 on the left side of the branch line and then from k_3 to $k_3 - i\infty$ on the right side. It is evident that the contribution to the integral F, which we denote $F^{(3)}$ is given by

$$F^{(3)} = \int_{k_3}^{k_3 - i\infty} \left[\frac{K_1}{K_1 + Z_2^{(+)}} - \frac{K_1}{K_1 + Z_2^{(-)}} \right] e^{-u_1(h-z)} \frac{\lambda}{u_1} H_0^{(2)}(\lambda\rho) d\lambda, \qquad (20)$$

where $Z_2^{(+)}$ and $Z_2^{(-)}$ are the values of Z_2 evaluated on the right and left portions of the branch line contributions. We observe immediately that, provided $|k_3\rho| \gg 1$, the Hankel function may be well approximated by the first term of its asymptotic expansion. That is,

$$H_0^{(2)}(\lambda\rho) \cong \left(\frac{2i}{\pi\lambda\rho}\right)^{1/2} \exp(-i\lambda\rho),$$

which indicates immediately that the major contribution to the integral $F^{(3)}$ is when λ is near k_3. This suggests that we introduce a new variable of integration s defined such that

$$\lambda = k_3 - is^2.$$

Thus, on the (+) side of the contour we note that

$$u_3 = (\lambda - k_3)^{1/2}(\lambda + k_3)^{1/2} = e^{-i\pi/4} s(2k_3 - is^2)^{1/2} \cong e^{-i\pi/4} s(2k_3)^{1/2},$$

whereas on the (−) side,

$$u_3 = -e^{-i\pi/4} s(2k_3 - is^2)^{1/2} \cong -e^{-i\pi/4} s(2k_3)^{1/2}.$$

The Hankel function in the integrand is thus

$$H_0^{(2)}(\lambda\rho) \cong \left(\frac{2i}{\pi k_3 \rho}\right)^{1/2} e^{-ik_3\rho} e^{-\rho s^2}, \tag{21}$$

which emphasizes the rapid decay as s becomes greater than zero. Elsewhere in the integrand for $F^{(3)}$, we note that

$$u_2 = (\lambda^2 - k_2^2)^{1/2} \cong (k_3^2 - k_2^2)^{1/2}$$

and

$$u_1 = (\lambda^2 - k_1^2)^{1/2} \cong (k_3^2 - k_1^2)^{1/2} = i(k_1^2 - k_3^2)^{1/2}.$$

Also, in order to achieve further simplification, we note that

$$\tanh u_2 h_2 \cong 1 - 2e^{-2u_2 h_2}$$

provided $|k_3 h_2| \gg 1$.

With the approximations indicated above, the integral $F^{(3)}$ is thus proportional to

$$\int_0^\infty s^2 e^{-\rho s^2} ds = \frac{\pi^{1/2}}{4\rho^{3/2}}, \tag{22}$$

provided the slowly varying factors are taken outside the integral. Thus, we find that

$$F^{(3)} \cong -i8 \frac{k_2^4 \exp(-ik_3\rho) \exp[-2(k_3^2-k_2^2)^{1/2}h_2 - (k_3^2-k_1^2)^{1/2}(h-z)]}{k_1^2(k_3^2-k_2^2)k_3\rho^2 \left[1 + \frac{(k_3^2-k_1^2)^{1/2}}{(k_3^2-k_2^2)^{1/2}} \frac{k_2^2}{k_1^2}\right]^2}. \tag{23}$$

The vertical electric field corresponding to this wave is

$$E_{1z}^{(3)} \cong \frac{Ids}{4\pi(\sigma_1 + i\epsilon_1\omega)} k_3^2 F^{(3)} \cong \frac{-2iIds\mu_0\omega}{\pi} \frac{k_2^4}{k_1^4} \frac{\exp(-ik_3\rho)}{k_3\rho^2}$$

$$\times \exp[-2(k_3^2-k_2^2)^{1/2}h_2] \exp[-(k_3^2-k_1^2)^{1/2}(h-z)]. \tag{24}$$

This corresponds to a heavily damped wave, since the arguments of the first two exponential factors have real parts somewhat greater than one.

Similar arguments apply to the branch line associated with k_1. The corresponding branch line contribution in this case leads to the result that $E_{1z}^{(1)} \propto \rho^{-2} \exp(-ik_1\rho)$, which is also very heavily damped, particularly in view of the relative largeness of $|k_1|$ compared with both $|k_2|$ and $|k_3|$.

Discussion of the Mode Sum

As indicated by (13) and (15), the actual field must be calculated as a sum of modes. Provided $|k_2\rho| \gg 1$, we must evaluate

$$W_{1,2} \cong \frac{(2\pi k_2\rho)^{1/2}}{k_2 h_2} e^{ik_2\rho} e^{-i\pi/4} \sum_j \delta_j S_j^{3/2} e^{-ik_2 S_j \rho} P_j^{(1,2)}, \tag{25}$$

where P_j is an appropriate "height gain" or "depth gain." For example,

$$P_j^{(1)} = \exp[-u_{1,j}(h-z)] C_j^2 (C_j^2 - \Delta_j^2)^{-1} \tag{26}$$

and
$$P_j^{(2)} = f_j(z) \exp[-u_{1,j}h]C_j^2(C_j^2 - \Delta_j^2)^{-1}, \qquad (27)$$

depending on whether the observer is in medium (1) or medium (2), respectively.

The summation in (25) is a sum of waveguide modes which have a complex propagation constant ik_2S_j in the radial direction. The modal equation (4) may be written in the form

$$\left(\frac{C-\Delta}{C+\Delta}\right) e^{\alpha C} = \exp\left[-2\pi i\left(j-\frac{1}{2}\right)\right] \exp[i2k_2h_2C] \qquad (28)$$

where $\Delta = (k_1/k_2)(1 - S^2k_1^2/k_2^2)^{1/2}$, $S = (1-C^2)^{1/2}$, and α is defined such that

$$R_l = -\exp(\alpha C), \text{ and } C = u_2/(ik_2).$$

Solutions of (28) yield the permissible values C_j. Thus, we note that the attenuation of the mode of order j is $-\text{Im } k_2S_j$ nepers per unit distance. For convenience, we shall assign the mode numbers j such that the imaginary parts of $-k_2S_1, -k_2S_2, -k_2S_3, -k_2S_4, \ldots$, are monotonically increasing.

For convenience, we now assume, for the modes of low attenuation, that $|\alpha C| \ll 1$ and $|\Delta/C| \ll 1$. The modal eq (28) is then expressible in the form

$$C(\alpha - i2k_2h_2) - 2\Delta/C = -2\pi i[j - (1/2)]. \qquad (29)$$

While this appears to be simple, we must remember that both α and Δ are functions of C. Fortunately, however, for the dominant modes, these are slowly varying and, for a given mode they may be regarded as a constant. For example, since for modes of low attenuation S is near unity, we may approximate

$$\Delta_j \cong \Delta \cong (k_1/k_2)(1 - k_1^2/k_2^2)^{1/2},$$

which is equal to the (normalized) surface impedance for a plane wave at grazing incidence on to the upper surface (i.e., at $z=0$).

The solutions of (29) are

$$C = C_j = \frac{2\pi\left(j-\frac{1}{2}\right) \pm \left[(2\pi)^2\left(j-\frac{1}{2}\right)^2 + 8\Delta i(2k_2h_2 + i\alpha)\right]^{1/2}}{2(2k_2h_2 + i\alpha)}. \qquad (30)$$

When the $+$ sign is taken before the radical, the solutions for C_j reduce to

$$C = C_j^{(0)} = \frac{2\pi\left(j-\frac{1}{2}\right)}{2k_2h_2 + i\alpha} \qquad (31)$$

under the condition $\Delta = 0$. Thus, regarding the second term within the radical of (30) as a perturbation, we readily find that

$$k_2S_j \cong k_2S^{(0)} - \frac{i\Delta}{h_2\left(1 + \frac{i\alpha}{2k_2h_2}\right)}, \qquad (32)$$

where
$$S_j^{(0)} = [1-(C_j^{(0)})^2]^{1/2}.$$

To illustrate the application of the above results, we assume that the lower medium has a conductivity which is not too great. First of all, we have the exact result

$$R_l = \frac{K_2-K_3}{K_2+K_3} = \frac{C-\frac{k_2}{k_3}\left(1-\frac{k_2^2}{k_3^2}S^2\right)^{1/2}}{C+\frac{k_2}{k_3}\left(1-\frac{k_2^2}{k_3^2}S^2\right)^{1/2}} = -e^{\alpha C}. \quad (33)$$

Clearly, this may be approximated by

$$R_l \cong -1 + \frac{2C}{\frac{k_2}{k_3}\left(1-\frac{k_2^2}{k_3^2}S^2\right)^{1/2}} \quad (34)$$

or
$$R_l \cong -1 + \alpha C,$$

where
$$\alpha = -2(k_3/k_2)(1-k_2^2 S^2/k_3^2)^{-1/2}.$$

The simplification indicated above is evidently valid for modes where $|\alpha C_j| \ll 1$, which means that the boundary is acting as a "magnetic wall," since the reflection coefficient R_l is near -1. Furthermore, for the modes of low attenuation, $|C_j|$ is small and, thus,

$$\alpha \cong -2/\Delta_l,$$

where
$$\Delta_l = (k_2/k_3)(1-k_2^2/k_3^2)^{1/2}$$

is a normalized surface impedance for plane waves at grazing incidence onto the lower interface (i.e., at $z = h_2$).

Setting
$$k_3^2/k_2^2 = K_r - \frac{i}{L}$$

where K_r and L are real, it is evident that

$$i\alpha \cong \frac{2}{[1+L^2(K_r-1)^2]^{1/2}}\left(K_r L^{1/2} - \frac{i}{L^{1/2}}\right)\left(\cos\left(\frac{\pi}{4}-\psi\right) - i\sin\left(\frac{\pi}{4}-\psi\right)\right) \quad (35)$$

where
$$\psi = \frac{1}{2}\arctan L(K_r-1).$$

This result may be used in (32) for calculating $k_2 S_j$.

Further simplification is now achieved by choosing the loss tangent δ in the waveguide region to be very small. Thus,

$$k_2 \cong |k_2|(1-i\delta/2) \quad \text{where } \delta \ll 1.$$

At the same time, the upper medium is taken to be well conducting. That is,

$$k_1 \cong |k_1|e^{i\pi/4}, \qquad \text{provided } \epsilon_1\omega/\sigma_1 \ll 1.$$

As a result, $|k_2| \cong N_2\omega/c_0$, where $N_2 = (\epsilon_2/\epsilon_0)^{1/2}$ is the refractive index of the waveguide medium, $\delta = \sigma_2/(\epsilon_2\omega)$, and $|k_1| \cong (\sigma_1\mu_0\omega)^{1/2}$. Under these conditions, we then find that

$$k_2 S_j \cong N_2(\omega/c_0)s_j + \frac{(c_j)^2(K_r L^{1/2} \cos - L^{-1/2} \sin)}{h_2[1 + L^2(K_r - 1)^2]^{1/2}} - \frac{|\Delta|}{2^{1/2}h_2}$$

$$- i\left[N_2(\omega/c_0)s_j\delta/2 + \frac{(c_j)^2(K_r L^{1/2} \sin + L^{-1/2} \cos)}{h_2[1 + L^2(K_r - 1)^2]^{1/2}} + \frac{|\Delta|}{2^{1/2}h_2}\right], \qquad (36)$$

where
$$|\Delta| \cong |k_1/k_2| \cong (\epsilon_2\omega/\sigma_1)^{1/2} = (\epsilon_0\omega/\sigma_1)^{1/2}N_2, \qquad K_r \cong \epsilon_3/\epsilon_2 \cong (\epsilon_3/\epsilon_0)N_2^{-2},$$

$$L \cong \epsilon_2\omega/\sigma_3 = (\epsilon_0\omega/\sigma_3)N_2^2, \qquad s_j = (1 - c_j^2)^{1/2}, \qquad c_j \cong \frac{\pi\left(j - \frac{1}{2}\right)}{|k_2|h_2} \cong \frac{\pi\left(j - \frac{1}{2}\right)c_0}{N_2\omega h_2}, \qquad c_0 = (\epsilon_0\mu_0)^{-1/2},$$

$$\sin = \sin\left(\frac{\pi}{4} - \psi\right), \text{ and } \cos = \cos\left(\frac{\pi}{4} - \psi\right).$$

If in the somewhat unlikely case that $\epsilon_2 = \epsilon_3$, then $K_r = 1$ and $\psi = 0$. Then (36) may be written in the somewhat simpler form,

$$k_2 S_j \cong N_2(\omega/c_0)s_j + \frac{(c_j)^2(L^{1/2} - L^{-1/2}) + |\Delta|}{2^{1/2}h_2} - i\left[N_2(\omega/c_0)s_j(\delta/2) + \frac{(c_j)^2(L^{1/2} + L^{-1/2}) + |\Delta|}{2^{1/2}h_2}\right]. \qquad (37)$$

When $N_2 = 1$ and $\delta = 0$, corresponding to a vacuum for region 2, the situation becomes fully analogous with the earth-ionosphere waveguide.
In the latter case, $L = \omega\nu/\omega_0^2$ where ω_0 and ν are the plasma frequency and collision frequency of the ideal sharply bounded ionosphere. In this situation, h_2 is the ionospheric reflecting height while σ_1 and ϵ_1 are the equivalent ground constants.

In order to illustrate the order of magnitude of some of the relevant quantities, we select an idealized earth crust waveguide with the following properties:

$$h_2 = 500 \text{ m}, \qquad \sigma_2 = 10^{-6} \text{ mho/m}, \qquad \epsilon_2 = N_2^2\epsilon_0 = 9\epsilon_0,$$

$$\sigma_1 = 10^{-1} \text{ mho/m}, \qquad \epsilon_1\omega/\sigma_1 \ll 1, \qquad \sigma_3 = 5 \times 10^{-4} \text{ mho/m}, \qquad \epsilon_3 = 9\epsilon_0,$$

and the operating frequency is $\omega/2\pi = 10^6$ c/s.

For the conditions stated, we find that

$$c_j = \left(j - \frac{1}{2}\right)\bigg/10,$$

which is relatively small for the low-order modes (i.e., $j = 1, 2,$ and 3). Thus, $s_j \cong 1 - (c_j^2/2) \cong 1$ for most cases of interest.

The attenuation is conveniently broken into three parts. Thus,

$$-\operatorname{Im} k_2 S_j = \alpha_w + \alpha_u + \alpha_l,$$

where α_w is the attenuation due to the ohmic losses within the waveguide, whereas α_u and α_l are the attenaution due to the losses associated with the upper and lower walls, respectively. For the present example, it is found that

$$\alpha_w \cong s_j |k_2| \delta/2 = \frac{60\pi}{N_2} \sigma_2 s_j = 0.063 \times s_j \text{ nepers/km},$$

$$\alpha_u \cong \left(\frac{\epsilon_0 \omega}{2\sigma_1}\right)^{1/2} \frac{N_2}{h_2} = 0.10 \text{ nepers/km},$$

$$\alpha_l \cong \frac{(c_j)^2(L^{1/2}+L^{-1/2})}{2^{1/2} h_2} \cong \left(j-\frac{1}{2}\right)^2 \times 0.0282 \text{ nepers/km}.$$

It is evident for the example chosen that all three components of loss are comparable to one another. However, for the lowest-order modes (e.g., $j=1$ and 2), the dissipation in the upper medium is the dominant loss mechanism. However, as the mode number increases, the loss in the lower medium becomes more important. For example, $\alpha_l > \alpha_u$ when the mode order $j>3$. On the other hand, α_w due to the dissipation within the waveguide does not depend significantly on mode number, at least, provided the modes are not near cutoff (i.e., c_j not near unity).

It is interesting to note from the above that α_l is approximately proportional to h_2^{-3}, whereas α_u is only proportional to h_2^{-1}. Thus, for thinner waveguides (i.e., $h_2 < 500$ m), α_l will become dominant, whereas the converse is true for thicker waveguides (i.e., $h_2 > 500$ m). Frequency, of course, also plays a role. It is evident (provided δ remains $\ll 1$) that α_w is, to a first order, independent of frequency. On the other hand, α_u is proportional to $\omega^{1/2}$, while α_l is roughly proportional to ω^{-2}.

Of special interest is the factor $L^{1/2}+L^{-1/2}$ in the expression for α_l. This has a broad minimum where $L=1$, which is the case of our example. Thus, if the conductivity σ_3 is less than or greater than 5×10^{-4}, the attenuation will be increased somewhat (provided, of course, all other parameters are kept fixed).

For the example chosen, we may also go back and compute the excitation factors for the various modes. However, for the modes of lowest attenuation, it appears that the factor δ_j is within a few percent of unity for the example chosen, provided $j<8$. Also, the quantity $S_j C_j^2 (C_j^2 - \Delta_j^2)^{-1}$ in (13) and (15) may be replaced by unity for all intents and purposes. On the other hand, the ratio $(k_2/k_1)^2$ appearing in (14) and (18) plays a key role. For the example considered above, its magnitude is 5×10^{-3}, which suggests that every effort should be made to locate both the source dipole and the observer within the waveguide. Otherwise, some other mechanism of excitation should be used such as a horizontal electric dipole. For example, this fact has recently been exploited by King, deBettencourt and Sandler, [1979] who made a very significant study of the general problem.

List of Principal Symbols

$\sigma_i, \epsilon_i, \mu_i$ electrical constants of three regions ($i=1, 2,$ and 3)
(ρ, ϕ, z) cylindrical coordinates
h_2 thickness of lossy dielectric slab
Ids current moment of source dipole
h height of source dipole above slab
Π_{1z} z-component of Hertz vector for region 1
$k_i^2 = -i\mu_i\omega(\sigma_i + i\epsilon_i\omega)$ is the square of the horizontal wave numbers for region i
$u_1 = (\lambda^2 - k_i^2)^{1/2}$ is the vertical propagation constant for region i
$K_i = \dfrac{u_i}{\sigma_i + i\epsilon_i\omega}$ is the vertically looking wave impedance for region i

F is a basic integral defined by (2)
F_p is the contribution to F from the poles
$F^{(3)}$ is the contribution to F from the branch line integration associated with the branch point at $\lambda = k_3$
Z_2 which is defined after (2), is the surface impedance for the upper layer
R_u and R_l are reflection coefficients
$H_0^{(2)}(\lambda\rho)$ Hankel function of second kind of order zero
j mode number
S_j normalized vertical wave number
C_j normalized vertical wave number
δ_j excitation factor defined by (8)
Δ_j normalized surface impedance for upper interface
E_{1z} vertical electric field in upper region
E_{1z}^p contribution to E_{1z} from poles (i.e., the waveguide modes)
$E_{1z}^{(1,3)}$ contribution to E_{1z} from the branch points at k_1 and k_3 (i.e., the lateral waves)
$W_{1,2}$ attenuation functions for the sum of modes
α defined by $R_l = -\exp(-\alpha C)$ and approximately given by (35)
δ loss tangent for lossy dielectric slab
N_2^2 dielectric constant of slab relative to free space
K_r and L real parameters defined by $k_3^2/k_2^2 = K_r - iL^{-1}$ and approximately given by $K_r \cong \epsilon_3/\epsilon_2$ and $L = \epsilon_2\omega/\sigma_3$
α_w, α_u, and α_l are components of the attenuation rates resulting from the losses within the waveguide, in the upper wall, and in the lower wall, respectively.

REFERENCES

HEACOCK, J.G., (1971), *The Structure and the Physical Properties of the Earth's Crust*, Monograph No. 14, American Geophysical Union, Washington, D.C.

KING, R.W.P., deBettencourt, J.T., and SANDLER, B.H., (1979), Lateral-wave propagation of electromagnetic waves in the lithosphere, *IEEE Trans.*, GE-17, 86-92.

WAIT, J.R., (1962), *Electromagnetic Waves in Stratified Media*, ch. 2, Pergamon Press, Oxford (or 2nd Edition, 1970).

WAIT, J.R., (1963), The possibility of guided electromagnetic waves in the earth's crust, *IEEE Trans.*, AP-11, 617-624.

WAIT, J.R., (1966), Electromagnetic propagation in an idealized earth crust waveguide, *Radio Science*, 1, 913-924.

Zvereva, Ye. V., Ryazantsev, SAMUYLOV, A.M., and SHAKHSUVAROV, D.N., (1975) Research on the propagation of electromagnetic waves in the earth's crust, *Rasprostraneniye Radiovoln*, Izd-vo "Nauka" Moscow, pp. 312-354.

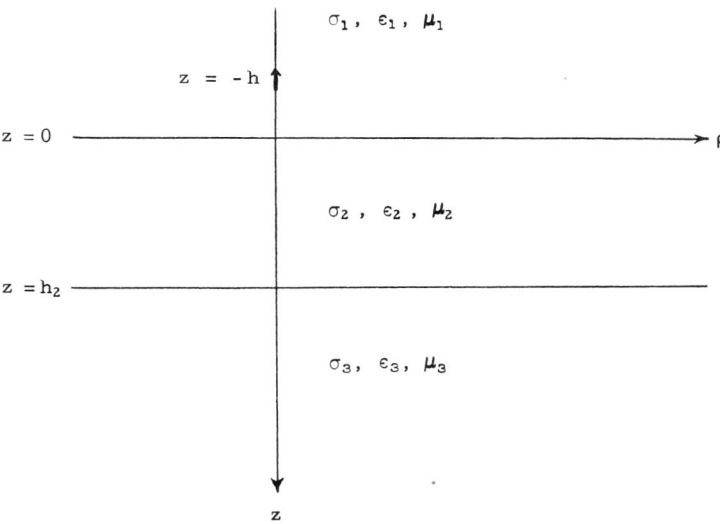

Fig. 1. The homogeneous lossy dielectric (subsurface) waveguide bounded by homogeneous conducting regions.

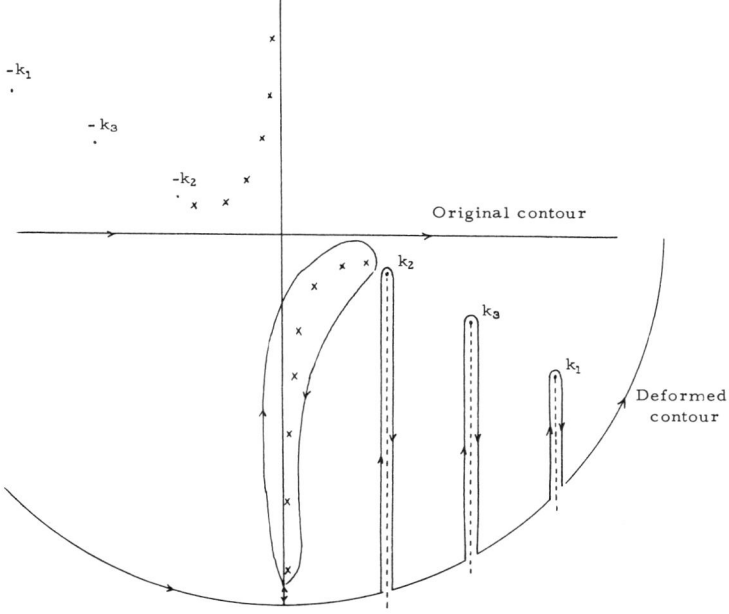

Fig. 2. The complex λ plane and the intergation contours.

15
Reflection from Inhomogeneous Media with Special Profiles

INTRODUCTION

Earlier we treated reflection from inhomogeneous planar structures from a stratified conductor by sub-dividing the medium into a number of parallel homogeneous layers. Such a method may be used for an approximation to a continuously varying conductivity profile. In fact, by taking a sufficiently large number of such layers of vanishing thickness, any desired degree of precision may be obtained. While this approach is direct, it is not very elegant. Furthermore, the calculations usually require a large-scale automatic computer. The alternative is to take some special form of conductivity variation, which allows the solutions to be expressed in terms of tabulated functions. This aspect of the subject is discussed briefly here. Particular attention is paid to profiles which lead to solutions in terms of Bessel functions.

GENERAL CONSIDERATIONS

For the moment, attention will be confined to perpendicular (or horizontal) polarization wherein the electric vector is always parallel to the plane of stratification (i.e. $z = 0$). Following the earlier work, the inhomogeneous medium will be taken to occupy the space $z > 0$. Unless otherwise stated, the region $z < 0$ is taken to be free space with electrical constants ε_0 and μ_0. The conductivity $\sigma(z)$ and dielectric constant $\varepsilon(z)$ are to be specified in some definite fashion for $z > 0$. For convenience, the permeability for $z > 0$ is also taken to be μ_0.

The electric field, which has only a y component, for the space $z < 0$, is given conveniently by [Wait, 1958]

$$E_y = E_0 \left(e^{-u_0 z} + R_\perp(\lambda) e^{u_0 z} \right) e^{-i\lambda x} \qquad (1)$$

where $\lambda = k_0 \sin \theta$, $u_0 = (\lambda^2 - k_0^2)^{1/2} = ik_0 \cos \theta$ and θ is the angle of incidence. As we have seen for the generally stratified half-space, the reflection coefficient has the form [Wait, 1958].

$$R_\perp(\lambda) = \frac{N_0 - Y_1}{N_0 + Y_1} \qquad (2)$$

where $N_0 = u_0/i\mu_0 \omega = \cos\theta/\eta_0$, $\eta_0 = \sqrt{(\mu_0/\varepsilon_0)} \cong 120\pi$. and Y_1 is the surface admittance at $z = 0$. In fact,

$$Y_1 = -\frac{H_{1x}}{E_{1y}}\bigg]_{z=0} \qquad (3)$$

The subscript 1 refers to the fact that the field is evaluated in the first layer of the multi-layer problem. Since we are dealing here with continuously stratified media, the subscript 1 may be dropped in what follows. Thus, in general,

(Reprinted, in part in revised form, from Electromagnetic Waves in Stratified Media, 2nd Edition, Pergamon Pres, New York, 1970)

Reflection from Inhomogeneous Media

$$R_\perp(\lambda) = \frac{N_0 - Y}{N_0 + Y} \quad (4)$$

where

$$Y = -\frac{H_x}{E_y}\bigg]_{z=0} \quad (5)$$

Alternately, we could write

$$R_\perp = -\frac{(\eta_0/\cos\theta) - Z}{(\eta_0/\cos\theta) + Z} \quad (6)$$

The problem of reflection of plane waves from a (planar) inhomogeneous media then boils down to finding an expression for the surface impedance evaluated at the interface.

The propagation constant $\gamma(z)$ is taken to have the following form

$$\gamma^2(z) = -k^2(z) = i\mu_0\omega[\sigma(z) + i\varepsilon(z)\omega] \quad (7)$$

Now remembering that the fields must vary everywhere as $e^{-i\lambda x}$ it is seen that Maxwell's equations in the inhomogeneous region can be written

$$i\mu_0\omega H_x = \frac{\partial E_y}{\partial z} \quad (8)$$

$$i\mu_0\omega H_z = i\lambda E_y \quad (9)$$

$$[\sigma(z) + i\varepsilon(z)\omega]E_y = \frac{\partial H_x}{\partial z} + i\lambda H_z \quad (10)$$

Combining these, it follows that

$$\frac{d^2 E_y}{dz^2} + (k^2(z) - \lambda^2)E_y = 0 \quad (11)$$

PROFILE WITH AN EXPONENTIAL TRANSITION

A special profile which also leads to Bessel functions is the exponential variation of the propagation constant [Wait, 1952]. For example, we let

$$-\gamma^2(z) = k^2(z) = (k_1^2 - k_2^2)e^{-\beta z} + k_2^2 \quad (12)$$

for $z > 0$ and $\text{Re } \beta > 0$. Thus

$$\lim_{z \to 0} k(z) = k_1$$

and

$$\lim_{z \to \infty} k(z) = k_2$$

Again assuming that the electric vector has only an E_y component, it is evident that Eq. (11) is now given by

$$\frac{d^2 E_y}{dz^2} + [(k_1^2 - k_2^2)e^{-\beta z} + (k_2^2 - k_0^2 \sin^2\theta)]E_y = 0 \quad (13)$$

To convert this into the required form, a new variable v is introduced as follows

$$v = \frac{2}{\beta} e^{-\beta z/2} (k_1^2 - k_2^2)^{1/2} \qquad (14)$$

Then, we find that

$$v^2 \frac{d^2 E_y}{dv^2} + v \frac{dE_y}{dv} + (v^2 - \nu^2) E_y = 0 \qquad (15)$$

provided

$$\nu^2 = -[k_2^2 - k_0^2 \sin^2 \theta] (2/\beta)^2. \qquad (16)$$

Solutions are then

$$E_y = \text{const } J_{\pm\nu}(v) e^{-i\lambda x} \qquad (17)$$

where

$$\nu = i[k_2^2 - k_0^2 \sin^2 \theta]^{1/2} (2/\beta) \qquad (18)$$

To select the proper index for the Bessel function, we consider the behavior for large values of z. For example, if $\beta z \gg 1$ it is seen that $v \ll 1$, and thus

$$J_{\pm\nu}(v) \cong \text{const} \times v^{\pm\nu}$$

$$\cong \text{const} \left[\frac{2}{\beta} (k_1^2 - k_2^2)^{1/2} \right]^{\pm\nu} \exp[\pm\nu(-\beta z/2)] \qquad (19)$$

It is now noted that Re $\nu > 0$ since k_2 has a finite negative imaginary part. Consequently, if fields are to be finite as $z \to \infty$, the upper sign must be chosen.

Now, from Maxwell's equations we see that

$$i\mu_0 \omega H_x = \frac{dE}{dv} \frac{dv}{dz}$$

$$= -e^{-\beta z/2} (k_1^2 - k_2^2)^{1/2} \frac{dE}{dv} \qquad (20)$$

Then

$$-\frac{E_y}{H_x} = \frac{i\mu_0 \omega J_\nu(v)}{e^{\beta z/2} (k_1^2 - k_2^2)^{1/2} J'_\nu(v)} \qquad (21)$$

An immediate check on this result is to allow β to approach infinity. Then, using the small argument approximation for $J_\nu(v)$, we see immediately that

$$-\frac{E_y}{H_x}\bigg|_{\beta \to \infty} = \eta_0 \left[\frac{k_2^2}{k_0^2} - \sin^2 \theta \right]^{1/2} \qquad (22)$$

which is the expected behavior for a homogeneous half-space of (complex) wave number k_2. Another interesting check on this result is to allow β to approach zero. Then we see that both ν and v become indefinitely large. Here, it is convenient to employ the following asymptotic approximation

$$J_{ip}(v) \cong (2\pi)^{-1/2} (p^2 + v^2)^{-1/4} \exp[-i\pi/4 + p\pi/2] \qquad (23)$$
$$\times \exp[i(p^2 + v^2)^{1/2} - ip \sinh^{-1}(p/v)]$$

where terms of order v^{-1} have been neglected. The above result is the first term of an expansion given by Bateman [1953] for purely imaginary order (i.e. p real). If the medium is a pure dielectric both k_1 and k_2 are real and ν is purely imaginary. Thus

$$\frac{J'_\nu(v)}{J_\nu(v)} \cong i \left[1 - \frac{\nu^2}{v^2}\right]^{1/2} \cong i \left[\frac{k_1^2 - k_0^2 \sin^2 \theta}{k_1^2 - k_2^2}\right]^{1/2} \qquad (24)$$

Thus

$$-\frac{E_y}{H_x}\bigg]_{\beta \to 0} = \eta_0 \left[\frac{k_1^2}{k_0^2} - \sin^2 \theta\right]^{1/2} \quad (25)$$

which again is the expected result.

The admittance at the boundary $z = 0$ is, of course,

$$Y = -\frac{H_x}{E_y}\bigg]_{z=0} = \frac{(k_1^2 - k_2^2)^{1/2} J_\nu(v_0)}{i\mu\omega J_\nu'(v_0)} \quad (26)$$

where

$$v_0 = (2/\beta)(k_1^2 - k_2^2)^{1/2}$$

The reflection coefficient is then given explicitly by

$$R_\perp = \frac{\cos\theta - Y\eta_0}{\cos\theta + Y\eta_0} \quad (27)$$

Other Exponential Profiles

A closely related profile is an exponentially increasing propagation constant. For example, if

$$-\gamma^2(z) = k^2(z) = k_1^2 e^{\beta z}, \quad \text{for } \beta > 0 \quad (28)$$

we readily find that

$$\frac{d^2 E_y}{dz^2} + (k_1^2 e^{\beta z} - k_0^2 \sin^2 \theta) E_y = 0 \quad (29)$$

Again, making the substitution

$$v = \frac{2}{\beta} e^{\beta z/2} k_1 \quad (30)$$

we then find

$$v^2 \frac{d^2 E_y}{dv^2} + v\frac{dE_y}{dv} + (v^2 - \nu^2) E_y = 0 \quad (31)$$

where

$$\nu^2 = k_0^2 (\sin^2\theta)(2/\beta)^2 \quad (32)$$

Independent solutions of this equation are still $J_{\pm\nu}(v)$, however, it is required that for z tending to $+\infty$, the field should be non-infinite. If the imaginary part of k_1 is negative, the required form is the Hankel function of the second kind $H_\nu^{(2)}(v)$ which is really a special linear combination of $J_{+\nu}$ and $J_{-\nu}$. Consequently,

$$-\frac{E_y}{H_x} = -\frac{i\mu_0\omega H_\nu^{(2)}(v)}{e^{\beta z/2} k_1 H_\nu^{(2)\prime}(v)} \quad (33)$$

which is in analogy to Eq. (21). In this particular case ν is taken as the positive or negative root of (32). However, for consistency, we shall write $\nu = k_0(\sin\theta)(2/\beta)$.

In certain physical problems the exponential increase may correspond to a conductivity which becomes indefinitely large. Thus, it is often more convenient to write

$$k^2(z) = -iB e^{\beta z} k_0^2 \quad (34)$$

where k_0^2 and B are essentially real when displacement currents can be neglected. Now, we choose the new variable to be

$$v = \frac{2k_0}{\beta} e^{i\pi/4} \sqrt{(B)} e^{\beta z/2} \tag{35}$$

Thus, we easily find that

$$\left(\frac{d^2}{dz^2} - ik_0^2 e^{\beta z} B - k_0^2 \sin^2\theta\right) E_y = 0 \tag{36}$$

becomes

$$\left[v^2 \frac{d^2}{dv^2} + v \frac{d}{dv} - (v^2 + \nu^2)\right] E_y = 0 \tag{37}$$

where $\nu = (2/\beta) k_0 \sin\theta$. This is the *modified* Bessel's equation of order ν. The appropriate solution is of the form

$$E_y = E_0 K_\nu(v) e^{-i\lambda x} \tag{38}$$

where $K_\nu(v)$ is the modified Bessel function of order ν and argument v. Therefore,

$$-\frac{H_x}{E_y} = -\frac{1}{\eta_0} \sqrt{(B)}\, e^{\beta z/2}\, e^{-i\pi/4} \frac{K'_\nu(v)}{K_\nu(v)} \tag{39}$$

and

$$Y = -\frac{H_x}{E_y}\bigg]_{z=0} = -\frac{1}{\eta_0} \sqrt{(B)}\, e^{-i\pi/4} \frac{K'_\nu(v_0)}{K_\nu(v_0)} \tag{40}$$

or

$$Y = -\frac{k(0)}{\mu_0 \omega} \frac{K'_\nu(v_0)}{K_\nu(v_0)} = \frac{k(0)}{\mu_0 \omega}\left[\frac{K_{\nu+1}(v_0)}{K_\nu(v_0)} - \frac{\nu}{v_0}\right] \tag{41}$$

where

$$v_0 = \frac{2k_0}{\beta} e^{i\pi/4} \sqrt{(B)} = i2k(0)/\beta.$$

One should note that in terms of the conductivity $\sigma(z)$, regard d as a function of of z,

$$v_0 = 2[i\sigma(0)\mu_0\omega]^{1/2}/\beta \tag{42}$$

Inserting the above expression for Y into Eq. (25) leads to an expression for the reflection coefficient at the interface $z = 0$ for a horizontally polarized plane wave incident in the free space region (i.e. $z < 0$). In the case of normal incidence (i.e. $\theta = 0$), it is seen that ν becomes zero in this particular problem. Thus, for *normal* incidence

$$Y = \frac{k(0)}{\mu_0 \omega} \frac{K_1(v_0)}{K_0(v_0)} \tag{43}$$

This result had been given previously [Wait, 1960].

The exponential profile is also well suited to a study of variable ionized media. A special feature of such propagation is the possibility of critical reflection. At normal incidence this occurs in the region where the dielectric constant is zero. The special exponential forms chosen above are not suitable for investigating this particular phenomenon. The form adopted is

$$k^2(z) = k_0^2[1 - b\, e^{\alpha z}] \tag{44}$$

which is assumed to hold in the whole region $-\infty < z < \infty$. When b and α are both real, it is seen that

$$k(z) = k_0 \quad \text{for} \quad z = -\infty$$

and

$$k(z) = 0 \quad \text{when} \quad z = \frac{1}{\alpha} \log(1/b) \tag{45}$$

For large positive values of z, $k^2(z)$ becomes an infinitely large *negative* real number.

The electric field (with only a y component) is taken to be parallel to the stratification. For large negative values of z, the field must be of the form

$$E_y = e^{-ikSx} e^{\pm ikCz}$$

where $C = \cos \theta$ and θ is again the angle of incidence. Now, in general, for any value of z

$$\frac{d^2 E_y}{dz^2} + k_0^2 [C^2 - b \, e^{\alpha z}] E_y = 0 \tag{46}$$

This equation is again reducible to Bessel's equation. Letting

$$v = (2ik_0 b^{1/2}/\alpha) \, e^{\alpha z/2} \tag{47}$$

and

$$\nu = 2ik_0 C/\alpha \tag{48}$$

the general solutions can again be made up in linear combinations of the functions $J_\nu(v)$ and $J_{-\nu}(v)$. The particular combination of these suitable for the present problem is the Hankel function of the first kind defined by

$$H_\nu^{(1)}(v) \cong \frac{J_{-\nu}(v) - e^{-i\pi\nu} J_\nu(v)}{i \sin \nu\pi} \tag{49}$$

This function, and only this function, has a non-infinite behavior as v tends to $+\infty$.

The admittance at any point in the medium, looking in the positive z direction, is then found to be

$$-\frac{H_x}{E_y} = \frac{\alpha v H_\nu^{(1)\prime}(v)}{2i \mu \omega H_\nu^{(1)}(v)} \tag{50}$$

For small values of v, corresponding to large negative values of z, it may be noted that

$$H_\nu^{(1)}(v) \sim \frac{1}{i \sin \nu\pi} \left[\frac{(v/2)^{-\nu}}{\Gamma(1 - \nu)} - \frac{e^{-i\pi\nu}(v/2)^\nu}{\Gamma(1 + \nu)} \right] \tag{51}$$

where each term in the square bracket is the leading term of the power series expansion of $J_{-\nu}(v)$ and $J_\nu(v)$, respectively. On replacing $(\alpha v/2)$ by $ik_0 C$ and changing the variable back to z, it then follows that (for $z \to -\infty$)

$$E_y \cong \text{const} \times [e^{-ik_0 Cz} + \hat{R} \, e^{ik_0 Cz}] e^{-ik_0 Sx} \tag{52}$$

where

$$\hat{R} = -\left(\frac{k_0}{\alpha}\right)^{2\nu} b^\nu \frac{\Gamma(1 - \nu)}{\Gamma(1 + \nu)} \tag{53}$$

It is clear that $\hat{R}$ has the nature of a reflection coefficient referred to in the plane $z = 0$. However, it is not at all the same as the reflection coefficient at an interface $z = 0$, which is free space for $z < 0$. In the latter case, the reflection coefficient would be given

$$R = \frac{C - Y\eta_0}{C + Y\eta_0} \tag{54}$$

where

$$Y = -\frac{H_x}{E_y}\bigg]_{z=0} = \frac{\alpha v_0 H_\nu^{(1)'}(v_0)}{2i\mu\omega H_\nu^{(1)}(v_0)} \quad (55)$$

where

$$v_0 = (2ik_0 b^{1/2}/\alpha) \quad (56)$$

In the case of an ionized gas $b = (1 - iZ)^{-1}$ where Z is a real quantity which is equal to the ratio of the collisional frequency ν to the angular frequency ω. We thus find Elias' [1930] and Budden's [1961] result that

$$|\hat{R}| = \exp\left[\frac{2k_0 C}{-\alpha} \tan^{-1} Z\right] \quad (57)$$

and

$$\arg \hat{R} = \pi + \frac{4k_0 C}{\alpha} \log\left(\frac{k_0}{\alpha}\right) - \frac{k_0 C}{\alpha} \log(1 + Z^2) + \\ + 2 \arg[\Gamma(1 - 2ik_0 C/\alpha)] \quad (58)$$

Following Budden [1961], it is very interesting to note that the phase integral method leads to an expression which is an excellent approximation for $\hat{R}$. The phase integral formula should read

$$\hat{R} \cong \left(\lim_{z_1 \to -\infty}\right) i \exp(i2k_0 C z_1) \exp\left[-i2 \int_{+z_1}^{z_0} [k^2(z) - k_0^2 S^2]^{1/2} dz\right] \quad (59)$$

where z_0 is defined by $k^2(z_0) = k_0^2 S^2$. It is noted that, for $z > z_0$, the integrand would be purely imaginary when $k(z)$ is real. The quantity z_1 can be conveniently imagined as a negative value of z sufficiently large that $k^2(z)$ can be replaced by k_0^2. Strictly in the case of an exponential profile defined by Eq. (44), the z_1 must be taken as $-\infty$. The appearance of the exponential factor $\exp(i2k_0 C z_1)$ on the right-hand side of the above equation means that $\hat{R}$ is referred to the plane $z = 0$. Then, on using

$$k^2(z) = k_0^2\left(1 - \frac{e^{\alpha z}}{1 - iZ}\right) \quad (60)$$

the integration in Eq. (75) is carried out to give

$$|\hat{R}| = \exp[-(2k_0 C/\alpha) \tan^{-1} Z] \quad (61)$$

and

$$\arg \hat{R} = \frac{\pi}{2} - \frac{4k_0 C}{\alpha}(\log 2C - 1) - \frac{k_0 C}{\alpha} \log(1 + Z^2) \quad (62)$$

As Budden comments, it is really quite remarkable that the phase integral formula gives the correct (i.e. exact) value of $|\hat{R}|$. However, the phase predicted by the phase integral method is only approximate. This is demonstrated by noting that $\arg \hat{R}$, given by Eq. (62), may be recovered from Eq. (58) since

$$2 \arg[\Gamma(1 - 2ik_0 C/\alpha)] \cong -\frac{\pi}{2} - \frac{4k_0 C}{\alpha}[\log(2k_0 C/\alpha) - 1] \quad (63)$$

when $2k_0 C/\alpha$ is somewhat greater than unity. The latter approximate relation is simply a consequence of using the first term in the Stirling formula for the gamma function.

Reflection from Inhomogeneous Media

LINEAR PROFILE

Another profile which leads to a convenient solution for horizontal polarization is the linear variation. For example, we let

$$k^2(z) = k_0^2[1 - \alpha z] \quad \text{for} \quad z > 0 \tag{64}$$
$$= k_0^2 \quad \text{for} \quad z < 0$$

and, again, we assume that the permeability is everywhere equal to μ_0. The equation for the electric field is thus

$$\frac{d^2 E_y}{dz^2} + k_0^2(C^2 - \alpha z)E_y = 0 \quad \text{for} \quad z > 0 \tag{65}$$

$$\frac{d^2 E_y}{dz^2} + k_0^2 C^2 E_y = 0 \quad \text{for} \quad z < 0 \tag{66}$$

As before, the electric vector is taken to have only a y component.

By making the substitution

$$t = (k_0^2 \alpha)^{1/3}(z - C^2/\alpha) \tag{67}$$

it is seen that

$$\frac{d^2 E_y}{dt^2} - t E_y = 0 \quad \text{for} \quad z > 0 \tag{68}$$

The parameter t should not be confused with time. Equation (68) is known as Stokes' differential equation and independent solutions are the 2 Airy functions $u(t)$ and $v(t)$. They may be defined in terms of definite integrals as follows [Miller, 1946].

$$u(t) = \frac{1}{\sqrt{\pi}} \int_0^\infty \{\exp(-\tfrac{1}{3}x^3 + tx) + \sin(\tfrac{1}{3}x^3 + tx)\}\, dx \tag{69}$$

$$v(t) = \frac{1}{\sqrt{\pi}} \int_0^\infty \cos(\tfrac{1}{3}x^3 + tx)\, dx \tag{70}$$

To choose the appropriate solution requires a knowledge of the asymptotic behavior for large positive or negative values of the argument. Following Miller [1946], we find that for $|t| \to \infty$, $|\arg t| < \pi/3$

$$u(t) \cong \frac{1}{t^{1/4}} \exp(\tfrac{2}{3} t^{3/2}) \quad \text{and} \tag{71}$$

$$v(t) \cong \frac{1}{2 t^{1/4}} \exp(-\tfrac{2}{3} t^{3/2}) \tag{72}$$

while for

$$|t| \to \infty, \quad |\arg(-t)| < 2\pi/3$$

$$u(t) \sim \frac{1}{(-t)^{1/4}} \cos\left[\tfrac{2}{3}(-t)^{3/2} + \frac{\pi}{4}\right] \tag{73}$$

and

$$v(t) \sim \frac{1}{(-t)^{1/4}} \sin\left[\tfrac{2}{3}(-t)^{3/2} + \frac{\pi}{4}\right] \tag{74}$$

Actually, these are the first terms of asymptotic expansions for large values of the arguments. It is also worth noting that Airy Functions are closely related to Bessel functions. This is indicated by the following identities

$$u(t) = \left(\frac{\pi t}{3}\right)^{1/2} [I_{-1/3}(\tfrac{2}{3}t^{3/2}) + I_{1/3}(\tfrac{2}{3}t^{3/2})] \tag{75}$$

$$u(-t) = \left(\frac{\pi t}{3}\right)^{1/2} [J_{-1/3}(\tfrac{2}{3}t^{3/2}) - J_{1/3}(\tfrac{2}{3}t^{3/2})] \tag{76}$$

$$v(t) = \frac{(\pi t)^{1/2}}{3} [I_{-1/3}(\tfrac{2}{3}t^{3/2}) - I_{1/3}(\tfrac{2}{3}t^{3/2})] \tag{77}$$

$$v(-t) = \frac{(\pi t)^{1/2}}{3} [J_{-1/3}(\tfrac{2}{3}t^{3/2}) + J_{1/3}(\tfrac{2}{3}t^{3/2})] \tag{78}$$

Further properties and relationships among Airy functions can be obtained from well-known relations in the theory of Bessel functions [Watson, 1944].

Now, if α is a real number greater than zero it is evident that, in view of Eq. (72), the desired solution is $v(t)$. In fact, this is still true if α is complex, provided $|\arg \alpha| < \pi/3$. Thus

$$E_y = E_0 v(t) \tag{79}$$

where E_0 is a constant. Furthermore,

$$i\mu\omega H_x = \frac{\partial E_y}{\partial z} = (k_0^2 \alpha)^{1/3} v'(t) \tag{80}$$

where the prime denotes a derivative with respect to t. The admittance of the wave looking in the positive direction is thus

$$-\frac{H_x}{E_y} = i\frac{k_0}{\mu_0 \omega}\left(\frac{\alpha}{k_0}\right)^{1/3}\frac{v'(t)}{v(t)} \tag{81}$$

for $z > 0$. The reflection coefficient at $z = 0$ is thus given by

$$R = \frac{C - \eta_0 Y}{C + \eta_0 Y} \tag{82}$$

where

$$Y = \frac{i}{\eta_0}\left(\frac{\alpha}{k_0}\right)^{1/3}\frac{v'(t_0)}{v(t_0)} \quad \text{or} \quad \frac{\eta_0 Y}{C} \cong i\frac{1}{(-t_0)^{1/2}}\frac{v'(t_0)}{v(t_0)} \tag{83}$$

where

$$\eta_0 = \mu_0 \omega / k_0 \quad \text{and} \quad t_0 = -(k_0/\alpha)^{2/3} C^2.$$

Thus

$$R = \frac{(-t_0)^{1/2} v(t_0) - iv'(t_0)}{(-t_0)^{1/2} v(t_0) + iv'(t_0)} \tag{84}$$

It is important to note that if α is real, t_0 is a negative real number which, in turn, indicates that the admittance Y is imaginary (i.e. it is a pure susceptance). Thus, in this case, $|R| = 1$ and the reflection is complete. Of course, there is a phase shift on reflection which is determined by the argument of Y. It may be noted that if $(-t_0)$ is greater than about 2, the asymptotic approximation for $v(t)$ given by Eq. (74) may be used. Thus

$$\frac{\eta_0 Y}{C} \simeq -i \cot\left[\tfrac{2}{3}(-t_0)^{3/2} + \frac{\pi}{4}\right] \tag{85}$$

and to this approximation the reflection coefficient is given by

$$R \cong i \exp[-i\tfrac{2}{3}(-t_0)^{3/2}]$$
$$\cong i \exp[-i\tfrac{2}{3}(k_0/\alpha)C^3] \tag{86}$$

It is of interest to compare this result with the "phase-integral method" of Eckersley [1932]. This would predict that the reflection coefficient should be given by

$$R = i \exp\left[-2ik_0 \int_0^{z_0} [N^2(z) - S^2]^{1/2}\, dz\right] \tag{87}$$

where $N(z) = k(z)/k_0$ is the refractive index which varies from 1 at $z = 0$ to a smaller value at z_0 such that $N^2(z_0) - S^2 = 0$. In this particular problem

$$N^2(z) - S^2 = C^2 - z\alpha \tag{88}$$

and $z_0 = C^2/\alpha$. Therefore, it readily follows that Eq. (86) is regained. It might be of interest to note that pure geometrical optics would not give the multiplier i. This can be associated with a $\pi/2$ phase advance at the caustic of the ray system. The phase integral approach is discussed in the following chapter.

EXTENSION TO VERTICAL POLARIZATION

In previous sections attention has been confined to "horizontal polarization" or parallel incidence. The other case of equal importance occurs when the magnetic vector is perpendicular to the plane of incidence. Then, since the electric vector is contained in the plane of incidence, this is aptly called "parallel polarization." Sometimes this case is described as vertical polarization since for near grazing incidence the electric vector is almost vertical relative to the (horizontal) stratification.

Again, unless otherwise stated, the region $z < 0$ is taken to be free space with electrical constants ε_0 and μ_0. For the region $z > 0$, the conductivity and dielectric constant are denoted $\sigma(z)$ and $\varepsilon(z)$, respectively, and, as before, the permeability for $z > 0$ is also taken to be μ_0.

The magnetic field, which is taken to have only a y component, for the space $z < 0$, is given by

$$H_y = H_0(e^{-u_0 z} + R_\parallel(\lambda) e^{u_0 z}) e^{-i\lambda x} \tag{89}$$

where $\lambda = k_0 \sin\theta$ and $u_0 = (\lambda^2 - k_0^2)^{1/2} = ik_0 \sin\theta$ as θ is the usual angle of incidence. As shown before, for the discretely stratified medium, the reflection coefficient is given by

$$R_\parallel(\lambda) = \frac{K_0 - Z_1}{K_0 + Z_1} \tag{90}$$

where

$$K_0 = \frac{u_0}{i\varepsilon_0 \omega} = \eta_0 \cos\theta, \tag{91}$$

and Z_1 is the surface admittance at $z = 0$ where

$$Z_1 = \frac{E_{1x}}{H_{1y}}\bigg]_{z=0} \tag{92}$$

Again, since we are now dealing with continuously stratified media, the subscript 1 may be dropped in what follows. Thus, in general,

$$R_{||}(\lambda) = \frac{K_o - Z}{K_o + Z} \quad (93)$$

where

$$Z = \frac{E_x}{H_y}\bigg]_{z=0}, \quad (94)$$

or

$$R_{||}(\lambda) = \frac{\eta_0 C - Z}{\eta_0 C + Z} \quad (95)$$

where $C = \cos\theta$.

Noting that the fields must vary as $\exp(-ik_0 Sx)$, Maxwell's equations, for the space $z > 0$, are given by

$$-i\mu_0\omega H_y = \frac{\partial E_x}{\partial z} + ik_0 S E_z \quad (96)$$

$$E_x = -\frac{1}{\sigma(z) + i\varepsilon(z)\omega} \frac{\partial H_y}{\partial z} \quad (97)$$

$$E_z = -\frac{ik_0 S}{\sigma(z) + i\varepsilon(z)\omega} H_y \quad (98)$$

From these we readily obtain

$$\frac{\partial^2 H_y}{\partial z^2} + [k^2(z) - k_0^2 S^2]H_y - \frac{1}{k^2(z)} \frac{\partial H_y}{\partial z} \frac{\partial k^2(z)}{\partial z} = 0 \quad (99)$$

where

$$k^2(z) = -i\mu_0\omega[\sigma(z) + i\varepsilon(z)\omega].$$

It can be seen that except for the latter term on the left-hand side, the equation for H_y is the same as Eq. (11) for E_y. The presence of this term for "vertical polarization" is a complicating feature. Often this term may be neglected when the medium is slowly varying such that derivatives with respect to z are small. In certain cases, however, such as where $k^2(z) - k_0^2 S^2$ is small, the latter term is very important. This means the physical processes are quite different for horizontal and vertical polarization when the medium is not slowly varying.

Exponential Profile With Vertical Polarization

For vertical polarization most of the special profiles mentioned above do not reduce simply to a standard differential equation. In general, numerical or iterative procedures are required. However, one case does reduce again to a form of Bessel's equation. Apparently, this was first pointed out by Galejs [1961], in a related problem. The special profile is

$$k^2(z) = -ik_0^2 B e^{\beta z} \quad (100)$$

for $\beta > 0$ and $z > 0$. When this is applied to a highly conducting medium where displacement currents are negligible, B is real.

For this exponential profile, the equation for H_y becomes

$$\frac{d^2 H_y}{dz^2} - k_0^2(S^2 + iB e^{\beta z})H_y - \beta\frac{dH_y}{dz} = 0 \quad (101)$$

Letting $u = -iB e^{\beta z}$, it readily follows that

$$\left[\frac{d^2}{du^2} + \left(\frac{k_0^2}{\beta^2 u} - \frac{k_0^2 S^2}{\beta^2 u^2}\right)\right]H_y = 0 \quad (102)$$

Reflection from Inhomogeneous Media

This equation has a solution of the form

$$H_y = H_0 v K_v(v) e^{-ikSx} \tag{103}$$

where H_0 is a constant, if

$$v = i(2k_0/\beta)u^{1/2} = (2k_0/\beta) e^{i\pi/4} B^{1/2} e^{\beta z/2} \tag{104}$$

and

$$v = [1 + (2k_0 S/\beta)^2]^{1/2} \tag{105}$$

It is noted that H_y is non-infinite as $z \to \infty$.

From Maxwell's equations

$$E_x = -\frac{1}{i\varepsilon_0 \omega u} \frac{\partial H_y}{\partial v} \frac{\partial v}{\partial z} \tag{106}$$

This leads quickly to

$$\frac{E_x}{H_y} = -\eta_0 \frac{e^{i\pi/4}}{B^{1/2} e^{\beta z/2}} \frac{\frac{d}{dv} v K_v(v)}{v K_v(v)} \tag{107}$$

It is interesting to compare this with the analogous formula for the admittance in the case of horizontal polarization. This is given by Eq. (55) where it can be seen that the argument v of the modified Bessel function is the same in the 2 cases. However, the argument v is defined differently.

Using the recurrence relation

$$K'_v = -K_{v-1} - \frac{v}{v} K_v \tag{108}$$

it now follows that

$$Z = \frac{E_x}{H_y}\bigg|_{z=0} = \eta_0 \left[\frac{1}{u_0^{1/2}} \frac{K_{v-1}(v_0)}{K_v(v_0)} - \frac{\beta}{2ik_0} \frac{1-v}{u_0} \right] \tag{109}$$

where

$$u_0^{1/2} = k(0)/k_0 \cong e^{-i\pi/4} (\sigma(0)/\varepsilon_0 \omega)^{1/2}$$

and where

$$v_0 = (2k_0/\beta) e^{i\pi/4} B^{1/2} \tag{110}$$

$$\cong 2[i\sigma(0)\mu_0\omega]^{1/2}/\beta$$

Here $\sigma(0)$ is the conductivity at $z = 0$. Also, it should be remembered that

$$v = [1 + (2k_0 S/\beta)^2]^{1/2} \tag{111}$$

Inserting the above expression into Eq. (95), leads to an exact expression for the reflection coefficient $R_\parallel$ for a plane wave incident from the free space region (i.e. $z < 0$). At normal incidence (i.e. $S = 0$), the impedance formula reduces to

$$Z = \frac{\mu_0 \omega}{k(0)} \frac{K_0(v)}{K_1(v_0)} \tag{112}$$

This is in agreement with Eq. (43) which is a check since the reflection coefficient at normal incidence is the same, regardless of the choice of coordinate axes.

It is important to note that Z given above is a good approximation for oblique incidence at vertical polarization provided $(2k_0 S/\beta)^2 \ll 1$.

As a further check on the general formula given above for Z, we allow β to tend to zero corresponding to a homogeneous conducting half-space. In this limiting situation both v and v_0 tend to infinity. Thus

$$\frac{\dfrac{d}{dv_0} v_0 K_\nu(v_0)}{v_0 K_\nu(v_0)}\bigg]_{\beta\to 0} \cong -\sqrt{\left(1+\frac{v^2}{v_0^2}\right)}\bigg]_{\beta\to 0} \cong -\sqrt{\left(1+\frac{S^2}{iB}\right)} \qquad (113)$$

Therefore

$$Z = \frac{\eta_0\, e^{i\pi/4}}{B^{1/2}} \sqrt{\left(1+\frac{S^2}{iB}\right)} \qquad (114)$$

This can be written

$$Z = \frac{\eta_0}{N^2} \sqrt{(N^2 - S^2)} \qquad (115)$$

where $N^2 = -iB$ is the square of the refractive index of the homogeneous half-space occupying $z > 0$.

POWER LAW PROFILE FOR NORMAL INCIDENCE

A profile of considerable generality is given as follows

$$k^2(z) = k_1^2 \left(1 + \frac{z}{a}\right)^{-\beta} \qquad (116)$$

where k_1, a and β are constants. Unfortunately, it does not seem possible to obtain closed-form solutions from Maxwell's equations in the general case of arbitrary values of a and β and oblique incidence. However, at normal incidence the fields in such a medium can again be expressed in terms of Bessel functions as first shown by Lahiri and Price [1939]. The analysis for this case is given in the following.

The electric vector is taken to have only a y component. Consequently, the magnetic vector has only an x component. One easily finds that

$$\left[\frac{d^2}{dz^2} + k_1^2 \left(1+\frac{z}{a}\right)^{-\beta}\right] E_y = 0 \qquad (117)$$

and

$$i\mu\omega H_x = \partial E_y/\partial z$$

As usual, μ is taken as a constant. Defining a new function by

$$E_y = \left(1+\frac{z}{a}\right)^{1/2} \psi \qquad (118)$$

it is seen that

$$\frac{\partial^2 \psi}{\partial z^2} + \frac{1}{a}\left(1+\frac{z}{a}\right)^{-1}\frac{\partial \psi}{\partial z} - \frac{1}{4a^2}\left(1+\frac{z}{a}\right)^{-2}\psi + k_1^2\left(1+\frac{z}{a}\right)^{-\beta}\psi = 0 \qquad (119)$$

Letting $\psi = Z_\nu(u)$, it is found that this is a Bessel function of order ν and argument u, if

$$\nu^2 = \left(\frac{1}{2-\beta}\right)^2 \qquad (120)$$

and

$$u = 2\nu(k_1 a)\left(1+\frac{z}{a}\right)^{(2-\beta)/2} \qquad (121)$$

The appropriate Bessel function to employ is dictated by the required behavior as z tends to $+\infty$. It is assumed that $\operatorname{Im} k_1 < 0$. For $\beta < 2$, it is seen that

Reflection from Inhomogeneous Media

$$E_y = E_0\left(1 + \frac{z}{a}\right)^{1/2} H^{(2)}_{1/(2-\beta)}\left[\frac{2}{2-\beta}(k_1 a)\left(1 + \frac{z}{a}\right)^{(2-\beta)/2}\right] \quad (122)$$

where E_0 is a constant and $H_1^{(2)}$ is the Hankel function of the second kind. On the other hand, for $\beta > 2$, it is found that

$$E_y = E_0\left(1 + \frac{z}{a}\right)^{1/2} J_{1/(\beta-2)}\left[\frac{2}{\beta - 2}(k_1 a)\left(1 + \frac{z}{a}\right)^{(2-\beta)/2}\right] \quad (123)$$

where J is the Bessel function of the first kind.

The special case $\beta = 2$ can be found from either of the above as limiting cases. This process requires the use of the appropriate asymptotic expansions for the Hankel and Bessel functions when the order and the argument are both indefinitely large. More simply, one considers this case separately. Thus, for $\beta = 2$,

$$\left[\frac{d^2}{dz^2} + k_1^2\left(1 + \frac{z}{a}\right)^{-2}\right] E_y = 0 \quad (124)$$

Letting

$$E_y = E_0\left(1 + \frac{z}{a}\right)^m \quad (125)$$

it follows that

$$\frac{d^2 E_y}{dz^2} = \frac{m(m-1)}{a^2}\left(1 + \frac{z}{a}\right)^{m-2} \quad (126)$$

Therefore

$$m(m-1) = -k_1^2 a^2 \quad (127)$$

or

$$2m = 1 \pm (1 - 4k_1^2 a^2)^{1/2}$$

If E_y is to be finite as z tends to infinity, it is clear that the $+$ sign must be rejected in the above radical.

Without further difficulty it is readily found that

$-E_y/H_x$

$$= -\frac{i\mu\omega}{k_1}\left(1 + \frac{z}{a}\right)^{\beta/2} \frac{H^{(2)}_{1/(2-\beta)}\left[\frac{2}{2-\beta}(k_1 a)\left(1 + \frac{z}{a}\right)^{(2-\beta)/2}\right]}{H^{(2)}_{1/(2-\beta)-1}\left[\frac{2}{2-\beta}(k_1 a)\left(1 + \frac{z}{a}\right)^{(2-\beta)/2}\right]}, \quad (\beta < 2) \quad (128)$$

$$= -\frac{i\mu\omega}{2k_1^2 a}[1 + (1 - 4k_1^2 a^2)^{1/2}]\left(1 + \frac{z}{a}\right), \quad (\beta = 2) \quad (129)$$

$$= -\frac{i\mu\omega}{k_1}\left(1 + \frac{z}{a}\right)^{\beta/2} \frac{J_{1/(\beta-2)}\left[\frac{2}{\beta - 2}(k_1 a)\left(1 + \frac{z}{a}\right)^{(2-\beta)/2}\right]}{J_{1/(\beta-2)+1}\left[\frac{2}{\beta - 2}(k_1 a)\left(1 + \frac{z}{a}\right)^{(2-\beta)/2}\right]}, \quad (\beta > 2) \quad (130)$$

If the region $z < 0$ is free space, then the reflection coefficient at $z = 0$, for a plane wave incident from $z = -\infty$, is

$$R = \frac{\eta_0 - Z}{\eta_0 + Z} \qquad (131)$$

where $Z = -E_y/H_x]_{z=0}$

$$= -\frac{i\mu\omega}{k_1} \frac{H^{(2)}_{1/(2-\beta)}\left(\frac{2k_1 a}{2-\beta}\right)}{H^{(2)}_{1/(2-\beta)-1}\left(\frac{2k_1 a}{2-\beta}\right)}, \qquad (\beta < 2) \qquad (132)$$

$$= -\frac{i\mu\omega}{2k_1^2 a}[1 + (1 - 4k_1^2 a^2)^{1/2}], \qquad (\beta = 2) \qquad (133)$$

$$= -\frac{i\mu\omega}{k_1} \frac{J_{1/(\beta-2)}\left[\frac{2k_1 a}{\beta-2}\right]}{J_{1/(\beta-2)+1}\left[\frac{2k_1 a}{\beta-2}\right]}, \qquad (\beta > 2) \qquad (134)$$

As a partial check on these results, it can be seen that all 3 of the forms for Z approach $\mu\omega/k_1$ as a tends to infinity. This would correspond to a homogeneous half-space with wave number k_1. Another check is to note that for $\beta = 0$,

$$Z = -\frac{i\mu\omega}{k_1} \frac{H^{(2)}_{1/2}\left[\frac{2k_1 a}{2-\beta}\right]}{H^{(2)}_{-1/2}\left[\frac{2k_1 a}{2-\beta}\right]} = \frac{\mu\omega}{k_1} \qquad (135)$$

which also corresponds to a homogeneous medium.

Exercise: Consider the inverse square profile defined by

$$\frac{k^2(z)}{k_0^2} = a^2 - \frac{b^2}{(z+z_0)^2}$$

Equation (11) now can be written in the form

$$\frac{d^2 E}{dz^2} + k_0^2\left[a_1^2 - \frac{b^2}{(z+z_0)^2}\right] E = 0$$

where $k_0 a_1^2 = k_0 a^2 - \lambda^2$ and where the subscript on the E has been dropped. Show that solutions of this equation are of the form

$$E = \text{const} \times \hat{z}^{1/2} V_\nu(\beta\hat{z}) \text{ where } \hat{z} = z + z_0$$

where

$$\beta^2 \hat{z}^2 V''_\nu(\beta\hat{z}) + \beta\hat{z} V'_\nu(\beta\hat{z}) + [k_0^2 a_1^2 \hat{z}^2 - (k_0^2 b^2 + \tfrac{1}{4})] V_\nu(\beta\hat{z}) = 0$$

where the primes indicate differentiations with respect to $\beta\hat{z}$. Show that V_ν is a Bessel function if

$$\beta = k_0 a_1 = k_0\sqrt{(a^2 - \sin^2\theta)}$$

and

$$\nu = (k_0^2 b^2 + \tfrac{1}{4})^{1/2}$$

A solution equivalent to this was developed by Rytov and Yudkevich [1946].

Assuming a slight loss in the medium, show that
$$Y = 1/Z = \left[\frac{H_\nu^{(2)\prime}(\beta z_0)}{H_\nu^{(2)}(\beta z_0)} + \frac{1}{2\beta z_0}\right]\frac{i\beta}{\mu_0\omega}$$
where
$$\beta = k_0\sqrt{(a^2 - \sin^2\theta)}$$
and
$$\nu = (k_0^2 b^2 + \tfrac{1}{4})^{1/2}$$

As a partial check on this result we can set $b = 0$, in which case
$$k(z) = k_0 a = \text{const.}$$

The Hankel functions are now of order $1/2$. These are known to be expressible in simple form. In fact
$$H_{1/2}^{(2)}(\beta z_0) = i\left(\frac{2}{\pi\beta z_0}\right)^{1/2} e^{-i\beta z_0}$$

On using this result in the above equation for Y, show that
$$Y = 1/Z = \frac{\beta}{\mu_0\omega} = \frac{1}{\eta_0}\sqrt{(a^2 - \sin^2\theta)}$$
which is the expected result.

REFERENCES

BATEMAN, H., (1953), *Higher Transcendental Functions*, Vol. II (edited by A. Erdelyi, et al.), McGraw-Hill, New York.

BREMMER, H., (1949) *Terrestrial Radio Waves*, Elsevier, New York and Amsterdam.

BUDDEN, K.G., (1961) *Radiowaves in the Ionosphere*, Cambridge University Press.

ECKERSLEY, T.L., (1932), Radio transmission problems treated by phase integral methods, *Proc. Roy. Soc.*, A, 136, 499.

ELIAS, G.J., (1930), Reflection of electromagnetic waves at ionized media with variable conductivity and dielectric constant, *Proc. I.R.E.*, 19, 891-907.

GALEJS, J., (1961), e.l.f. waves in the presence of exponential ionospheric conductivity profiles, *Trans. I.R.E.*, AP-9, 554-562.

LAHIRI, B.N., and PRICE, A.T., (1939), Electromagnetic induction in non-uniform conductors, and the determination of the conductivity of the earth from terrestrial magnetic variations, *Phil. Trans. Roy. Soc.*, London, Ser. A., 237, 507-540.

MILLER, J.C.P., (1946), *The Airy Integral, Giving Tables of Solutions of the Differential Equation $y'' = xy$*, Cambridge University Press.

RYTOV, S.M., and YUDEVICH, F.S., (1946), Electromagnetic wave reflection from a layer with a negative dielectric constant, *J. Expl. Theor. Phys.*, **10**, 285. (In Russian).

WAIT, J.R., (1952), Reflection of electromagnetic waves obliquely from an inhomogeneous medium, *J. Appl. Phys.*, **23**, 1403-1404.

WAIT, J.R., (1958), Transmission and reflection of electromagnetic waves in presence of stratified media, *J. Res. Nat. Bur. Stand.*, **61**, No. 3, 205-232.

WAIT, J.R., (1960), Terrestrial propagation of v.l.f. radio waves - a theoretical investigation, *J. Res. Nat. Bur. Stand.*, **64D**, 153-203.

WATSON, G.N., (1944), *Theory of Bessel Functions*, Cambridge University Press, 2nd Ed.

OTHER REFERENCES

BOSSY, L., and DEVUYST, A., (1959), Relations entre les champs électriques et magnétique d'une de peroid trés longue induits dans un milieu de conductivité variable, *Geofis. Pur. Appl., Milano*, **44**, Pt. III, 119-134.

BREKHOVSKIKH, L.M., (1960), *Waves in Layered Media*, Academic Press, New York.

BURMAN, R., (1964), Some approximate formulas concerning the reflection of electromagnetic waves from a stratified semi-infinite medium, *Radio Science*, **68D**, 1215-1218.

BURMAN, R., and GOULD, R.N., (1965), The reflection of waves in a generalized Epstein profile, *Can. J. Phys.*, **43**, 921-934.

FÖRSTERLING, K., (1931), Über die Ausbreitung des Lichtes in inhomogenen Medien, *Ann. Phys.*, (Series 5), **11**, 1-39.

GALEJS, J., (1972), *Terrestrial Propagation of Long Electromagnetic Waves*, Chap. 6, Pergamon Press, Oxford.

GRAY, MARION, (1934), Mutual impedance of grounded wires lying on the surface of the earth when the conductivity varies exponentially with depth, *Physics*, 5, 76-80.

JOHLER, J.R., and HARPER, J.D., Jr., (1962), Reflection and transmission of radio waves at a continuously stratified plasma with arbitrary magnetic induction, *J. Res. Nat. Bur. Stand.*, 66D, (Radio Prop) No. 1, 81-100.

PEKERIS, C.L., (1946), Theory of propagation of sound in a half-space of variable sound velocity, *J. Acous. Soc. Amer.*, 18, 295-315.

SHMOYS., J., (1956), Long range propagation of low frequency radio waves between the earth and the ionosphere, *Proc. I.R.E.*, 44, 163-170.

STANLEY, J.P., (1950), The absorption of long and very long waves in the ionosphere, *J. Atmos. Terr. Phys.*, 1, 65.

TAYLOR, L.S., (1961), Electromagnetic propagation in an exponential ionization density, *Trans. I.R.E.*, AP-9, 483-487.

WAIT, J.R., (1962), On the propagation of v.l.f. and e.l.f. radio waves when the ionosphere is not sharply bounded, *J. Res. Nat. Bur. Stand.*, 66D, (Radio Prop.) No. 1, 53-62.

WESTCOTT, B.S., (1969), Exact solutions for vertically polarized electromagnetic waves, *Proc. Camb. Phil. Soc.*, 66, 675-684.

WESTCOTT, B.S., (1970), Soluble profiles for inhomogeneous gyrational media, *Quart. Journ. Mech. and Appl. Math.*, 23(Pt. 3), 431-440.

WESTCOTT, B.S., (1970), Oblique refexion of electromagnetic waves from isotropic media with limiting Epstein-type stratifications, *Proc. Camb. Phil. Soc.*, 68 (Pt. 3), 765-771.

Addition to Note No. 15

EXERCISE: Consider the square or parabolic profile given by

$$k^2(z) = k_o^2[1 - (z/z_1)^2] \quad \text{for} \quad z > 0$$
$$\qquad\qquad - k_o^2 \quad \text{for} \quad z < 0$$

Then (11), for $z > 0$, can be written

$$\frac{d^2E}{dz^2} + \left[k_o^2 - \left(\frac{k}{z_1}\right)^2 z^2\right]E = 0$$

in the case of normal incidence (i.e., $S = 1$). By changing variable to $\zeta = (2k_o/z_1)^{1/2} z$ show that the governing equation is

$$\frac{d^2E}{d\zeta^2} + \left(n + \frac{1}{2} - \frac{1}{4}\zeta^2\right)E = 0$$

where $n = (k_o z_1 - 1)/2$. Solutions that vanish as ζ tends to infinity are the parabolic cylinder functions denoted by $D_n(\zeta)$ in the notation of Whittaker and Watson (1950). Show that

$$\frac{dE/dz}{k_o E} = \frac{\eta_o Y}{i} = \frac{1}{\left(n + \frac{1}{2}\right)^{1/2}} \frac{D'_n(0)}{D_n(0)}$$

where

$$D_n(0) = \Gamma\left(\frac{1}{2}\right) 2^{\frac{n}{2}} \Big/ \Gamma\left(\frac{1}{2} - \frac{n}{2}\right)$$

and

$$D'_n(0) = \Gamma\left(-\frac{1}{2}\right) 2^{\frac{n}{2} - \frac{1}{2}} \Big/ \Gamma\left(-\frac{n}{2}\right)$$

or

$$D'_n(0)/D_n(0) = -2^{1/2} \Gamma\left(\frac{1}{2} - \frac{n}{2}\right) \Big/ \Gamma\left(-\frac{n}{2}\right)$$

Further show that

$$\eta_o Y = \frac{2i}{(k_o z_1)^{1/2}} \frac{\Gamma\left(\frac{1}{4}(k_o z_1 + 3)\right)}{\Gamma\left(\frac{1}{4}(k_o z_1 + 1)\right)} \tan\left[\frac{\pi}{4}(k_o z_1 - 1)\right]$$

Then show that, in the phase integral approximation,

$$\eta_o Y \simeq i \tan \phi/2$$

where
$$\phi \simeq (\pi/2)kz_1 - (\pi/2).$$

Compare this with the asymptotic limit of the exact solution when $|kz_1| \to \infty$. An equivalent problem was considered originally by Hartree (1931). Closely related work was done by another pioneer, T.L. Eckersley (1931), who pointed out the close analogy with the Schrodinger equation used in the quantum theory of the harmonic oscillator.

REFERENCES

ECKERSLEY, T.L., (1931), On the connection between ray theory of electric waves and dynamics, *Proc. Roy. Soc.*, A **132**, 38-98, London.

HARTREE, D.R., (1931), Optical and equivalent paths in a stratified medium, *Proc. Roy. Soc.*, A **131**, 428-450, London.

WHITTAKER, E.T., and WATSON, G.N., (1950), *A Course of Modern Analysis*, 4th Ed., Cambridge at the University Press, Sec. 16.5.

16
Approximate Methods for Inhomogeneous Media

Introduction and the Conventional WKB Method

In the previous chapter, the propagation in an inhomogeneous medium was treated by adopting special profiles for a continuously stratified medium. Here, approximate methods, which are not restricted to the form of the profile, are investigated.

To illustrate the central idea, the one dimensional wave equation is considered

$$\frac{d^2\psi}{dz^2} + k^2(z)\psi = 0 \tag{1}$$

where $\psi(z)$ and $k(z)$ are scalar quantities. Such an equation is appropriate for plane waves propagating normal to the stratification. To seek an approximate solution of this equation, we set

$$\psi(z) = \Lambda(z)\, e^{-i\phi(z)} \tag{2}$$

where Λ and ϕ can be regarded as the amplitude and phase, respectively.

Substituting this into the wave equation, it follows that

$$\Lambda'' - 2i\phi'\Lambda' - i\phi''\Lambda + [k^2(z) - (\phi')^2]\Lambda = 0 \tag{3}$$

where the prime indicates a derivative with respect to z. An approximate solution of this latter equation is now obtained by using the following physical arguments.

The dimension l is taken to be the distance over which quantities Λ and ϕ vary by a significant amount. Thus, a slowly varying medium can be defined as one in which $kl \gg 1$.

Order of magnitude estimates can now be made for the various terms in Eq. (3). Thus

$$\Lambda'' \sim \Lambda/l^2$$

is of second order of smallness and for a first crude approximation it may be replaced by zero. Then $\Lambda \exp(-i\phi)$ is an approximate solution of Eq. (3) if

$$\phi' = \pm k(z) \quad \text{or} \quad \phi = \pm \int^z k(z)\, dz \tag{4}$$

and

$$\frac{\Lambda'}{\Lambda} = -\frac{k'}{2k} \quad \text{or} \quad \Lambda = \frac{\text{const}}{\sqrt{[k(z)]}} \tag{5}$$

Then, the general solution is

$$\psi(z) \cong [1/k(z)]^{1/2}\left\{a \exp\left[-i\int_{z_0}^z k(z)\, dz\right] + b \exp\left[+i\int_{z_0}^z k(z)\, dz\right]\right\} \tag{6}$$

where a, b, and z_0 are arbitrary constants. This result is usually known as the WKB solution where the letters stand for Wentzel, Kramers and Brillouin,

(Reprinted in part from J.R. Wait, Electromagnetic Waves in Stratified Media, Pergamon Press, 2nd Ed., 1970)

who used the method extensively in quantum mechanics in the mid-1920's. Sometimes it is called the WKBJ method where the last letter is for Sir Harold Jeffreys [1923], who described the method in a definitive paper. In most applications the lower limit of integration, z_0, is set equal to zero since, in effect, this changes only the arbitrary constants a and b.

The physical meaning of Eq. (6) is fairly clear if $k(z)$ is real. The exponent $\int_{z_0}^{z} k(z)\,dz$ is the change of phase of a wave which has travelled from the arbitrary point z_0 to the point z. Obviously, the first exponential term in Eq. (6), which is preceded by an $-i$, corresponds to a wave propagating in the positive z direction. Conversely, the second exponential term, which is preceded by an $+i$, corresponds to a wave travelling in the negative direction.

The presence of the factor $[1/k(z)]^{1/2}$, in Eq. (6), can be reconciled as the basis of energy flow. For example, if ψ is the electric field component E_y, then the magnetic field component, H_x, must be proportional to $\partial\psi/\partial z$ or $k(z)\psi$. Then if the time average of the Poynting vector, in the z direction, is $(1/2)\,\mathrm{Re}\,E_y H_x^*$, which is constant for both upgoing and downcoming waves.

Actually, Eq. (6) is still valid for complex $k(z)$, corresponding to a medium with losses, but the physical interpretation is not as simple.

WKB METHOD FOR OBLIQUE INCIDENCE

The above derivation can be readily extended to oblique incidence. For example, if the electric vector has only a y component, E_y, it is necessary to start with the equation

$$\left[\frac{d^2}{dz^2} + q^2(z)\right] E_y = 0 \tag{7}$$

where $q^2(z) = k^2(z) - \lambda^2$. Here, λ, which is a constant, describes the x variation of the fields (i.e. $\partial/\partial x = -i\lambda$). For example, if a plane wave was incident from $z = -\infty$ and if $k(z) \to k_0$ as $z \to -\infty$, then $\lambda = k_0 S = k_0 \sin\theta_0$ where θ_0 is the conventional angle of incidence. Consequently, all field quantities must vary as $\exp(-ik_0 S x)$ or $\exp(-i\lambda x)$. It thus follows that the WKB solution of Eq. (7) is given by

$$E_y = [1/q(z)]^{1/2}\left\{A_0 \exp\left[-i\int_0^z q(z)\,dz\right] + B_0 \exp\left[i\int_0^z q(z)\,dz\right]\right\} e^{-ik_0 S x} \tag{8}$$

when A_0 and B_0 are also arbitrary constants.

The WKB approximation for the magnetic field components are now found from Maxwell's curl equation

$$-i\mu_0 \omega H = \mathrm{curl}\, E \tag{9}$$

where μ_0, the magnetic permeability, is assumed constant everywhere. Thus

$$i\mu_0 \omega H_x = \frac{\partial E_y}{\partial z} \tag{10}$$

or

$$\mu_0 \omega H_x = \left\{-A_0 q^{1/2} \exp\left[-i\int_0^z q\,dz\right] + B_0 q^{1/2} \exp\left[+i\int_0^z q\,dz\right]\right.$$
$$\left. + \frac{i}{2}\frac{1}{q^{3/2}}\frac{dq}{dz}\left[A_0 \exp\left[-i\int_0^z q\,dz\right] + B_0 \exp\left[+i\int_0^z q\,dz\right]\right]\right\} e^{-ik_0 S x} \tag{11}$$

where the z dependence on q is understood. Since it has already been assumed that the medium is slowly varying, the term containing dq/dz in the above equation can be neglected. The remaining field component is then obtained from

$$-i\mu_0\omega H_z = \frac{\partial E_y}{\partial x} \tag{12}$$

or simply

$$\mu_0\omega H_z = k_0 S E_y \tag{13}$$

GENERALIZATION OF WKB METHOD

To investigate the true significance of the WKB method and to extend its usefulness, one should obtain correction terms. Only in this way, can one place a precise meaning on the term "slowly varying." Furthermore, following the ideas of Bremmer [1949], it is possible to show that the WKB approximation is indeed the first term of a rigorous series solution. The treatment used here is more general than Bremmer's. The form of the WKB approximate solution for E_y, given above, suggests that we write

$$E_y = \left\{ \frac{A(z)}{[q(z)]^{1/2}} \exp\left[-i\int_0^z q(z)\,dz\right] + \frac{B(z)}{[q(z)]^{1/2}} \exp\left[i\int_0^z q(z)\,dz\right] \right\} e^{-ik_0 S x} \tag{14}$$

and

$$\mu_0\omega H_x = \left\{ -A(z)[q(z)]^{1/2} \exp\left[-i\int_0^z q(z)\,dz\right] + B(z)[q(z)]^{1/2} \exp\left[+i\int_0^z q(z)\,dz\right] \right\} e^{-ik_0 S x} \tag{15}$$

and

$$\mu_0\omega H_z = k_0 S E_y$$

The factors A and B are now allowed to be functions of z of a form yet to be determined. It is evident that the conventional WKB approximation is obtained when A and B are replaced by the constants A_0 and B_0.

Now, the fields must satisfy the Maxwellian equations

$$-q^2 E_y = i\mu_0\omega\, \partial H_x/\partial z \tag{16}$$

$$\partial E_y/\partial z = i\mu_0\omega H_x \tag{17}$$

On substituting E_y and H_x given by Eqs. (14) and (15) into the latter pair of equations, one easily finds that A and B must satisfy

$$\frac{dA}{dz} = \frac{B}{2q}\frac{dq}{dz} \exp\left[+2i\int_0^z q\,dz\right] \tag{18}$$

and

$$\frac{dB}{dz} = \frac{A}{2q}\frac{dq}{dz} \exp\left[-2i\int_0^z q\,dz\right] \tag{19}$$

where the z dependence of A, B, and q is understood.

For convenience of physical discussion, the latter 2 equations can be written in the form

$$\frac{dA(z)}{dz} = \varepsilon P(z) B(z) \tag{20}$$

$$\frac{dB(z)}{dz} = \varepsilon Q(z) A(z) \tag{21}$$

where ε is a small (constant) parameter. It is now assumed that a solution can be written as power series of the form

$$A = A_0 + \varepsilon A_1 + \varepsilon^2 A_2 + \cdots \tag{22}$$

and

$$B = B_0 + \varepsilon B_1 + \varepsilon^2 B_2 + \cdots \tag{23}$$

where A_0 and B_0 are constants. It is evident that if these are substituted into Eqs. (20) and (21), the successive approximations are related by

$$\frac{d}{dz} A_{m+1}(z) = P(z) B_m(z) \tag{24}$$

$$\frac{d}{dz} B_{m+1}(z) = Q(z) A_m(z) \tag{25}$$

It is now clear that the first terms of the series correspond to the conventional WKB solution, whereas the succeeding correction terms arise from internal reflection within the medium. For example, the εB_1 term corresponds to the first-order reflected wave propagating in the negative z direction resulting from a wave, A_0, propagating in the positive z direction. Thus

$$B_1(z) = \int_\infty^z Q(z) A_0 \, dz \tag{26}$$

where the lower limit is chosen so that εB_1 vanishes as z tends to $+\infty$. In general,

$$B_{m+1}(z) = \int_\infty^z Q(z) A_m(z) \, dz \tag{27}$$

On the basis of identical reasoning, εA_1 is the first-order reflected wave, propagating in the positive z direction, which arises from an incident wave, B_0, propagating in the negative direction. Thus

$$A_1(z) = \int_{-\infty}^z P(z) B_0 \, dz \tag{28}$$

or, in general,

$$A_{m+1}(z) = \int_{-\infty}^z P(z) B_m(z) \, dz \tag{29}$$

As an explicit example, the second-order reflected waves for the incidence wave A_0 are written as a double integral of the form

$$A_2(z) = \int_{-\infty}^z P(z') \int_\infty^{z'} Q(z'') A_0 \, dz'' \, dz' \tag{30}$$

where the primes have been used to distinguish between the 2 variables of integration.

Generalized WKB Method for Vertical Polarization

The formalism developed above can be readily extended to vertical polarization where the magnetic vector has only a component H_y (perpendicular to the plane of incidence). It is instructive to outline the derivation for this case also.

The field components are written in the following form

$$H_y = \left\{ \frac{N(z)A^*(z)}{[q(z)]^{1/2}} \exp\left[-i\int_0^z q(z)\,dz\right] + \frac{N(z)B^*(z)}{[q(z)]^{1/2}} \exp\left[+i\int_0^z q(z)\,dz\right] \right\}$$
$$\times \exp(-ik_0 Sx) \quad (31)$$

$$\varepsilon_0 \omega E_x = \left\{ \frac{A^*(z)[q(z)]^{1/2}}{N(z)} \exp\left[-i\int_0^z q(z)\,dz\right] \right. \quad (32)$$
$$\left. - \frac{B^*(z)[q(z)]^{1/2}}{N(z)} \exp\left[+i\int_0^z q(z)\,dz\right] \right\} \exp(-ik_0 Sx)$$

and

$$\varepsilon_0 \omega E_z = -\frac{k_0 S}{N^2(z)} H_y \quad (33)$$

where $q(z) = k_0[N^2(z) - S^2]^{1/2}$, as before. It is clear that if the new functions, A^* and B^* are replaced by constants, A_0^* and B_0^*, the solutions correspond to the conventional WKB form. In this case, upgoing and downcoming waves are uncoupled and the normal wave impedances are

$$\pm \eta_0 [N^2(z) - S^2]^{1/2}/N^2(z)$$

in accordance with expected behavior.

To find the corrected form of the WKB solution, Eqs. (31) and (32) are substituted into the following Maxwellian equations

$$-\frac{q^2(z)}{N^2(z)} H_y = -i\varepsilon_0 \omega \frac{\partial E_x}{\partial z} \quad (34)$$

$$\frac{1}{N^2(z)} \frac{\partial H_y}{\partial z} = -i\varepsilon_0 \omega E_x \quad (35)$$

This gives rise to the following coupled differential equations

$$\frac{dA^*}{dz} = \left[\frac{1}{2q^*} \frac{dq^*}{dz} \exp\left[+2i\int_0^z q\,dz\right]\right] B^* \quad (36)$$

and

$$\frac{dB^*}{dz} = \left[\frac{1}{2q^*} \frac{dq^*}{dz} \exp\left[-2i\int_0^z q\,dz\right]\right] A^* \quad (37)$$

where $q^* = q/N^2$. The z dependence of the quantities is understood. These equations have the same form as the pair given by Eqs. (18) and (19) and, therefore, the method of successive approximations is also valid here.

Relation to Geometrical Optics

We are now in a position to discuss the validity of geometrical optics which corresponds to the retention of only the first term of the series expansions for A and B. Except for special circumstances, it can be expected that the neglect of partial or internal reflections is justified if εB_1 is small compared with A_0. It is not difficult to see that, for horizontal polarization

$$\frac{\varepsilon B_1}{A_0} = \frac{1}{2} \int_\infty^z \left[\frac{1}{q}\frac{dq}{dz}\right] \exp\left[-2i\int_0^z q\,dz\right] dz \quad (38)$$

On the other hand, for vertical polarization, the first square bracket term in the above is replaced by $(1/q^*)(dq^*/dz)$.

Now, it is noted that the integrand in the integral is a rapidly oscillating function with a period approximately equal to π/q. Consequently, an order of magnitude estimate of the integral is simply

Approximate Methods for Inhomogeneous Media

$$\left|\frac{1}{q}\frac{dq}{dz} \times \frac{1}{q}\right|$$

Thus, for horizontal polarization, the condition for the validity of geometrical optics is

$$\left|\frac{d}{dz}\frac{1}{q}\right| \ll 1$$

In terms of the refractive index $N(z)$ and the angle $\theta(z)$ of the ray†, this condition is

$$\frac{d}{dz}\frac{1}{N(z)\cos\theta(z)} \ll \lambda_0^{-1} \tag{39}$$

In the case of vertical polarization, the order of magnitude estimate for $\varepsilon B_1/A_0$ is

$$\left|\frac{1}{q^*}\frac{dq^*}{dz} \times \frac{1}{q}\right|$$

Thus, the condition for the validity of geometrical optics is

$$\left|\frac{1}{qq^*}\frac{dq^*}{dz}\right| \ll 1 \tag{40}$$

In terms of the angle $\theta(z)$ and the refractive index $N(z)$, the restriction is

$$\left|\frac{1}{N^2(z)}\frac{d}{dz}\left[\frac{N(z)}{\cos\theta(z)}\right]\right| \ll \lambda_0^{-1} \tag{41}$$

which is equivalent to

$$\left|\frac{1}{\cos^2\theta(z)}\frac{d}{dz}\left[\frac{\cos\theta(z)}{N(z)}\right]\right| \ll \lambda_0^{-1} \tag{42}$$

The latter form was quoted by Brekhovskikh [1960] who, following Bremmer [1949], treated only the case for vertical polarization.

It is quite apparent that internal reflections are particularly important at near grazing angles where $\cos\theta(z)$ is small and is, itself, changing. Indeed, this is true for both polarizations.

APPLICATION TO TROPOSPHERIC PROPAGATION

An interesting application of the preceding development is to tropospheric propagation of radio waves. It is now known that large horizontal layers may exist in the troposphere. Order-of-magnitude estimates of the parameters of these layers have been given by du Castel et al. [1960], as follows

horizontal extent—kilometers
vertical extent—tens of meters
height (above ground)—hundreds of meters

The maximum change of the refractive index is of the order of 10^{-5}.

The reflection coefficient for a plane wave incident on a troposphere layer extending from z_1 to z_2 can be well approximated by

$$r \cong \frac{\varepsilon B_1}{A_0} \cong \frac{1}{2}\int_{z_2}^{z_1}\left[\frac{1}{q}\frac{dq}{dz}\right]\exp\left[-2i\int_0^z q\,dz\right]dz \tag{43}$$

† (Note that $\theta(z)$ may be defined by $N(z)\sin\theta(z) = S$ where λ_0 is the free space wavelength).

which is valid for horizontal polarization. Now, in the troposphere, the refractive index can be written as

$$N(z) = 1 + \delta N(z) \tag{44}$$

when $|\delta N| \ll 1$. Then, since

$$q = k_0(N^2 - S^2)^{1/2} \cong k_0(C^2 + 2\delta N)^{1/2} \tag{45}$$

one finds that

$$q \cong k_0 C$$

and

$$\frac{dq}{dz} \cong \frac{k_0}{C}\frac{d}{dz}(\delta N) = \frac{k_0}{C}\frac{dN}{dz}$$

Here, C is the usual cosine of the angle of incidence as the latter approximations require that $C^2 \gg \delta N$. Thus, the reflection coefficient is given by

$$r \cong \frac{1}{2C^2}\int_{z_2}^{z_1}\left[\frac{d}{dz}\delta N\right]e^{-2ik_0 Cz}\,dz \tag{46}$$

This result has been obtained and used by many previous authors [Schelkunoff, 1948; Friis et al. 1957; du Castel et al. 1960]. They have obtained it by starting with a discontinuous model. The idea of their derivation is given here, briefly.

The Fresnel reflection coefficient at horizontal polarization, for a homogeneous half-space of constant refraction index, $1 + \delta N$, is given by

$$r = \frac{C - [2\delta N + (\delta N)^2 + C^2]^{1/2}}{C + [2\delta N + (\delta N)^2 + C^2]^{1/2}} \tag{47}$$

Now, if $|\delta N| \ll 1$ and $C^2 \gg \delta N$, this can be well approximated by

$$r \cong -\frac{\delta N}{2C^2} \tag{48}$$

Then, using somewhat heuristic reasoning, the reflection coefficient, dr, resulting from a change, $d(\delta N)/dz$, within a layer is given by

$$dr = -\frac{1}{2C^2}\frac{d(\delta N)}{dz}e^{-i2k_0 Cz}\,dz \tag{49}$$

where the phase $2k_0 Cz$ accounts for the two-way path traversed by the ray. Then it is supposed that the total reflection coefficient is obtained by integrating over the vertical extent of the layer. Thus

$$r = \frac{1}{2C^2}\int_{z_2}^{z_1}\left[\frac{d(\delta N)}{dz}\right]e^{-i2k_0 Cz}\,dz \tag{50}$$

where $z_2 > z_1$. This result is identical to Eq. (46) derived from the WKB series solution. It should be emphasized that this simple formula neglects all internal reflections. It is not just the total change δN in a layer, but the derivative $d(\delta N)/dz$ which must be small to justify the use of such a simple formula.

It is not difficult to show that the reflection coefficient for a tropospheric layer at vertical polarization is essentially the same as for horizontal polarization. This follows from the approximations

Approximate Methods for Inhomogeneous Media

$$q^* = q^2/N^2(z) \cong k_0 C$$

and

$$dq^*/dz \cong (k_0/C)\, d(\delta N)/dz$$

when

$$|\delta N| \ll 1 \quad \text{and} \quad C^2 \gg |\delta N|$$

The integral in Eq. (46) can be readily evaluated in closed form when certain analytical forms for the derivative of the refractive index are assumed, such as discussed by duCastel, Misme and Voge (1960).

Exercise: Derive the following expressions for the reflection coefficient r for the profiles indicated (assume $dN(z)/dz = 0$ outside the layer):

(1) Linear layer,

$$dN(z)/dz = g \quad \text{for } -\tfrac{h}{2} < z < \tfrac{h}{2}$$

$$r = \frac{gh}{2C^2} \frac{\sin \chi}{\chi} \tag{51}$$

where $\chi = k_0 Ch$;

(2) Parabolic layer,

$$dN(z)/dz = g\left(1 - \frac{|z|}{h}\right) \quad \text{for } -h < z < h$$

$$r = \frac{gh}{2C^2}\left(\frac{\sin \chi}{\chi}\right)^2 \tag{52}$$

(3) Sinusoidal layer,

$$dN(z)/dz = \frac{\pi g}{4}\cos\left(\frac{\pi z}{2h}\right) \quad \text{for } -h < z < h$$

$$r = \frac{gh}{2C^2}\frac{\cos 2\chi}{1 - \left(\frac{4\chi}{\pi}\right)^2} \tag{53}$$

(4) Hyperbolic layer,

$$dN(z)/dz = \frac{g}{2}\cosh^{-2}\left(\frac{z}{h}\right) \quad \text{for } -\infty < z < \infty$$

$$r = \frac{gh}{2C^2}\frac{\pi\chi}{\sinh(\pi\chi)} \tag{54}$$

It is seen that for each of these cases the total change of the refractive index across the layer is gh. In each case, as χ approaches zero, $|r|$ tends to $gh/2C^2$ which is the value appropriate to a sharp discontinuity. It is apparent that when χ is large the reflection coefficient is greatly reduced in magnitude.

The Phase Integral Approach

When the medium is slowly varying, the WKB method and its extended form are very convenient. However, when the quantity $(1/q)(dq/dz)$ is not small, the method fails. In particular, if q becomes zero at some point in the medium some further modification is required as pointed out by Langer [1937] and Budden [1961]. The approach used here is adapted from their work.

To illustrate the method, $N^2(z)$ is allowed to be a real and a smooth monotonically decreasing function of z. For $z \to -\infty$, $N(z)$ approaches unity and therefore $q(z)$ approaches $k_0 C$. At $z = z_0$, $q(z_0) = 0$ or $N(z_0) = S$.

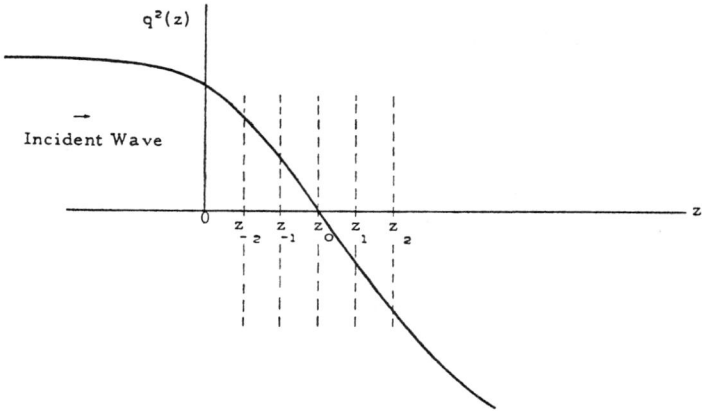

FIG. 1. Sketch of $q^2(z) = k_0^2[N^2(z) - S^2]$

If attention is restricted to horizontal polarization, the electric field E_y is a solution of

$$\frac{d^2 E_y}{dz^2} + q^2(z) E_y = 0 \tag{55}$$

The medium *outside* the interval $z_{-1} \leq z \leq z_1$ is assumed to be sufficiently slowly varying that the usual WKB solutions are valid. Now, in the region $z < z_{-1}$, the appropriate form of the WKB solution is

$$E_y = \frac{P}{q^{1/2}} \left[\exp\left(-i \int_0^z q \, dz\right) + R \exp\left(i \int_0^z q \, dz\right) \right] e^{-ik_0 Sx} \tag{56}$$

where P and R are arbitrary constants. It is noted that if $z \to -\infty$, this could be written

$$E_y = \frac{P}{(k_0 C)^{1/2}} [e^{-ik_0 Cz} + R \, e^{ik_0 Cz}] e^{-ik_0 Sx} \tag{57}$$

which is the superposition of 2 plane waves. The constant R can thus be regarded as a reflection coefficient which has yet to be determined.

In the region $z > z_1$, the WKB solution has the form

$$E_y = \frac{T}{(-q^2)^{1/4}} \exp\left[-\int_{z_0}^z (-q^2)^{1/2} \, dz\right] \tag{58}$$

where now

$$iq, \quad \text{or} \quad (-q^2)^{1/2}$$

is essentially a real quantity. T is another unknown constant which is also to be determined.

For the whole region extending from z_{-2} to z_2, which includes the portion z_{-1} to z_1, it is assumed that the profile can be approximated by

$$q^2(z) \cong -k_0^2 \alpha(z - z_0) \tag{59}$$

where α is a constant. Thus E_y, in this region, satisfies

$$\frac{d^2 E_y}{dz^2} - k_0^2 \alpha(z - z_0) E_y = 0 \tag{60}$$

Introducing the new variable

$$t = (k_0^2 \alpha)^{1/3}(z - z_0) \tag{61}$$

it readily follows that

$$\frac{d^2 E_y}{dt^2} - t E_y = 0 \tag{62}$$

As pointed out in the previous chapter, solutions of this equation are the Airy functions $u(t)$ and $v(t)$. On inspection of the asymptotic forms it is found that only $v(t)$ gives rise to an exponentially damped wave for large positive values of z. Specifically, the asymptotic forms for $v(t)$ may be written

$$v(t) \cong \frac{1}{(-t)^{1/4}} \frac{\exp\left[\frac{2i}{3}(-t)^{3/2}\right] \exp\left(\frac{i\pi}{4}\right) - \exp\left[-\frac{2i}{3}(-t)^{3/2}\right] \exp\left(-\frac{i\pi}{4}\right)}{2i} \tag{63}$$

when t is a large negative number, while

$$v(t) \cong \frac{1}{2t^{1/4}} \exp(-\tfrac{2}{3} t^{3/2}) \tag{64}$$

when t is a large positive number. These can be rewritten in terms of q and z. Thus

$$v(t) \cong \frac{(k_0^2 \alpha)^{1/6}}{2iq^{1/2}} \left[\exp\left(-i \int_{z_*}^{z} q \, dz\right) \exp\left(\frac{i\pi}{4}\right) - \exp\left(i \int_{z_0}^{z} q \, dz\right) \exp\left(-\frac{i\pi}{4}\right)\right] \tag{65}$$

when $(z_0 - z) \gg (k_0^2 \alpha)^{-1/3}$, and

$$v(t) \cong \frac{(k_0^2 \alpha)^{1/6}}{2(-q^2)^{1/4}} \exp\left[-\int_{z_0}^{z}(-q^2)^{1/2} \, dz\right] \tag{66}$$

when

$$(z - z_0) \gg (k_0^2 \alpha)^{-1/3}$$

It is immediately apparent that these are the WKB forms. It is assumed that they are valid in the respective intervals, z_{-2} to z_{-1} and z_1 to z_2. Thus Eqs. (65) and (66) should be proportional to Eqs. (56) and (58), respectively, in the intervals of common validity. Therefore, it is easily found that

$$R = i \exp\left[-2i \int_{0}^{z_0} q(z) \, dz\right] \tag{67}$$

and

$$T = e^{i\pi/4} \exp\left[-i \int_{0}^{z_0} q \, dz\right] \tag{68}$$

which completes the solution.

A Generalization of the Phase Integral Method

It is evident that the successive steps in the previous derivation are equivalent, more or less, to using the following substitutions

$$t = \left[i\frac{3}{2}\int_{z_0}^{z} q(z)\, dz\right]^{2/3} \tag{69}$$

when $z < z_0$, and

$$t = \left[\frac{3}{2}\int_{z_0}^{z} [-q^2(z)]^{1/2}\, dz\right]^{2/3} \tag{70}$$

when $z > z_0$. Here, z_0 is defined by the condition $q(z_0) = 0$. Thus, the electric field is given by

$$E \cong E_0 v(t) t^{1/4}/q^{1/2} \tag{71}$$

throughout the whole range of z. This can be verified by observing that the WKB solutions given by Eqs. (56) and (58) are regained when $-t$ and t are respectively large compared with unity. However strictly this is only true when $q^2(z)$ is a linear function of z in the transition region. Also, it is important that no additional zeroes or other singularities of $q^2(z)$ are near the principal zero z_0. Implicit in this latter statement is the requirement that the curve $q^2(z)$ vs. z has a small curvature at $z = z_0$.

The substitutions (69) and (70), used in conjunction with (71), are sometimes called the "extended WKB approximation." It has been used by Langer [1937], Pekeris [1946], Bremmer [1960], and possibly others.

In the preceding discussion of the phase integral method and its relation to the extended WKB approximation, the parameter $q^2(z)$ was regarded as a positive (or negative) real quantity. The root z_0 was then always also a real quantity and corresponded to the "level of reflection." If the square of the refractive index $N^2(z)$ is complex, it is evident that $q^2(z)$ is also complex and, consequently, the root z_0 is complex. Provided $q^2(z)$ is still approximated by a linear function of $(z - z_0)$ in the vicinity of z_0, the previous arguments still hold. Furthermore, the reflection coefficient still has the form

$$R = i \exp\left[-2i \int_0^{z_0} q(z)\, dz\right] \tag{72}$$

Here, the contour of integration extending from the origin to z_0 may be chosen in any convenient way. It must be remembered, of course, that account must be taken of any singularities that are crossed in the formation of the contour. An interesting discussion of the physical significance of the contour used in the phase integral formula has been given by Budden[1961].

Phase Integral for Vertical Polarization

The phase integral method or "extended" WKB approximation has been discussed only in relation to horizontal polarization. In considering vertical polarization, at oblique incidence, where the magnetic field has only a y component, H_y, the differential equation is

$$\frac{d^2 H_y}{dz^2} + q^2(z) H_y - \frac{1}{N^2(z)} \frac{dN^2(z)}{dz} \frac{dH_y}{dz} = 0 \tag{73}$$

where

$$q^2(z) = k_0^2[N^2(z) - S^2]$$

This is identical to Eq. (115) discussed in the previous chapter. By introducing a new function $\Phi(z)$, defined by

Approximate Methods for Inhomogeneous Media

$$\Phi(z) = H_y/N(z) \tag{74}$$

the differential equation may be written in the form

$$\left[\frac{d^2}{dz^2} + Q^2(z)\right]\Phi(z) = 0 \tag{75}$$

where

$$Q^2 = q^2 + \frac{1}{2N^2}\frac{d^2N^2}{dz^2} - \frac{3}{4N^4}\left(\frac{dN^2}{dz}\right)^2 \tag{76}$$

The previous arguments for horizontal polarization may now be carried over without modification *provided* $Q^2(z)$ is an approximate linear function in the vicinity of its principal zero $\hat{z}_0$ defined by $Q^2(\hat{z}_0) = 0$. In this case, the reflection coefficient $R_\parallel$ defined in the usual way for vertical polarization, is given by the phase integral

$$R \cong i \exp\left[-2i \int_0^{\hat{z}_0} Q(z)\,dz\right] \tag{77}$$

It is of interest to examine briefly the significance of the phase integral formula for vertical polarization when $q^2(z)$, rather than $Q^2(z)$, is a linear function of z in the vicinity of the reflection level. For example, if

$$N^2(z) = 1 - \frac{z-a}{b} \tag{78}$$

where a and b are constants, it is seen that

$$q^2(z) = k_0^2\left(C^2 - \frac{z-a}{b}\right) \tag{79}$$

and

$$Q^2(z) = q^2(z) - \frac{3}{4(z-a-b)^2} \tag{80}$$

Now, since $q^2(z) = 0$ when $z = z_0 = a + bC^2$, it is desirable to express $Q^2(z)$ as a Taylor expansion about z_0. Thus

$$Q^2(z) = -\left\{\frac{3}{4b^2S^2} + (z-z_0)\left(\frac{k_0^2}{b} + \frac{3}{2b^3S^6}\right) + (z-z_0)^2\left(\frac{9}{4b^4S^6}\right) + \ldots\right\} \tag{81}$$

It is readily apparent that $Q^2(z)$ has a zero near $z = z_0$ if $k_0 b \gg 1$ and S is not small compared with unity. Therefore, at sufficiently high frequencies and oblique incidence, $Q(z)$ may simply be replaced by $q(z)$ in the phase integral formula given above. When these conditions are not met, the phase integral method for vertical polarization fails.

For additional remarks concerning the validity of the phase integral method at vertical polarization, the reader is referred to Budden's [1961] treatise.

RAPIDLY VARYING TRANSITION REGION

Introduction

When the properties of the medium change rapidly the WKB and the phase integral methods are unsuitable. For example, the theoretical treatment of v.l.f. radio waves* from the ionosphere requires a different approach. Here, although precise information is not yet available, the refractive index changes significantly in a distance small compared with a wavelength [Wait, 1960, 1962]. A first, and often a very good approximation, is to represent the lower edge by a sharply bounded homogeneous medium of constant refractive index N_1. Then, for a vertically polarized plane wave incident from

---
* v.l.f. = very low frequency (the range 3–30 Hz).

below the reflection coefficient, R, is given by the well-known Fresnel coefficient

$$R = \frac{N_1^2 C_0 - (N_1^2 - S_0^2)^{1/2}}{N_1^2 C_0 + (N_1^2 - S_0^2)^{1/2}} \qquad (82)$$

where $C_0 = (1 - S_0^2)^{1/2} = \cos\theta_0$ in terms of the angle of incidence θ_0. (To simplify the discussion, the earth's magnetic field has been neglected.) We are now interested in obtaining a correction which will account for the finite thickness of a transition region between the free space below and the isotropic ionosphere above.

To start with, we shall assume an inhomogeneously stratified medium. With respect to the usual rectangular coordinate system (x, y, z), the refractive index is assumed to be only a function of z. As z tends to positive and negative infinity the refractive index is assumed to approach constant values. That is

$$N(z)]_{z \to +\infty} = N_1$$

$$N(z)]_{z \to -\infty} = 1$$

At $z = -\infty$, a plane wave is incident at an angle θ_0 with respect to the positive z direction.

Differential Equation for the Reflection Coefficient

Without any loss of generality, the H vector is taken to have only a y component. Thus, for a time factor, $e^{i\omega t}$, Maxwell's equations are given by

$$\frac{\partial E_x}{\partial z} - \frac{\partial E_z}{\partial x} = -i\mu\omega H_y$$

$$i\varepsilon\omega E_x = -\frac{\partial H_y}{\partial z}, \quad \text{and} \quad i\varepsilon\omega E_z = \frac{\partial H_y}{\partial x} \qquad (83)$$

where μ and ε are the magnetic permeability and permittivity of the medium. For sake of generality, both μ and ε can be regarded as functions of z. Of course, in applications to the ionosphere μ can be replaced by its free space value μ_0. The retention of a variable μ in the theory permits one to readily adapt the results to arbitrary polarization. Furthermore, the analogy in acoustics is readily brought out.

Now the sum of the incident and reflected waves is defined in the following manner. The sole magnetic field component is written

$$H_y = [A(z) + B(z)]\exp[-i(\varepsilon\mu)^{1/2}\omega Sx] \qquad (84)$$

while the 2 electric field components are written

$$E_x = [A(z) - B(z)]C(\mu/\varepsilon)^{1/2} \exp[-i(\varepsilon\mu)^{1/2}\omega Sx] \qquad (85)$$

and

$$E_z = -[A(z) + B(z)]S(\mu/\varepsilon)^{1/2} \exp[-i(\varepsilon\mu)^{1/2}\omega Sx] \qquad (86)$$

where C and S are also functions of z.

In this analysis, $k = (\varepsilon_0\mu_0)^{1/2}\omega$ where ε_0 and μ_0 are the constants of free space, while $A(z)$ and $B(z)$ are not yet defined. Since these field components are to satisfy Maxwell's equations, it is required that

$$(\varepsilon\mu)^{1/2}S = \text{constant}$$

and since $S = \sin\theta_0$, $\varepsilon = \varepsilon_0$, $\mu = \mu_0$, at $z = -\infty$, it follows that

$$(\varepsilon\mu)^{1/2}S = (\varepsilon_0\mu_0)^{1/2} \sin\theta_0 \qquad (87)$$

which is just a statement of Snell's law. Furthermore,

$$C^2 + S^2 = 1$$

Again, as a consequence of Maxwell's equations, $A(z)$ and $B(z)$ must satisfy

$$\frac{dA}{dz} + i\delta A + \Gamma(A - B) = 0 \tag{88}$$

$$\frac{dB}{dz} - i\delta B + \Gamma(B - A) = 0 \tag{89}$$

where

$$\delta = (\varepsilon\mu)^{1/2}\omega C$$

and

$$\Gamma = \frac{\varepsilon}{2\delta}\frac{d}{dz}\left(\frac{\delta}{\varepsilon}\right)$$

Equations (88) and (89) are easily combined into a single equation for the ratio $B/A = R(z)$. Thus

$$\frac{dR}{dz} = 2i\delta R + \Gamma(1 - R^2) \tag{90}$$

where $R(z)$ is, by definition, a reflection coefficient. Results, more or less equivalent to (90) have been given by Budden [1961] and Brekhovskikh [1960].

Iterative Solution

To obtain a solution, $R(z)$ is written in terms of a new function $v(z)$. Thus

$$R = \frac{g(z)v(z) - g_1}{g(z)v(z) + g_1} \tag{91}$$

where

$$g(z) = \frac{K}{N^2}(N^2 - S_0^2)^{1/2}, \quad K = \mu/\mu_0, \quad S_0 = \sin\theta_0$$

and

$$g_1 = \lim_{z \to \infty} g(z) = \frac{K_1}{N_1^2}(N_1^2 - S_0^2)^{1/2}, \quad K_1 = \mu_1/\mu_0$$

Now, since $\lim_{z \to \infty} R(z) = 0$, it follows that $\lim_{z \to \infty} v(z) = 1$. The differential equation for $v(z)$ is obtained by substituting (91) into (90). This can be written in the relatively simple form

$$-\frac{dv}{dz} = \frac{ikN^2 g_1}{K}\left(1 - \frac{g^2}{g_1^2}v^2\right) \tag{92}$$

where $k = (\varepsilon_0\mu_0)^{1/2}\omega$ and N, K and g are functions of z. Using a method of successive approximations, the solution can be expressed as an ascending series in powers of k. For example, the zero'th approximation is to replace the right-hand side of (92) by zero, thus v is a constant which must be unity to satisfy the limiting condition at $z \to \infty$. The first approximation is obtained by replacing v^2 on the right-hand side by unity. Thus

$$v = 1 + ikg_1 \int_z^\infty \frac{N^2}{K}\left(1 - \frac{g^2}{g_1^2}\right)dz \tag{93}$$

where the limits of the integration are chosen so that v satisfies the limiting condition at $z = \infty$. The second approximation is then obtained by substituting the latter result for v into the right-hand side of (92). In general, the

n'th approximation, v_n, can be found from the $(n - 1)$'th approximation, v_{n-1}, by using

$$v_n = 1 + ikg_1 \int_z^\infty \frac{N^2}{K}\left(1 - \frac{g^2}{g_1^2} v_{n-1}^2\right) dz \qquad (94)$$

Some Simple Extensions of the Solution

While these results have been developed with specific reference to an incident wave with the electric vector in the plane of incidence, the results are also applicable to the other polarization. If the magnetic vector of the incident wave is in the plane of incidence (i.e. horizontal polarization), the results are still valid if the following transformations are made,

$$H_y \to E_y, \quad E_x \to -H_x, \quad E_z \to -H_z, \quad \mu \to \varepsilon, \text{ and } \varepsilon \to \mu$$

Thus, the formula for the reflection coefficient R, given by (91), is still valid if K is replaced by $\varepsilon/\varepsilon_0$ and N is not changed.

There is also a well defined acoustic analogy to the problem being discussed. In this case $N(z) = c_0/c(z)$ where $c(z)$ is the velocity of sound and c_0 is the limiting value of c at $z = -\infty$. Thus the velocity is varying from a constant value c_0 to a differing constant value c_1 at $z = +\infty$. Also, $N^2/K = \rho(z)/\rho_0$ where $\rho(z)$ is the density and ρ_0 is its limiting value at $z = \infty$. The component H_y is then analogous to the acoustic pressure and E_x and E_z are analogous to the x and z components, respectively, of the particle velocity.

Discussion of the Form of the Solution

It is interesting to note that for the zero'th approximation (corresponding to $v = 1$), the reflection coefficient may be written

$$R = \frac{C_0 - (K_1/N_1^2)(N_1^2 - S_0^2)^{1/2}}{C_0 + (K_1/N_1^2)(N_1^2 - S_0^2)^{1/2}} \quad \text{where } C_0 = \cos\theta_0 \text{ and } S_0 = \sin\theta_0 \qquad (95)$$

This is the Fresnel reflection coefficient for the reflection of a plane wave at oblique incidence from a sharply bounded and homogeneous medium. Thus the higher terms in the ascending k series account for the "gradualness" of the boundary.

In the general case, the reflection coefficient may be written

$$R(z_0) = \frac{C_0 - (K_1/N_1^2)(N_1^2 - S_0^2)^{1/2}/v(z_0)}{C_0 + (K_1/N_1^2)(N_1^2 - S_0^2)^{1/2}/v(z_0)} \qquad (96)$$

where z_0 is some convenient level, below which $K(z)$ and $N(z)$ may be regarded as unity. Thus the total field in the region $z < z_0$ can be written

$$H_y = H_0 \, e^{-ikxS_0}[e^{-ik(z-z_0)C_0} + R(z_0) \, e^{+ik(z+z_0)C_0}] \qquad (97)$$

where H_0 is the value of the incident wave at the fictitious interface, $z = z_0$. In the first approximation, neglecting terms in k^3 and higher,

$$v(z_0) \cong 1 + ikg_1 \int_{z_0}^\infty \frac{N^2}{K}\left(1 - \frac{g^2}{g_1^2}\right) dz$$

$$+ 2k^2 \int_{z_0}^\infty \frac{N^2}{K} g^2 \left[\int_z^\infty \frac{N^2}{K}\left(1 - \frac{g^2}{g_1^2}\right) dz\right] dz \qquad (98)$$

Succeeding terms quickly become more complicated. It can be seen that the integrands contain the factor $1 - (g/g_1)^2$ in each of these terms. The presence of this factor permits one to replace the upper limit of each of these integrals by z_1, where z_1 is the level, above which $N(z)$ and $K(z)$ may be replaced by

N_1 and K_1. Thus the transition region may be defined as the interval $z_0 < z < z_1$. On this basis, it is apparent that the n'th term in the series for $v(z_0)$ is the order of $[k(z_1 - z_0)]^n$. Consequently, the series converges rapidly when the electrical thickness of the transition layer is small.

As an interesting check, the transition is replaced by a homogeneous slab. Thus,

$$N(z) = \hat{N} \quad \text{and} \quad K(z) = \hat{K} \quad \text{when} \quad z_0 < z < z_1$$

In this case, the reflection coefficient is given exactly by [Wait, 1958]

$$R = \frac{C_0 v(z_0) - (N_1^2 - S_0^2)^{1/2}(K_1/N_1^2)}{C_0 v(z_0) + (N_1^2 - S_0^2)^{1/2}(K_1/N_1^2)} \tag{99}$$

provided

$$v(z_0) = \frac{1 + \dfrac{K_1 \hat{N}^2 (N_1^2 - S_0^2)^{1/2}}{\hat{K} N_1^2 (\hat{N}^2 - S_0^2)^{1/2}} \tanh[ik(z_1 - z_0)\sqrt{(\hat{N}^2 - S_0^2)}]}{1 + \dfrac{\hat{K} N_1^2 (\hat{N}^2 - S_0^2)^{1/2}}{K_1 \hat{N}^2 (N_1^2 - S_0^2)^{1/2}} \tanh[ik(z_1 - z_0)\sqrt{(\hat{N}^2 - S_0^2)}]} \tag{100}$$

If $|k(z_1 - z_0)\sqrt{(\hat{N}^2 - S_0)}| \ll 1$, it is seen that to a first order,

$$v(z_0) \cong 1 + \frac{\hat{N}^2 (N_1^2 - S_0^2)^{1/2}}{N_1^2}\left[1 - \frac{\hat{K}^2 N_1^4}{K_1^2 \hat{N}^4} \frac{\hat{N}^2 - S_0^2}{N_1^2 - S_0^2}\right] ik(z_1 - z_0) \frac{K_1}{\hat{K}} \tag{101}$$

which is consistent with the first 2 terms of the series given by (93).

For some applications to v.l.f. propagation, it is desirable to express the reflection coefficient in the following form

$$R = \frac{\beta C_0 - 1}{\beta C_0 + 1} \tag{102a}$$

where

$$\beta = \frac{N_1^2 v(z_0)}{K_1 (N_1^2 - S_0^2)^{1/2}} \tag{102b}$$

Then, if βC_0 is regarded as a small parameter, the following expansion results

$$R = -e^{-2\beta C_0} + \tfrac{2}{3}(\beta C_0)^3 - \tfrac{4}{3}(\beta C_0)^4 + \cdots \tag{103a}$$

Thus if $|\beta C_0|^3 \ll 1$,

$$R \cong -e^{-2\beta C_0} \tag{103b}$$

which is a convenient form when the incidence is highly oblique and the frequency is not too low. On the other hand, if βC_0 is regarded as a large parameter, it is convenient to use the expansion

$$R = \exp\left(-\frac{2}{\beta C_0}\right) - \frac{2}{3}\frac{1}{(\beta C_0)^3} + \frac{4}{3}\frac{1}{(\beta C_0)^4} + \cdots \tag{104}$$

which can be approximated by the first term if $|\beta C_0|^3 \gg 1$.

REFERENCES

BREKHOVSKIKH, L., (1960), *Waves in Layered Media*, Academic Press, New York and London.

BREMMER, H., (1949), The propagation of electromagnetic waves through a stratified medium and its WKB approximation for oblique incidence, *Physica*, 15, 593-608.

BREMMER, H., (1960), On the theory of wave propagation through a concentrically stratified troposphere with a smooth profile, *J. Res. Nat. Bur. Stand.*, 64D, (Radio Prop.), No. 5, 467-482.

BUDDEN, K.G., (1961), *Radio Waves in the Ionosphere*, Cambridge University Press, Cambridge.

DU CASTEL, F., MISME, P., and VOGE, J., (1960), Sur le rôle des phénomènes de réflexion dans la propagation lointaine des ondes ultracourtes, in *Electromagnetic Wave Propagation*, 670-683, Academic Press, London and New York.

FRIIS, H.T., CRAWFORD, A.B., and HOGG, D.C., (1957), A reflection theory for propagation beyond the horizon, *Bell Syst. Tech. J.*, 36, 627-644.

JEFFREYS, H., (1923), On certain approximate solutions of linear differential equations of the second order, *Proc. Lond. Math. Soc.*, 23, 428.

LANGER, R.E., (1937), On the connection formulas and the solution of the wave equation, *Phys. Rev.*, 51, 669-676.

PEKERIS, C.L., (1946), Theory of propagation of sound in a half-space of variable sound velocity under conditions of formation of a shadow zone, *J. Acoust. Soc. Amer.*, 18, 295-315.

SCHELKUNOFF, S.A., (1948), *Applied Mathematics for Engineers and Scientists*, p. 212, Van Nostrand, New York.

WAIT, J.R., (1958), Transmission and reflection of electromagnetic waves in the presence of stratified media, *J. Res. Nat. Bur. Stand.*, 61, No. 3, 205-232.

WAIT, J.R., (1962), On the propagation of v.l.f. and e.l.f. radio waves when the ionosphere is not sharply bounded, *J. Res. Nat. Bur. Stand.*, 65D, (Radio Prop.), No. 1, 53-62.

OTHER REFERENCES

BAHAR, E., (1967), Generalized WKB method with application to problems of propagation in non-homogeneous media, *J. Math. Phys.*, 9, 1735-1746.

BAHAR, E., and AGRAWAL, B.S., (1976), Vertically polarized waves in inhomogeneous media with critical coupling regions, energy conservation and reciprocity, *Radio Science*, 11, 885-896.

BUDDEN, K.G., and TERRY P.D., (1971), Radio ray tracing in complex space, *Proc. Roy. Soc. London*, A.321, 275-301.

BUDDEN, K.G., (1975), Phase integral methods for studying the effect of the ionosphere on radio propagation, *Phil. Trans. Roy. Soc. London*, A.280, 111-130.

BUDDEN, K.G., and SMITH, M.S., (1976), Phase memory and additional memory in W.K.B. solutions for wave propagation in stratified media, *Proc. Roy. Soc. London*, A.350, 27-46.

BURMAN, R., (1965), A note concerning the reflection of waves in inhomogeneous layers with asymmetric profiles, *Radio Science*, 69D, 701-703.

GAGE, K.S., and GREEN, J.L., (1978), Evidence for specular reflection from monostatic VHF radar observations of the stratosphere, *Radio Science*, 13, 991-1001. [Equ. (10) in this paper for the reflection coefficient for a thin layer of thickness t and constant gradient is wrong; instead it applies to a semi-infinite linear layer with the same gradient].

HEADING, J., (1962), *An Introduction to Phase-Integral Methods*, Methuen and Co. Ltd., London.

NORTHOVER, F.H., (1962), The reflection of electromagnetic waves from thin ionized gaseous layers, *J. Res. Nat. Bur. Stand.*, 65D (Radio Prop.), No. 1, 73-80.

RYDBECK, O.E.H., (1961), On the coupling of waves in inhomogeneous systems, Report No. 56, Research Lab. of Electronics, Chalmers University, Gothenburg, Sweden.

THAYER, G.D., (1970), Radio reflectivity of tropospheric layers, *Radio Science*, 5, 1293-1299.

WAIT, J.R., (1960), Terrestrial propagation of v.l.f. radio waves, *J. Res. Nat. Bur. Stand.*, 64D (Radio Prop.), No. 2, 153-204.

WAIT, J.R., (1964), A note on VHF reflection from a tropospheric layer, *Radio Science*, 68D, 847-848. (Discusses an error in the earlier analysis by Bean, Frank and Lane who neglect the effect of the finite vertical extent of the layer.)

WAIT, J.R., and JACKSON, C.M., (1964), Influence of the refractive index profile in V.H.F. reflection from a tropospheric layer, *Trans. IEEE*, AP-12, 512-513.

17
High Frequency Electromagnetic Coupling Between Small Loops Over an Inhomogeneous Half-Space

Introduction

Insulated electric-current loops have considerable value as antennas for electromagnetic probing of an inhomogeneous environment. In particular, for time-harmonic currents, measurement of the mutual impedance between two loops over a range of frequencies or locations is a common geophysical-survey technique (Ward, 1967a and b; Grant and West, 1965). Usually, the objective is to describe the variation of the earth's electrical parameters with depth or to recognize the presence of lateral inhomogeneities. In both cases, the experimental data may be compared with curves computed for the mutual impedance of two loops over a stratified half-space. Therefore, successful interpretation of the data depends upon the availability of theoretical solutions for various profiles of horizontal stratification.

A basic and relevant source model is a loop that is oriented with its axis perpendicular to the earth-air interface. The loop may be located on or above the earth's surface. Solutions for the time-harmonic fields can be formulated for many analytical profiles of earth stratification, but the resulting integrals are often difficult to evaluate; zero radius (i.e., a magnetic dipole) is a special case which is frequently assumed in the literature. An exact, closed-form solution for the fields and mutual coupling of vertical magnetic dipoles on the surface of a homogeneous, conducting half-space has been given by Wait (1952a). The loop of finite radius and the elevated magnetic dipole were also treated, using a quasi-static approximation valid when the separation distance is small compared with the free-space wavelength (Wait, 1954, 1955). The solution for the finite loop indicated that the magnetic dipole approximation is accurate to within one percent, when the source-receiver separation is greater than ten-loop radii.

The important case of a magnetic dipole over a layered earth was formulated by Wait (1951). Approximate closed-form solutions were given for the two-layered earth, when the conductivity contrast of the two layers is small (Wait, 1952b), and when the vertical or horizontal separation of the source and receiver is large compared with a skin depth in the upper layer (Wait, 1958). For all of these solutions the quasi-static approximation was used. The integrals derived in the above references, in more general situations, must be evaluated by numerical integration. Extensive tabulations of the basic integrals T_0, T_1, and T_2 defined by Wait (1958) were compiled by the USGS and published by Frischknecht (1967).

The solution for a loop of finite radius over an n-layered half-space was formulated by Morrison et al (1969) in connection with a time-domain problem. Using this formulation Ryu et al (1970) computed the fields by careful numerical integration and published the results as an interesting and comprehensive collection of curves.

The layered half-space profiles are adequate for many purposes. For this reason and because of the difficulty of obtaining results in a usable form, profiles exhibiting smooth transitions have received little attention in the literature. A formulation for the fields of a vertical magnetic dipole over a layered half-space containing a linear transition layer was given recently by Mallick and Roy (1971). However, neither numerical integration nor the identification of simple approximate cases was discussed.

In this note we consider an exponentially stratified model with vertical magnetic-dipole excitation. Deep within the half-space, the conductivity and permittivity are essentially constant, but their values decrease smoothly as the surface is approached. This profile could be a useful model for a water table in the ground, for a frozen surface, or possibly for the lunar crust. In these cases, the electrical properties of the surface layers are inhomogeneous, while the deeper regions are more nearly homogeneous.

We first give an integral formulation for the fields of a finite loop located over an exponential half-space. Then we consider the limiting case of a magnetic dipole placed on the interface. The resulting integral is

*Reprinted in part from: J.R. Wait and J.A. Fuller, *Geophysics*, 37, 997–1004, 1972.

evaluated by making an approximation that is valid when the numerical distance to the receiver is much greater than one. Finally, we give simple expressions for the mutual impedance between two small loops and between a small loop and a horizontal electric dipole, where all of these antennas are located on the surface. The effect of the exponential stratification is represented by a single factor which modifies the formulas for a homogeneous half-space (Wait, 1952a). A tabular comparison of the mutual impedances computed by numerical integration and by the approximate formulas is provided.

Formulation

A loop of radius a is located at a height h above a conducting half-space, and a cylindrical coordinate system (ρ, ϕ, z) is established as shown in Figure 1. The z axis, which coincides with the loop axis, is directed into the earth, and the $z=0$ plane represents the earth-air interface. The loop carries an assumed uniformly distributed current $Ie^{i\omega t}$.

Using Dirac delta functions for cylindrical coordinates, the source may be represented by a ϕ-directed current density,

$$J_\phi = I(a/\rho)\delta(\rho-a)\delta(z+h). \tag{1}$$

The three nonzero field components are related by the following:

$$i\omega\mu_0 H_\rho = \frac{\partial E_\phi}{\partial z}, \tag{2a}$$

$$i\omega\mu_0 H_z = -\frac{1}{\rho}\frac{\partial}{\partial \rho}(\rho E_\phi), \tag{2b}$$

and

$$\frac{\partial}{\partial \rho}\frac{1}{\rho}\frac{\partial}{\partial \rho}(\rho E_\phi) + \frac{\partial^2 E_\phi}{\partial z^2} + k^2(z)E_\phi = i\omega\mu_0 J_\phi. \tag{2c}$$

In the air ($z<0$), we have $k^2(z)=k_0^2=\omega^2\mu_0\varepsilon_0$, while in the earth ($z>0$), the assumed exponential profile is

$$k^2(z) = (k_1^2 - k_2^2)\exp(-z/l) + k_2^2, \tag{3}$$

where $k_1^2 = -i\omega\mu_0(\sigma_1 + i\omega\varepsilon_1)$, $k_2^2 = -i\omega\mu_0(\sigma_2 + i\omega\varepsilon_2)$, and l is a "depth constant." Therefore, the conductivity and dielectric constant at the surface are σ_1 and ε_1; these quantities increase exponentially toward the larger values σ_2 and ε_2 deep within the earth. The general form of this variation is shown in Figure 2.

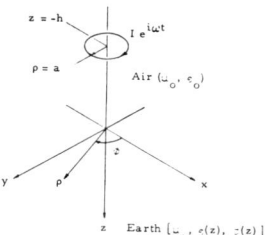

Fig. 1. Current loop with vertical axis over a plane stratified earth.

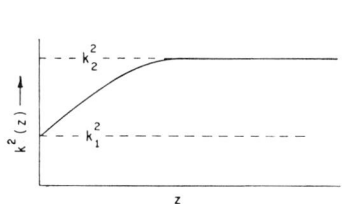
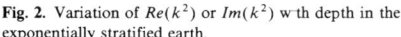

Fig. 2. Variation of $Re(k^2)$ or $Im(k^2)$ with depth in the exponentially stratified earth.

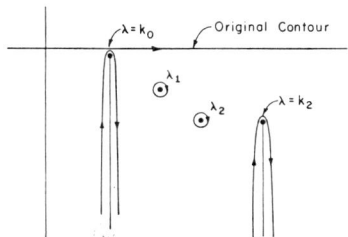

Fig. 3. Transform inversion contours in the complex λ plane.

It is sufficient for the present purpose to obtain solutions in the free-space region ($z<0$). Following the usual transform procedures, the ρ dependence of (2c) is eliminated. The appropriate Fourier-Bessel transform is defined as follows:

$$E(\lambda) = \int_0^\infty E_\phi(\rho, z) J_1(\lambda\rho) \rho \, d\rho \quad \text{and} \tag{4a}$$

$$E_\phi(\rho, z) = \int_0^\infty E(\lambda) J_1(\lambda\rho) \lambda \, d\lambda. \tag{4b}$$

Solutions of the transformed differential equation are of the form

$$E(\lambda) = A(\lambda) e^{u_0 z} + B(\lambda) e^{-u_0 z}, \tag{5}$$

where $u_0^2 = \lambda^2 - k_0^2$. To facilitate the asymptotic evaluation of the inverse transform discussed later, the branch cut chosen for $u_0 = (u_0^2)^{1/2}$ is a sligh modification of the principal cut. When mapped into the λ plane, this modified cut appears as the cut originating from $\lambda = k_0$ in Figure 3. With this definition of the root, $Re(u_0) \geq 0$ for λ on the inversion contour, although the inequality is not always valid when λ is analytically continued into the complex plane. Provided that $A(\lambda)$ and $B(\lambda)$ are properly determined for real λ, analytic continuation into the cut plane is permissible, regardless of the values that u_0 may then assume.

The coefficients $A(\lambda)$ and $B(\lambda)$ are determined by requiring outgoing waves at $z = -\infty$, by requiring an appropriate discontinuity in the horizontal magnetic field across the plane of the source ($z = -h$), and by introducing a reflection coefficient to account for the earth-air interface. The reflection coefficient for a perpendicularly polarized wave incident on an exponentially graded half-space has been given by Wait (1962). The reflection coefficient for a plane wave incident on the half-space at an angle θ can be utilized in the transform solution here by setting $\lambda = k_0 \sin\theta$ and allowing complex angles of incidence. The reflection coefficient is $R(\lambda) = (N_0 - Y)/(N_0 + Y)$, where $N_0 = u_0/i\omega\mu_0$ and Y is the surface admittance of the half-space,

$$Y = \frac{(k_1^2 - k_2^2)^{1/2}}{i\omega\mu_0} \cdot \frac{J_\nu'(v_0)}{J_\nu(v_0)}. \tag{6}$$

Here, $v_0 = (2l)(k_1^2 - k_2^2)^{1/2}$ is the root with a positive real part, and $\nu = (2l)(\lambda^2 - k_2^2)^{1/2}$ is defined by a cut similar to that for u_0. For convenience, we define $u_2^2 = \lambda^2 - k_2^2$. Then the cut for u_2 is the one originating from $\lambda = k_2$ in Figure 3.

When l approaches zero, the parameters of the deep earth extend uniformly to the surface, and thus $Y = Y_\infty = u_2/(i\omega\mu_0)$. The effect of surface stratification can be introduced by defining a factor Q, which modifies Y_∞ for finite l; accordingly, let $Y = Y_\infty/Q$, where

$$\frac{1}{Q} = \frac{v_0 J_\nu'(v_0)}{\nu J_\nu(v_0)}. \tag{7}$$

For small l it can be shown that Q approaches unity. It follows that

$$R(\lambda) = \frac{u_0 - u_2/Q}{u_0 + u_2/Q}. \tag{8}$$

The solution in the region $-h < z < 0$ is

$$E(\lambda) = \frac{-i\omega\mu_0 aI}{2} \cdot \frac{J_1(\lambda a)}{u_0} \{\exp[-u_0(z+h)] + R(\lambda)\exp[u_0(z-h)]\}. \tag{9}$$

The formulation of problems such as this obviously is quite straightforward, when $R(\lambda)$ is known for the half-space of interest. This quantity has been derived for a number of special profiles by Wait (1962, Chapter 3). For the solution to be useful, quantitative information must be given for the inverse transform of (9).

For the asymptotic evaluation to be considered here, it is convenient to deal with an electric-vector potential, rather than with the electric field. Only a single component, F_z, of the vector potential is required. The electric field is given by $E_\phi = \partial F_z/\partial \rho$, and the magnetic components then follow from (2a) and (2b). Since $J_0'(x) = -J_1(x)$, we have

$$F_z = -\int_0^\infty E(\lambda) J_0(\lambda\rho) d\lambda. \tag{10}$$

The case of a magnetic dipole having moment $\pi a^2 I$ is obtained by letting $a \to 0$ and replacing $J_1(\lambda a)$ by the small argument limit $\lambda a/2$ in (9). The integrand can be rearranged to identify the primary contribution from the source and its image in a perfectly conducting half-space. This is accomplished by letting $F_z = F + \Delta F$, where

$$F = C\int_0^\infty [e^{-u_0(h+z)} - e^{-u_0(h-z)}] J_0(\lambda\rho)\frac{\lambda}{u_0} d\lambda, \tag{11a}$$

$$\Delta F = C\int_0^\infty e^{-u_0(h-z)} J_0(\lambda\rho)\frac{2\lambda}{u_0 + u_2/Q} d\lambda, \tag{11b}$$

and $C = i\omega\mu_0\pi a^2 I/4\pi$. Now, F has the form of a Sommerfeld-type integral, with the result that

$$F = C\left\{\frac{\exp[-ik_0\sqrt{\rho^2 + (h+z)^2}]}{\sqrt{\rho^2 + (h+z)^2}} - \frac{\exp[-ik_0\sqrt{\rho^2 + (h-z)^2}]}{\sqrt{\rho^2 + (h-z)^2}}\right\}. \tag{12}$$

Evaluation of the Potential Integral

Here we assume that the transmitter and receiver are located on the earth's surface, so that $F_z = \Delta F$. In this case, the convergence of (11b) is particularly slow (since $h - z = 0$), but it can be transferred into a more suitable representation by deforming the contour of integration. It then becomes apparent that an approximate evaluation can be obtained if the distance separating the antennas is sufficiently large.

The integration limits in (11b) can be changed to $[-\infty, +\infty]$ by using the identity (valid for $x \geq 0$), $J_0(x) = \frac{1}{2}[H_0^{(2)}(x) - H_0^{(2)}(-x)]$. Setting $\Omega = \Delta F/C$, it follows that

$$\Omega = \int_{-\infty}^\infty \frac{\lambda}{u_0(\lambda) + u_2(\lambda)/Q(\lambda)} H_0^{(2)}(\lambda\rho) d\lambda. \tag{13}$$

Table 1-A. Mutual impedance of VMD-HED antenna pair ($\rho=50$ m). $\varepsilon_1/\varepsilon_0=5$, $\sigma_1=10^{-5}$ mho/m, $\varepsilon_2/\varepsilon_0=10$, and $\sigma_2=10^{-3}$ mho/m.

f(hz)	$|P_m|$	Z_1^a/Z_{01} (ohms)	Z_1/Z_{01} (ohms)
		$l=10$ m	
10^5	6.5	4.335 ($-$ 79.57°)	5.487 ($-$ 43.71°)
10^6	4.4	0.7898($-$103.82°)	0.9578($-$ 92.94°)
10^7	21.1	0.5080 (112.10°)	0.5104 (112.15°)
10^8	209.0	0.5001 (115.09°)	0.5001 (115.09°)
		$l=50$ m	
10^5	3.3	8.727 ($-$ 70.79°)	25.33 ($-$ 20.64°)
10^6	2.0	1.758 ($-$127.63°)	2.334 ($-$141.59°)
10^7	20.9	0.5112 (105.21°)	0.5117 (105.18°)
10^8	209.0	0.5001 (114.40°)	0.5001 (114.40°)

Table 1-B. Mutual impedance of VMD-HED antenna pair ($\rho=100$ m). $\varepsilon_1/\varepsilon_0=5$, $\sigma_1=10^{-5}$ mho/m, $\varepsilon_2/\varepsilon_0=10$, and $\sigma_2=10^{-3}$ mho/m.

f(hz)	$|F_m|$	Z_1^a/Z_{01} (ohms)	Z_1/Z_{01} (ohms)
		$l=10$ m	
10^5	13.2	1.090 ($-$ 79.57°)	1.089 ($-$ 63.76°)
10^6	8.8	0.3461($-$120.20°)	0.3507($-$116.74°)
10^7	42.2	0.5027($-$120.11°)	0.5038($-$120.06°)
10^8	418.0	0.5000($-$128.25°)	0.5000($-$128.25°)
		$l=50$ m	
10^5	6.6	2.194 ($-$ 70.79°)	3.000 ($-$ 32.67°)
10^6	4.0	0.7701($-$144.01°)	0.8900($-$143.58°)
10^7	41.8	0.5059($-$127.01°)	0.5062($-$127.02°)
10^8	415.0	0.5001($-$128.93°)	0.5001($-$128.93°)

Table 2-A. Mutual impedance of VMD-VMD antenna pair ($\rho=50$ m). $\varepsilon_1/\varepsilon_0=5$, $\sigma_1=10^{-5}$ mho/m, $\varepsilon_2/\varepsilon_0=10$, and $\sigma_2=10^{-3}$ mho/m.

f(hz)	$|P_m|$	Z_2^a/Z_{02} (ohms)	Z_2/Z_{02} (ohms)
		$l=10$ m	
10^5	6.5	12.99 ($-$ 79.57°)	19.87 ($-$ 35.27°)
10^6	4.4	2.037 ($-$101.74°)	2.979 ($-$ 87.62°)
10^7	21.1	5.204 ($-$163.65°)	5.229 ($-$163.64°)
10^8	209.0	52.39 ($-$155.46°)	52.39 ($-$155.46°)
		$l=50$ m	
10^5	3.3	26.15 ($-$ 70.79°)	107.2 ($-$ 20.06°)
10^6	2.0	4.533 ($-$125.55°)	6.311 ($-$149.42°)
10^7	20.9	5.237 ($-$170.55°)	5.242 ($-$170.59°)
10^8	209.0	52.40 ($-$156.14°)	52.40 ($-$156.14°)

Table 2-B. Mutual impedance of VMD-VMD antenna pair ($\rho=100$ m). $\varepsilon_1/\varepsilon_0=5$, $\sigma_1=10^{-5}$ mho/m, $\varepsilon_2/\varepsilon_0=10$, and $\sigma_2=10^{-3}$ mho/m.

f(hz)	$|P_m|$	Z_2^a/Z_{02} (ohms)	Z_2/Z_{02} (ohms)
		$l=10$ m	
10^5	13.2	3.253 ($-$ 79.56°)	3.424 ($-$ 56.24°)
10^6	8.8	0.6938($-$ 91.10°)	0.7500($-$ 86.02°)
10^7	42.2	10.48 ($-$ 32.88°)	10.50 ($-$ 32.84°)
10^8	418.0	104.8 ($-$ 38.52°)	104.8 ($-$ 38.52°)
		$l=50$ m	
10^5	6.6	6.549 ($-$ 70.79°)	11.63 ($-$ 24.58°)
10^6	4.0	1.544 ($-$114.91°)	1.906 ($-$122.38°)
10^7	41.8	10.54 ($-$ 39.78°)	10.55 ($-$ 39.79°)
10^8	418.0	104.8 ($-$ 39.20°)	104.8 ($-$ 39.20°)

The integration contour now can be deformed to infinity in the lower half of the λ plane, but it must pass around the two branch cuts and a discrete set of poles, as shown in Figure 3. There is no contribution from the contour at infinity. This suggests that we write

$$\Omega = \Omega^{(0)} + \Omega^{(2)} + \Sigma_j \Omega_j, \qquad (14)$$

where $\Omega^{(0)}$ represents the integral around the cut at $\lambda = k_0$, $\Omega^{(2)}$ is the integral around the cut at $\lambda = k_2$, and Ω_j is the contribution from a pole at $\lambda = \lambda_j$. Physically, the two branch-line integrals account for a continuous spectrum of radiation in the homogeneous regions above the interface and deep within the earth, respectively. The pole contributions, of course, can be identified with wave-guide modes in the stratified region of the earth.

When the source and receiver are located above or on the surface of the earth, and when they are separated horizontally by a distance which is several wavelengths in the earth, the contributions to Ω which depend on propagation through the lossy ground should be negligible. In this case it is sufficient to evaluate $\Omega^{(0)}$.

In the λ plane the branch line from k_0 can be described by $\lambda = k_0 - is^2$, where s is a real variable. In the integrand of $\Omega^{(0)}$, u_0 is the only function which changes sign, when the branch cut is crossed. If $q = (2k_0)^{1/2} \exp[-i\pi/4]$, then on the right side of the cut,

$$u_0 = u_0^+(s) = q \cdot |s| \cdot (1 - s^2/q^2)^{1/2}, \qquad (15)$$

and on the left side $u_0 = -u_0^+(s)$. The branch-line integral is given by

$$\Omega^{(0)} = -i4 \int_0^\infty \frac{(q^2 s^2 - s^4)^{1/2}}{u_0^2 - u_2^2/Q^2} H_0^{(2)}(k_0 \rho - is^2 \rho)(k_0 - is^2) s \, ds. \qquad (16)$$

Noting the rapid attenuation provided by the Hankel function as s increases, we approximate the denominator for small s as follows. First, we put $u_2/Q \simeq (k_0^2 - k_2^2)^{1/2}/Q_0$, where

$$\frac{1}{Q_0} = \frac{v_0 J_\mu'(v_0)}{\mu J_\mu(v_0)}, \qquad (17)$$

with the order $\mu = 2l(k_0^2 - k_2^2)^{1/2}$. Since $\nu = 2l(k_0^2 - k_2^2 + q^2 s^2 - s^4)^{1/2}$, (17) is valid when $s^2 \ll |(k_0^2 - k_2^2)/q^2|$. Then we observe that

$$|u_0^2| = |q^2 s^2 - s^4| \ll |(k_0^2 - k_2^2)/Q_0^2|,$$

if

$$s^2 \ll |(k_0^2 - k_2^2)/(qQ_0)^2|.$$

For the cases considered here, this second condition on s is the more restrictive (i.e., $|Q_0| \geqslant 1$). If the argument of the Hankel function becomes sufficiently large before that condition is violated, then the denominator in (16) may be approximated by a constant without serious error. Therefore, we require that $|k_0 \rho - is^2 \rho| \gg 1$, when $s^2 \geqslant |(k_0^2 - k_2^2)/(qQ_0)^2|$. At grazing incidence with a horizontally polarized electric field, the numerical distance is defined by

$$p_m = \rho(k_0^2 - k_2^2)/(qQ_0)^2, \qquad (18)$$

in accordance with the concept introduced by Sommerfeld (Wait, 1962, p. 40). When $|p_m| \gg 1$, the approximations are justified, and the following integral remains:

$$\Omega^{(0)} \simeq \frac{i4 Q_0^2}{k_0^2 - k_2^2} \int_0^\infty H_0^{(2)}(k_0 \rho - is^2 \rho)(q^2 s^2 - s^4)^{1/2}(k_0 - is^2) s \, ds. \qquad (19)$$

In order to evaluate (19), we utilize the following:

$$P(\alpha) = \frac{\partial^2}{\partial \alpha^2}\left[\frac{\exp(-ik_0 R)}{R}\right] = \int_0^\infty \exp(-\alpha u_0) J_0(\lambda \rho) u_0 \lambda \, d\lambda$$

$$= \{-ik_0 R - (1-(ik_0\alpha)^2) + 3ik_0\alpha^2/R + 3\alpha^2/R^2\}\exp(-ik_0 R)/R^3. \tag{20}$$

Here, as before, $u_0 = (\lambda^2 - k_0^2)^{1/2}$, and $R = (\rho^2 + \alpha^2)^{1/2}$. By a procedure similar to that applied to (11b), the contour of integration is deformed onto the (single) branch cut in the lower half plane. In this case, there are no poles, and the following representation is exact:

$$P(\alpha) = -i2\int_0^\infty \cosh(\alpha u_0^+) H_0^{(2)}(k_0\rho - is^2\rho)(q^2 s^2 - s^4)^{1/2}(k_0 - is^2) s \, ds. \tag{21}$$

Setting $\alpha = 0$, we obtain the required integral,

$$F(0) = (-1 - ik_0\rho)\exp(-ik_0\rho)/\rho^3. \tag{22}$$

Using the above result, we find that the electric and magnetic field components in the interface are

$$E_\phi \simeq \frac{i\omega\mu_3\pi a^2 I Q_0^2}{2\pi(k_2^2 - k_0^2)\rho^4}(3 + 3ik_0\rho + (ik_0\rho)^2)\exp(-ik_0\rho), \tag{23a}$$

and

$$H_z \simeq \frac{\pi a^2 I Q_0^2}{2\pi(k_2^2 - k_4^2)\rho^5}(9 + 9ik_0\rho + 4(ik_0\rho)^2 + (ik_0\rho)^3)\exp(-ik_0\rho). \tag{23b}$$

These are valid under the condition $|\mathit{F}_m| \gg 1$. As required, these expressions yield the fields on a homogeneous half-space ($Q_0 = 1$) (Wait, 1952a), under the condition that $|k\rho| \gg 1$, where k is the complex wavenumber for the half-space.

Mutual Coupling Formulas

Two commonly used antenna configurations are considered here. In both cases, the transmitter and the receiver are located on the earth's surface, and the transmitter is a vertical magnetic dipole (VMD) having a moment $\pi a^2 I$. In the first case (indicated by a subscript 1), the receiving antenna is a horizontal electric dipole (HED) of length dl. The angle between the electric dipole and the line joining the two antennas is defined as ψ. In the second case (indicated by a subscript 2), the receiving antenna is a vertical magnetic dipole, represented by a small loop of radius b. Simple expressions are derived for the mutual impedances of these two antenna configurations.

The emf's induced in the receiving electric and magnetic dipoles are $V_1 = E_\phi \, dl \cdot \sin\psi$ and $V_2 = -i\omega\mu_0\pi b^2 H_z$, respectively. If the transmitting loop carries a current I, the mutual impedances for the two cases are $Z_1 = V_1/I$ and $Z_2 = V_2/I$. Using (23a) and (23b), the following approximate normalized impedances are obtained:

$$Z_1^a/Z_{01} = \frac{2Q_0^2}{k_0^2\rho^2(1 - k_2^2/k_0^2)}[3 + 3(ik_0\rho) + (ik_0\rho)^2]\exp(-ik_0\rho), \tag{24a}$$

and

$$Z_2^a/Z_{02} = \frac{2Q_0^2}{k_0^2\rho^2(1 - k_2^2/k_0^2)}[9 + 9(ik_0\rho) + 4(ik_0\rho)^2 + (ik_0\rho)^3]\exp(-ik_0\rho). \tag{24b}$$

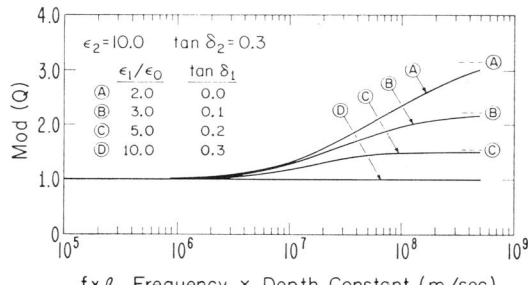

Fig. 4(a). Modulus of the correction factor Q for typical earth parameters in the HF range.

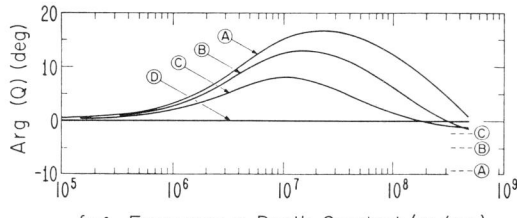

Fig. 4(b). Phase of the correction factor Q for typical earth parameters in the HF range.

This normalization is now common in the literature; it refers to the near-field mutual impedances, when the antennas are in free space,

$$Z_{01} = -\frac{i\omega\mu_0(\pi a^2)\,dl}{4\pi\rho^2}\sin\psi \tag{25a}$$

and

$$Z_{02} = \frac{i\omega\mu_0(\pi a^2)(\pi b^2)}{4\pi\rho^3}. \tag{25b}$$

As a check on the validity of the approximations used to derive (24), impedances have been computed by numerical integration of field equations derived directly from (16). A comparison of the results is presented in phasor form in Tables 1 and 2; the column headed $Z_n^a(n=1$ or $2)$ contains impedances based on (24a) or (24b), while Z_n denotes the more accurate quantities computed by numerical integration.[1] The magnitude of numerical distance is also tabulated. The approximate formulas appear to be useful, when $|p_m|>10$.

[1] In all of the computations discussed here, Q was computed by making the substitution,

$$J'_\nu(v_0)/J_\nu(v_0) = \nu/v_0 - J_{\nu+1}(v_0)/J_\nu(v_0),$$

and then using the continued fraction representation for the ratio, $J_{\nu+1}(v_0)/J_\nu(v_0)$ (Watson, 1944, p. 153).

In applications of (24), values for the stratification factor Q_0 must be available readily. A sample set of curves is plotted in Figure 4, where the abscissa is a normalized frequency scale. The several sets of earth parameters shown are typical of the HF range.

Concluding Remarks

We have obtained a general formulation for the electromagnetic field of a small transmitting loop located on or over the surface of an inhomogeneous plane earth. The inhomogeneity consists of a surface layer with a smooth exponential transition in its electrical properties. In this fashion we may account for such conditions as a frozen soil or a permafrost region above a more highly conducting subsoil. The analogous situation for the lunar crust also may be encompassed in the present development.

We show that the surface inhomogeneity or transition region can be accounted for by a factor multiplying the previously known exact formulas for the mutual impedances of coplanar small loops. The illustrative examples suggest that such results should be useful in the interpretation of measured mutual impedances in terms of the conductivity and dielectric constant structure of the surface and subsoil regions at HF.

References

FRISCHKNECHT, F.C., (1967), Fields about an oscillating magnetic dipole over a two-layer earth, and application to ground and airborne electromagnetic surveys, *Quart. Colo. Scho. Mines*, Vol. 62, p. 1–326.

FULLER, J.A. and WAIT, J.R., (1972), High frequency electromagnetic coupling between small loops over an inhomogeneous ground, *Geophys.*, Vol. 37, p. 997–1004.

GRANT, F.A., and WEST, G.F., (1965), *Interpretation Theory in Applied Geophysics*, New York, McGraw-Hill Book Company, Inc., p. 444–546.

MALLICK, K., and ROY, A., (1971), Vertical magnetic dipole over a transitional earth, *Geophys. Prosp.*, Vol. 19, p. 388–394.

MORRISON, H.F., PHILLIPS, R.J., and O'BRIEN, D.P., (1969), Quantitative interpretation of transient electromagnetic fields over a layered half-space, *Geophys. Prosp.*, Vol. 17, p. 82–101.

RYU, J., MORRISON, H.F., and WARD, S.H., (1970), Electromagnetic fields about a loop source of current, *Geophys.*, Vol. 35, p. 862–896.

WAIT, J.R., (1951), The magnetic dipole over a horizontally stratified earth, *Can. J. Phys.*, Vol. 29, p. 577–592.

WAIT, J.R., (1952a), Current-carrying wire loops in a simple inhomogeneous region, *J. Appl. Phys.*, Vol. 23, p. 497–498.

WAIT, J.R., (1952b), Mutual inductance of circuits on a two-layer earth, *Can. J. Phys.*, Vol. 30, p. 450–452.

WAIT, J.R., (1954), Mutual coupling of loops lying on the ground, *Geophys.*, Vol. 19, p. 290–296.

WAIT, J.R., (1955), Mutual electromagnetic coupling of loops over a homogeneous ground, *Geophys.*, Vol. 20, p. 630–637.

WAIT, J.R., (1958), Induction by an oscillating magnetic dipole over a two-layer ground, *Appl. Sci. Res.*, Sec. B, Vol. 7, p. 73–80.

WAIT, J.R., (1962), *Electromagnetic Waves in Stratified Media*, Oxford, Pergamon Press.

WARD, S.H., (1967a), Electromagnetic theory for geophysical applications, *Mining Geophysics*, Vol. 2, Tulsa, SEG, p. 10–196.

WATSON, G.N., (1944), *Theory of Bessel Functions*, Cambridge, Cambridge University Press.

18
Reflection of VLF Radio Waves from an Inhomogeneous Isotropic Ionosphere

PART I   EXPONENTIAL MODEL

Introduction

In the study of VLF radio wave propagation it is often assumed that the ionosphere can be regarded as a sharply bounded medium on its underside. This step may be considered reasonable in view of the long wavelength and the relatively rapid change of the electron density of the D region. Furthermore, experimental data are often in accord with calculations based on this model. The general agreement is particularly good at highly oblique incidence provided that appropriate corrections are made for earth curvature. Nevertheless, there are many occasions when the sharply bounded model appears to be inadequate. Therefore, it is worthwhile to consider a more realistic model of the lower ionosphere.

On examining much of the recent literature on the characteristics of the lower D layer, it appears that the effective dielectric constant of the medium can be well approximated by an exponential function. Then, to within this approximation, the relative permittivity may be written in the form

$$K(z) = K_0 \left(1 - i \frac{1}{L} \exp \beta z \right), \qquad (1)$$

where K_0 is a reference permittivity, and L and β are constants. The level $z=0$ may be defined as the reference level and thus

$$K(0) = K_0 \left(1 - i \frac{1}{L}\right), \qquad (2)$$

which is a familiar form.

Under the assumptions that the angular frequency ω is much less than both the collision frequency ν and the plasma frequency ω_0, it is known that [Wait, 1962]

$$K_0 = 1 \quad \text{and} \quad L = \frac{\omega \nu}{\omega_0^2}, \qquad (3)$$

provided that the earth's magnetic field can be neglected. The latter assumption is strictly valid only if ν is somewhat greater than the gyrofrequency ω_H. In actual cases, however, the isotropic assumption is useful even when ν is of the order of ω_H, provided the magnetic field is not transverse to the direction of propagation [Crombie, 1961; Johler and Harper, 1962]. For the purposes of the present paper, the influence of the terrestrial magnetic field is not considered.

The constant β in the exponent of the equation for $K(z)$ is a measure of the sharpness of the gradient. For example, when $\beta = 1$ km^{-1}, it means that the ratio ω_0^2/ν or N/ν increases by a factor 2.71 for each km of vertical height. The best available information on expected values of β can be found from observations of pulse cross modulation [Barrington et al., 1962] and from various rocket measurements [Kane, 1962]. From these results it appears that β is of the order of $\frac{1}{2}$ km^{-1} for quiet daytime conditions although it may differ by a factor of two or more at certain times. In the present study it appears to be desirable to allow β to vary from 0.3 to 3.0 to encompass most cases of interest.

*Reprinted in part from: J.R. Wait and L.C. Walters, *Jour. Res. NBS.*, 67D, Radio Propagation, Parts I, II and III, 1963.

Reflection of VLF Radio Waves

BACKGROUND THEORY

The propagation of electromagnetic waves in a medium whose permittivity $\epsilon(z)$ varies in an exponential manner has been studied extensively. The usual case considered is for horizontal polarization where the electric vector is parallel to the stratification. For example, if the electric vector has only a y component E_y, it is a simple matter to show that [Wait, 1962]

$$\left[\frac{d^2}{dz^2}+k^2(K(z)-S^2)\right]E_y=0, \tag{4}$$

when the field varies in the x direction according to $\exp(-ikSx)$, where S is a dimensionless constant. Here $k=\sqrt{\epsilon_0\mu_0}\,\omega$ is the free-space wave number and ϵ_0 and μ_0 are the permittivity and permeability of free space, respectively.

At a sufficiently large negative value of z, the relative permittivity $K(z)$ becomes unity, and thus, E_y satisfies

$$\left(\frac{d^2}{dz^2}+k^2C^2\right)E_y=0, \tag{5}$$

where $C^2=1-S^2$. The general solution of this equation has the form

$$E_y=E_0\left(e^{-ikCz}+R_h e^{+ikCz}\right)e^{-ikSx}, \tag{6}$$

when E_0 and R_h are constants. Recognizing that the solution of (4) must reduce to (5), for z tending to $-\infty$, it may be shown that [Wait, 1962]

$$R_h=-\left(\frac{k}{\beta}\right)^{2\nu}\frac{1}{(-iL)^{\nu_0}}\frac{(-\nu_0)!}{(\nu_0)!}, \tag{7}$$

where $\nu_0=2ikC/\beta$.

The term $E_0 e^{-ikCz}e^{-ikSx}$ may be regarded as an incident wave whose direction of propagation makes angle θ with the z axis where $C=\cos\theta$ or $S=\sin\theta$. Then $E_0 R e^{+ikCz}e^{-ikSx}$ can be interpreted as a reflected wave and R can be defined as the reflection coefficient of the exponential layer. It should be stressed that this reflection coefficient, while referred to the level $z=0$, is only valid in the free space region at large negative values of z.

The reflection coefficient R_h, given by (7), may be conveniently written in the form

$$R_h=\exp\left(-\frac{2\pi^2}{\lambda_0\beta}C\right)\exp i\Phi, \tag{8}$$

where

$$\Phi=\frac{8\pi}{\lambda_0\beta}C\log\left(\frac{2\pi}{\lambda_0\beta L^{1/2}}\right)\pm\pi+2\arg\left[\left(-i\frac{4\pi C}{\lambda_0\beta}\right)!\right],$$

where $\lambda_0=2\pi/k$ is the free-space wavelength. Thus, the amplitude of the reflection coefficient is given by the very simple form

$$|R_h|=\exp\left(-\frac{2\pi^2}{\lambda_0\beta}C\right), \tag{9}$$

which is independent of L. On the other hand, the phase factor Φ is relatively complicated since it involves the factorial function of imaginary argument. Nevertheless, (8), for the complex value of R_h can be used to obtain quantitative results for exponential-type layers.

Extension to Vertical Polarization

In the VLF radio problem one is mainly interested in vertical polarization such that the magnetic vector is parallel to the stratification. For example, if the magnetic field has only a y component, H_y, and assuming again that the fields vary in the x direction according to $\exp(-ikSx)$, it follows that [Wait, 1962]

$$\left[\frac{d^2}{dz^2} - \frac{1}{K(z)}\frac{dK(z)}{dz}\frac{d}{dz} + k^2(K-S^2)\right]H_y = 0. \tag{10}$$

Intrinsically, this is a more complicated form than (4) for horizontal polarization.

In the region of large negative z the (10) reduces to the elementary form

$$\left(\frac{d^2}{dz^2} + k^2 C\right)H_y = 0, \tag{11}$$

and thus

$$H_y = H_0\left(e^{-ikCz} + R_v e^{+ikCz}\right)e^{-ikSx}, \tag{12}$$

where H_0 is a constant and R_v is, by definition, the reflection coefficient for vertical polarization.

Unfortunately, for the exponentially varying permittivity of the form defined by (1), it does not appear to be possible to obtain a closed form expression for R_v. For this reason, the required quantitative results were obtained by a numerical method. Essentially, the procedure is to replace the continuous $K(z)$ profile by a finite number of steps. The situation is illustrated in figure 1 where the function

$$\frac{1}{L(z)} = \frac{1}{L}\exp\beta z, \tag{13}$$

is shown plotted versus z along with its step approximation. Thus, between the limits $z = -z_0$ and $z = +T$, the medium is divided into M homogeneous layers of width $h_1, h_2 \ldots h_m, \ldots h_{M-1}, h_M$. The value of $L(z)$ in each of these slabs is then replaced by a constant value L_m. The method is really equivalent to the usual method [Budden, 1961] of numerically integrating the basic differential equations.

The problem now boils down to finding the wave impedance Z_1 at $z = -z_0$ in terms of the properties of all the individual slabs. Here $Z_1 = E_x/H_y$ which is to be evaluated at $z = -z_0$. The reflection coefficient R_v, which is referred to the level $z = 0$, is then found from

$$R_v = \frac{C - Z_1/\eta_0}{C + Z_1/\eta_0}\exp\left[-i\frac{4\pi}{\lambda_0}Cz\right], \tag{14}$$

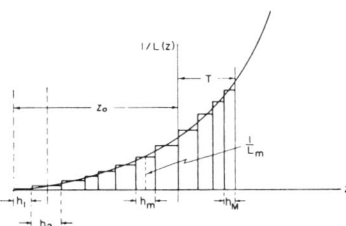

Fig. 1. The step approximation to an exponential profile.

Reflection of VLF Radio Waves

for $z = -z_0$, and where $\eta_0 = \sqrt{\mu_0/\varepsilon_0} = 20\pi$. Here z_0 is chosen sufficiently large that $1/L(z)$ may be regarded as zero. (Strictly speaking, $z_0 \to \infty$.)

Iterative Process

From the theory of wave propagation in stratified media [e.g., Wait, 1962], it is known that Z_1 can be obtained by a series of iterative processes. Thus

$$Z_1 = K_1 Q_1, \tag{15}$$

where

$$Q_1 = \frac{K_{2,1} Q_2 + \tanh D_1}{1 + K_{2,1} Q_2 \tanh D_1} \tag{16}$$

$$Q_2 = \frac{K_{3,2} Q_3 + \tanh D_2}{1 + K_{3,2} Q_3 \tanh D_2}, \tag{17}$$

$$\cdots$$

$$Q_m = \frac{K_{m+1,m} Q_{m+1} + \tanh D_m}{1 + K_{m+1,m} Q_{m+1} \tanh D_m}, \tag{18}$$

and so on. The various factors are defined below. The parameter D_m which is a measure of the thickness of an individual slab is defined by

$$D_m = \frac{i 2\pi h_m}{\lambda_0} N_m \left(1 - \frac{S^2}{N_m^2}\right)^{1/2}, \tag{19}$$

where N_m is the refractive index of the mth slab given by[1]

$$N_m = \left(1 - \frac{i}{L_m}\right)^{1/2}.$$

The quantity $K_{m+1,m}$ is the ratio of the wave impedances of layer $m+1$ and layer m. Thus

$$K_{m+1,m} = \frac{K_{m+1}}{K_m}, \tag{20}$$

where

$$K_m = \frac{\eta_0}{N_m} \left(1 - \frac{S^2}{N_m^2}\right)^{1/2},$$

is the wave impedance in the mth layer. Alternatively,

$$K_{m+1,m} = \frac{\left(1 - \dfrac{i}{L_m}\right)}{\left(1 - \dfrac{i}{L_{m+1}}\right)} \left[\frac{C^2 - \dfrac{i}{L_{m+1}}}{C^2 - \dfrac{i}{L_m}}\right]^{1/2}. \tag{21}$$

---

[1] The square roots are *defined* such that $D_m = (i2\pi h_m/\lambda_0) N_m$ if $N_m \to \infty$, while $D_m = (i2\pi h_m/\lambda_0) C$ if $N_m \to 1$. Furthermore, $N_m \simeq (1/L_m)^{1/2} e^{-i\pi/4}$ if $|N_m| \gg 1$ and $N_m = 1$ if $L_m \to \infty$.

Finally, because the quantities Q_m are given in terms of Q_{m+1} it is necessary to know the initial value Q_{M+1} at $z=T$. for example,

$$Q_M = \frac{K_{M+1,M} Q_{M+1} + \tanh D_M}{1 + K_{M+1,M}[\tanh D_M] Q_{M+1}}, \tag{22}$$

and similarly, Q_{M-1} is expressed in terms of Q_M. The process is continued until Q_1 is obtained.

In some problems of this kind it is not necessary to know precisely the value of Q_{M+1}. However, the economy of the calculations is greatly improved if a good starting value of Q_{M+1} is known. Recognizing that the wave impedance at $z=T$ is given by

$$Z_{M+1} = K_{m+1} \; Q_{m+1} \tag{23}$$

and utilizing the condition $|N_{M+1}| \gg 1$, it follows from previous work [Wait, 1962] that

$$Q_{M+1} \simeq \frac{K_0\left[\sqrt{i}\,\frac{2\pi}{\lambda_0} N_{M+1} \frac{2}{\beta}\right]}{K_1\left[\sqrt{i}\,\frac{2\pi}{\lambda_0} N_{M+1} \frac{2}{\beta}\right]}, \tag{24}$$

where K_0 and K_1 are modified Bessel functions.

In a practical sequence of calculations it is necessary to choose T large enough that the final result for R_v is insensitive to further changes in T. For example, if $\beta = 1$ km^{-1} and $\lambda_0 = 15$ km, it was found that $T = 4$ km was sufficiently large to obtain four-figure accuracy in R_v. Furthermore, for some values of β and λ_0, it was found that R_v approaches a constant as z_0 was increased to 20 km. In general, for smaller values of β it was necessary to increase T and z_0 to larger values in order to achieve stability of the final results. The width of the steps, h_m, must also be chosen sufficiently small to achieve an adequate simulation of the smooth profile. Generally, it was found that $|D_m| < 10^{-2}$ was a satisfactory criterion.

Discussion of Reflection Coefficient Calculations

Using the basic definition of the reflection coefficient R_v, given by (14), the amplitude and phase of R_v were calculated for a wide range of conditions. While, strictly speaking, R_v is defined as the limit for $z = -z_0$, where $z_0 \to \infty$, it is of interest to first demonstrate how R_v approaches this limit. An example of such calculations is shown in figures 2a and 2b. Here $C = 0.1$, corresponding to highly oblique incidence, and $\lambda_0 = 15$ km, or $f = 20$ kHz. Various values of β are shown on the curves. It is immediately apparent that for sufficiently negative values of z (i.e., sufficiently far below the layer), both the amplitude and phase of R_v approach a limit.

Actually, the curves in figures 2a and 2b have more than just a mathematical interest. Physically, the R_v, corresponding to $z = -z_0$ where z_0 is finite, is the reflection coefficient for a permittivity profile of the form

$$K(z) = K_0\left[1 - \frac{i}{L} e^{\beta z}\right] \quad \text{for} \quad z > -z_0$$

$$= 0 \quad \text{for} \quad z < -z_0. \tag{25}$$

Therefore, these curves describe reflection from a sharply bounded ionosphere which behaves exponentially above its lower edge. Of course, as z_0 tends to ∞, the discontinuous profile becomes a continuous exponential. It is rather interesting to observe that as the discontinuity is moved from above the reference level ($z=0$) to below, the reflection coefficient passes through a minimum. This is related to a Brewster angle phenomenon since to the right of this minimum in figure 2a the reflection is metalliclike while, to the left, the reflection is dielectriclike.

Reflection of VLF Radio Waves

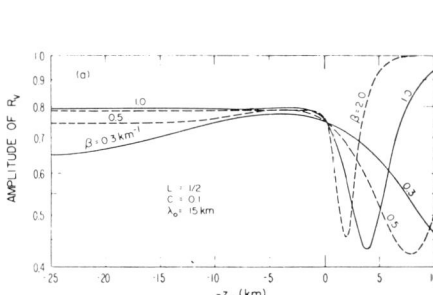 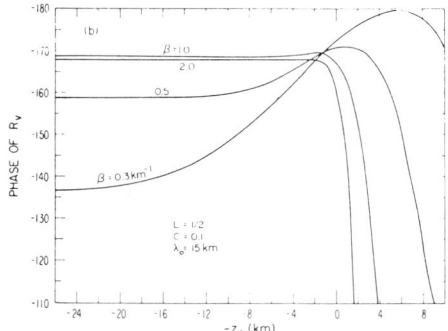

Fig. 2a. The reflection coefficient as a function of height for an exponential profile and a fixed angle of incidence.

Fig. 2b. The reflection coefficient as a function of height for an exponential profile and a fixed angle of incidence.

For the remainder of the present paper, attention will be confined to the limiting value of R_v far below the reference level $z=0$ (i.e., $z \cong -z_0 \cong -\infty$). On the other hand, it should be emphasized that R_v is always referred to the level $z=0$.

The magnitude of the reflection coefficients R_v and R_h are shown plotted in figure 3 for vertical and horizontal polarization, respectively. The values of $|R_v|$ were obtained from (14) using the multislab model described. The values of $|R_h|$ were calculated directly from (9) and, on this log-linear scale they are merely straight lines. As an important check, the value of $|R_h|$ was also obtained from a multislab model using the same procedure as described for vertical polarization. Within four-figure accuracy, the values of $|R_h|$ obtained by the two methods were the same.

A number of significant features are evident in figure 3. In the first place, $|R_v|$ exhibits a Brewster angle phenomenon, provided β is sufficiently large. Here the reflection process is dielectriclike at grazing angles and is more or less metalliclike at normal incidence. Of course, $|R_h|$ does not exhibit this phenomenon. However, for very small values of β, corresponding to a relatively slowly varying medium, the curves for $|R_h|$ and $|R_v|$ become rather close to one another. This is consistent with the optical behavior of waves in an inhomogeneous medium. In fact, by a direct application of the phase integral method of Eckerskey, the equation for $|R_h|$ is found also to be applicable to $|R_v|$. The use of the phase integral method for such applications is necessarily restricted to slowly varying media [Wait, 1962]. It is apparent, that for VLF radio waves, where β is of the order of 0.5, the phase integral method is inapplicable to vertical polarization.

Another interesting feature of figure 3 is the near linear dependence of all the curves for small values of C. Fortunately, it is just these values which are important in the long-distance propagation of VLF radio waves. The linearity of the phase curves for R_v are indicated in figure 4 for the same conditions. Here, it is evident that they all approach $-180°$ at grazing incidence. The phase curves for R_h exhibit a similar property but they are not shown here since they have only an academic interest at VLF.

The variation of the reflection coefficient R_v as a function of the gradient parameter β is illustrated in figures 5a and 5b at oblique incidence. It is rather remarkable that $|R_v|$ is relatively insensitive to β if it is in the range from 0.7 to 3.0. Furthermore, it appears that $|R_v|$ has a broad maximum for β approximately equal to 1.2.

The general behavior of the amplitude and phase of the reflection coefficients at highly oblique incidence suggests that, if R_v is written in the form

$$R_v = -\exp(\alpha C), \qquad (26)$$

the function α should be almost independent of C. Writing

$$\alpha = \alpha_1 + i\alpha_2,$$

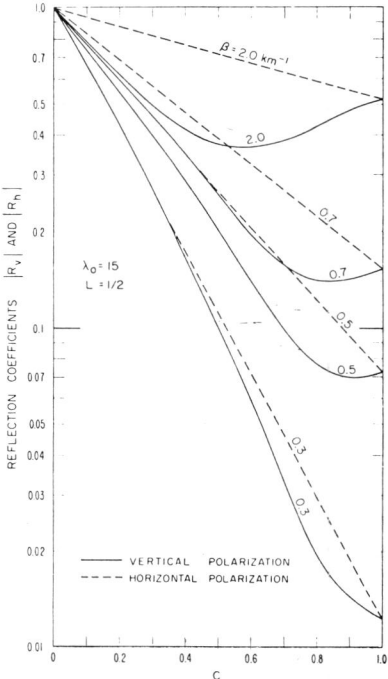

Fig. 3. The amplitude of the reflection coefficients for vertical and horizontal polarization as a function of C which is the cosine of the angle of incidence.

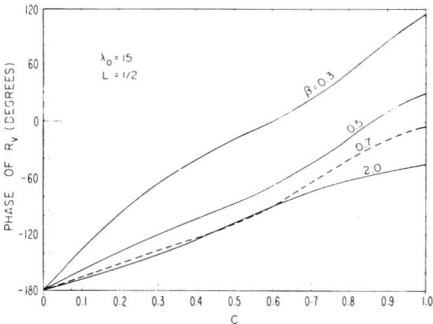

Fig. 4. The phase of the reflection coefficient for vertical polarization as a function of C.

Reflection of VLF Radio Waves

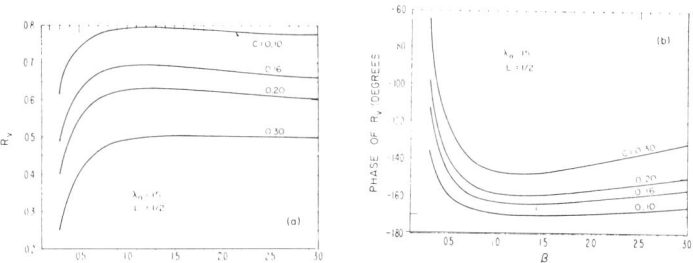

Fig. 5. The reflection coefficient as a function of the gradient parameter β for an exponential profile.

where α_1 and α_2 are real, it is a simple matter to compute the complex coefficient α from the numerical data of R_v. The results are shown in figures 6a and 6b where $(-\alpha_1)$ and α_2, respectively, are plotted as a function of C in the range 0.05 to 0.30. It is apparent that over this range of C, the coefficient α can be regarded essentially as a complex constant.

For all the results given in the foregoing figures, it has been assumed that $L=1/2$ and $\lambda_0=15$. Actually, the results for the magnitude of R_v and R_h do not depend on L. In fact, $|R_v|$ and $|R_h|$ are identical for $\beta\lambda_0=$ constant. Furthermore, the phase of the reflection coefficients is also simply related although the situation is slightly complicated by the choice of the reference level where L is required to have a special value. In fact, it is convenient to choose the reference level (i.e., $z=0$) so that

$$L = \frac{\omega v}{\omega_0^2} = \frac{1}{2} \times \frac{15}{\lambda_0}. \qquad (27)$$

In this way, the results can be readily compared as a function of frequency or wavelength. At the reference level $z=0$, the effective conductivity parameter ω_r or ω_0^2/v has the value 2.51×10^5.

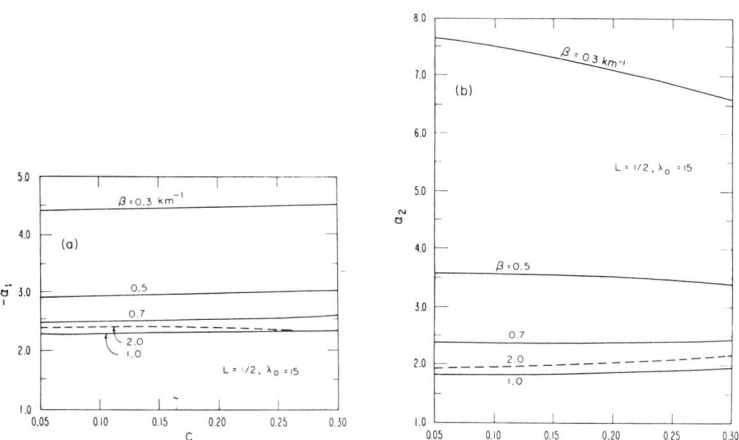

Fig. 6. The real and imaginary parts of the function α defined by $R_v = -\exp(\alpha C)$ for $\lambda_0=15$ and various values of β.

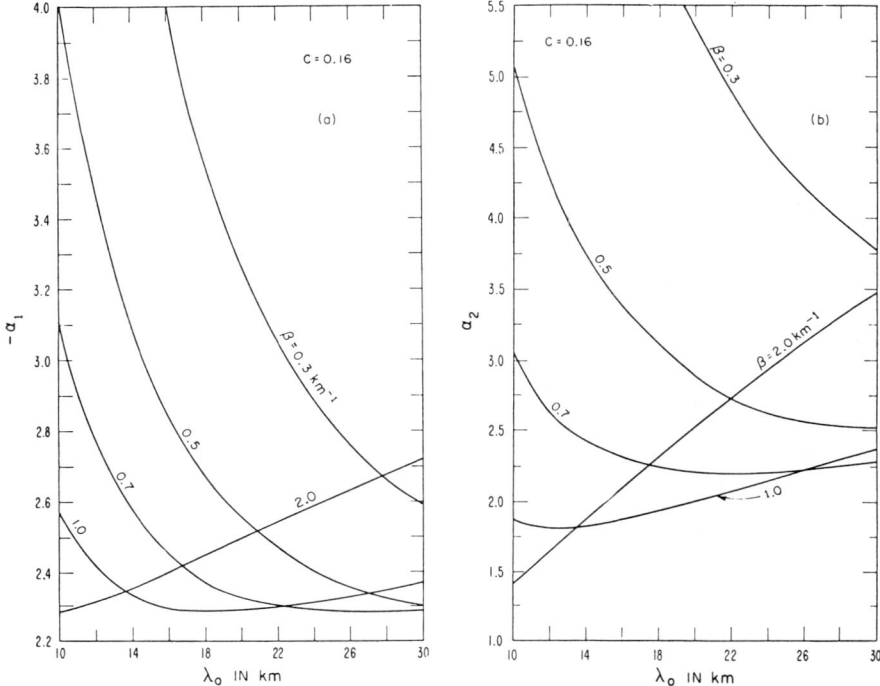

Fig. 7. The real and imaginary parts of the function α plotted as a function of the wavelength λ_0 for $C=0.16$ and various values of β.

For wavelengths other than $\lambda_0 = 15$, it is also found that R_v may be approximated by the function $-\exp(\alpha C)$ where α is approximately a complex constant. To illustrate the wavelength dependence, $-\alpha_1$ and α_2 are plotted as a function of λ_0 in figures 7a and 7b, respectively. For these curves, C is chosen to be 0.16; the corresponding curves for other values of C in the range 0.10 to 0.20 are almost indistinguishable. The ordinate in figure 7a is simply related to the magnitude of the reflection coefficient and, thus, small values of $-\alpha_1$ are associated with high reflection coefficients. It is apparent that for small values of β corresponding to a diffuse layer, the reflection coefficient becomes very small for the shorter wavelength. On the other hand, for a rapidly varying layer, corresponding to large values of β, the reflection coefficient decreases with increasing wavelength. For intermediate values of β, the reflection coefficient has a minimum in the wavelength range between 10 and 30 km. For example, when $\beta = 1.0$, the optimum wavelength is about 17 km or approximately 18 kHz.

The curves for α_2 in figure 7b also have a particular significance. Noting that

$$\arg R_v = -\pi + \alpha_2 C, \tag{28}$$

it is apparent that $\alpha_2 C$ is the phase shift resulting from the imperfect reflecting properties of the exponential layer. To attach a physical meaning to this term it is often desirable to imagine the reflection taking place at a height Δz_1 *below* the level $z=0$. In this case, Δz_1 is chosen so that the arg of R_v is always $-\pi$. Clearly,

$$\alpha_2 = 2k\Delta z_1 = 4\pi\Delta z_1/\lambda_0, \tag{29}$$

or

$$\Delta z_1 = \alpha_2 \lambda_0/(4\pi). \tag{30}$$

For example, at $\lambda_0 = 15$ km (i.e., 20 kHz) and $\beta = 1$, the effective height of reflection is depressed by approximately 3 km. For smaller values of β, it is seen from figure 7b that Δz_1 may be much greater.

The results, for the coefficients α_1 and α_2, given here may be introduced into the waveguide mode theory. In this way, attenuation rates and phase velocities of the modes may be obtained. [Wait, 1962].

Here we have adopted a numerical method to deduce the complex reflection coefficient for an unperturbed exponential profile. Under some cases the results can be obtained by using a phase integral approach. After this research was completed Booker, Fejer and Lee (1968) uncovered an interesting theorem that permitted the reflection coefficient for vertical polarization to be obtained from the results for horizontal polarization viz a complex height transformation. The method was also used by Booker and Crain (1968) to obtain a simple method to estimate reflection loss at loss frequencies.

PART II   PERTURBED EXPONENTIAL MODEL

Introduction

In Part I, referred to as (I), oblique reflection of (VLF) radio waves from a continuously stratified ionized medium was considered. The profile of the effective conductivity was taken to be exponential in form. Actually, this is a fairly good representation of the actual D layer of the ionosphere under daytime conditions.

It is the purpose of the present paper to consider profiles which are no longer exponential in form. Since the objective is to gain insight into the mechanism of reflection from perturbed layers, a number of idealizations are made. First, it is assumed, under quiescent conditions, that the ionospheric conductivity varies exponentially with height. Then the idealized perturbation is assumed to have a Gaussian form. Again, for sake of simplicity, the earth's magnetic field is neglected as in (I). This is well justified when considering effects which result from ionization in the lowest ionosphere.

Description of the Profile

The notation follows that used in (I) as closely as possible. Thus the undisturbed profile, as a function of height z, is defined by the conductivity parameter $1/L(z)$ where

$$\frac{1}{L(z)} = \frac{1}{L}\exp(\beta z), \tag{31}$$

and L is a constant, β is a gradient parameter and z is the height above the reference level $z=0$. Under the isotropic assumption, it is known that [Wait, 1962]

$$L = \frac{\omega(\nu + i\omega)}{\omega_0^2}, \tag{32}$$

in terms of the angular frequency ω, collision frequency ν, and plasma frequency ω_0. At VLF, $\nu \gg \omega$, and therefore

$$L \simeq \frac{\omega}{\omega_r} \quad \text{where } \omega_r = \frac{\omega_0^2}{\nu}, \tag{33}$$

to within a very good approximation.

In general, it is seen that $1/L(z)$ is proportional to $N(z)/\nu(z)$ where $N(z)$ and $\nu(z)$ are the electron density and collision frequency regarded as a function of height. The constant β, in the exponent, is a measure of the sharpness of the gradient. For example, $\beta=1$ km^{-1} means that the ratio of $\omega_r(z)$ or $N(z)/\nu(z)$ increases by 2.71 for each km of vertical height. From the recent work of Barrington et al. [1962], Kane [1962], and Belrose [1963], it appears that β for an undisturbed ionosphere may be in the range from 0.2 to 0.8. If the level from about 60 km to 70 km is considered, it appears that $\beta=0.3$ typifies many of these daytime D-layer profiles. A detailed study of the influence of changing β is to be found in (I).

Having specified our undisturbed profile, we now wish to introduce the perturbation. It is assumed that the collision frequency profile is unchanged whereas the ionization is to be increased by an amount $\Delta N(z)$ where

$$\Delta N(z) = \Delta N_0 \exp\left[-\left(\frac{z-F}{D}\right)^2\right], \tag{34}$$

and ΔN_0, F and D are constants. Clearly, the maximum value of $\Delta N(z)$ is ΔN_0 which is located at $z=F$. Furthermore, the thickness of this layer is $2D$ which is the vertical distance between the levels where $\Delta N(z)$ drops to $\Delta N_0/e$.

In order to estimate correctly the influence of this Gaussian shaped layer, it is necessary to assume something about the collision frequency profile. A careful study of the recent literature indicates that an exponential variation of $\nu(z)$ with height z is not unreasonable. The form chosen here is

$$\nu(z) = \nu_0 \exp\left(-\frac{\beta}{2}z\right), \tag{35}$$

where $\beta = 0.3$ km^{-1}. Therefore, the resulting conductivity perturbation has the form

$$\frac{\Delta N(z)}{\nu(z)} = \frac{\Delta N_0}{\nu_0} \exp\left(\frac{\beta}{2}z\right) \exp\left[-\left(\frac{z-F}{D}\right)^2\right]. \tag{36}$$

The complete profile, under these idealized conditions, is given by

$$\frac{1}{L(z)} = \frac{1}{L(0)}\left[\exp(\beta z) + A \exp\left(\frac{\beta}{2}z\right) \exp\left[-\left(\frac{z-F}{D}\right)^2\right]\right], \tag{37}$$

where the right-hand side is proportional to the effective conductivity of the medium as a function of height above (or below) the reference level at $z=0$. The coefficient A defines the strength of the perturbation. In fact,

$$A = \frac{\Delta N_0}{N_0},$$

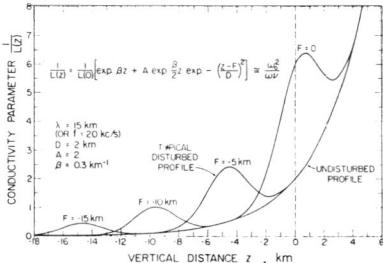

Fig. 8. The undisturbed and disturbed conductivity profiles.

Reflection of VLF Radio Waves

where N_0 is the electron density of the undisturbed profile at the reference level $z=0$. As in (I), $L(0)=7.5/\lambda$ where λ is the wavelength in kilometers.

It is admitted that other ways to define a pertubation in the profile may be preferable. Here the electron density anomaly, for a given value of A, does not change with its vertical location F. Consequently, we may anticipate that the influence of this type of perturbation will be diminished at sufficiently low heights because of the increasing collision frequency. However, we shall see the problem is not quite this simple as other factors come into play.

A sketch of the profiles used is given in figure 8. The undisturbed profile is the exponential form, while the disturbed profiles have the superimposed Gaussian "bump." The location of the "bump" for five typical profiles is specified by the appropriate value of F.

Results of the Calculations

The method used to calculate the reflection coefficient R has been described in detail in (I). The quantities considered are the amplitude $|R|$ and the phase of R for a vertically polarized plane wave incident at an angle whose cosine is C. The reflection coefficient is evaluated in the free space region corresponding to $z \to -\infty$. However, it is important to remember that the *phase is referred to the level $z=0$*.

The plan of the calculations is to vary the value of one parameter while keeping the others constant. To obtain a complete understanding of the various phenomena, an enormous number of calculations is needed. In order to keep the problem within reasonable bounds and to reduce the expense of the computation, only a limited number of cases was considered. These results are shown in graphical form in figures 9 to 13. In all cases $\beta = 0.3$ km^{-1}.

In figure 9a the amplitude of the reflection coefficient is plotted as a function of F for $\lambda=15$ km ($f=20$ kHz), $A=2$, $D=2$ km, and C values varying from 0.05 to 0.4. Small values of C here correspond to angles

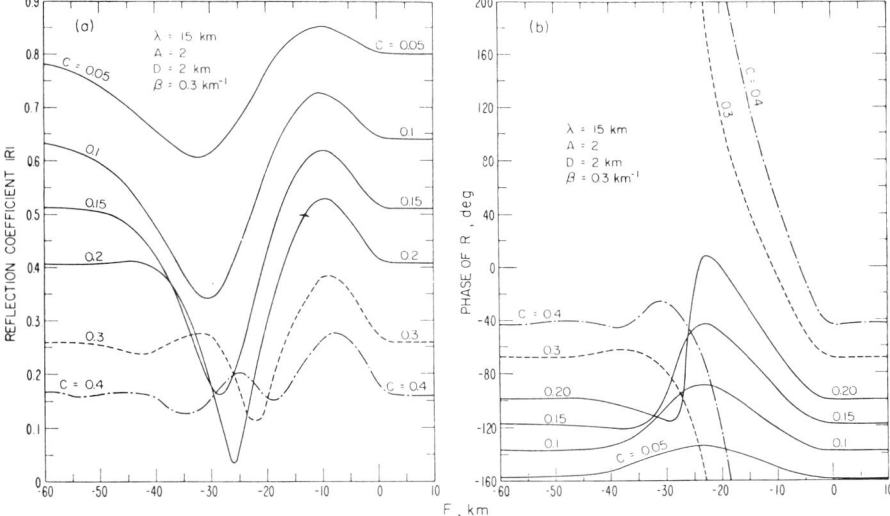

Figs. 9a and b. The reflection coefficient as a function of the vertical location, F, of the Gaussian perturbation, for various angles of incidence.

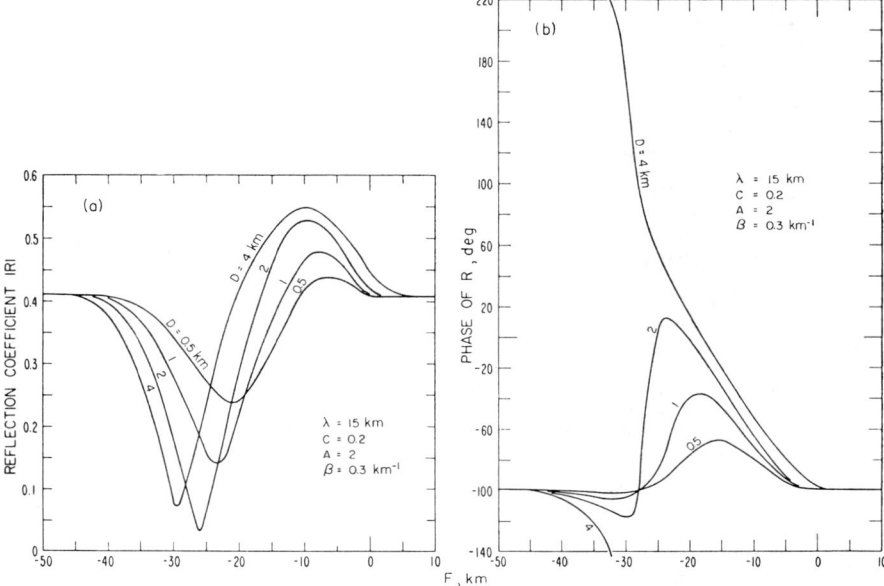

Figs. 10a and b. The reflection coefficient, as a function of F, for various widths of the Gaussian perturbation when $C=0.2$ (i.e., angle of incidence is 78°).

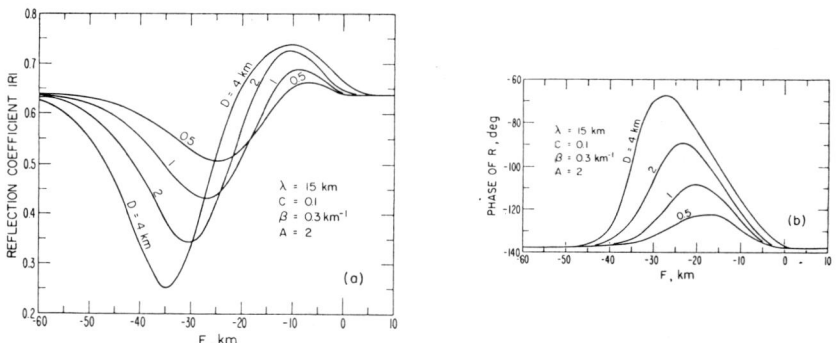

Figs. 11a and b. The reflection coefficient, as a function of F, for various widths of the Gaussian perturbation when $C=0.1$ (i.e., angle of incidence is 84°).

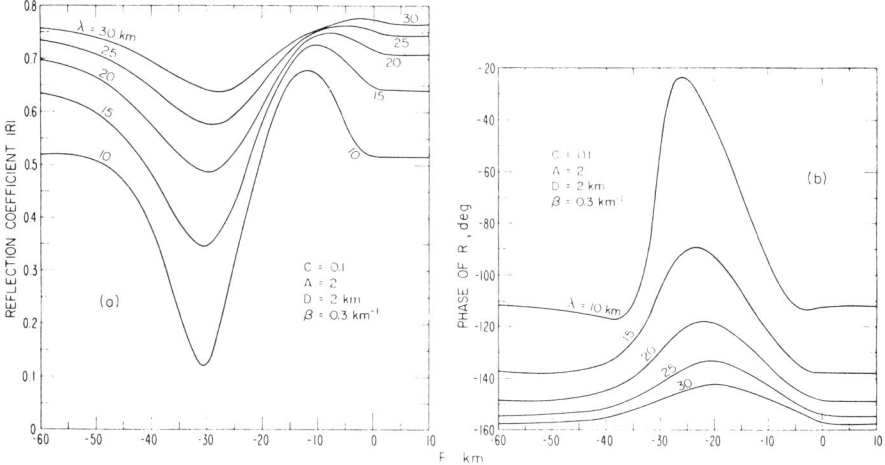

Figs. 12a and b. The reflection coefficient, as a function of F, for various wavelengths (from 10 to 30 km).

near grazing. For long distance propagation of VLF radio waves, values of C near 0.1 are most important. For this case, it is interesting to note, when F is near or above zero, that $|R|$ takes the same value as for the undisturbed profile. As the "bump" or perturbed layer is lowered, the reflection coefficient first increases then decreases. Eventually, as the "bump" is brought down to very low heights, $|R|$ returns to its undisturbed value. The other curves for highly oblique incidence have a similar behavior. Thus, the "bump" may either improve or degrade the reflection. Presumably, at the lower heights the Gaussian layer is acting as an absorber

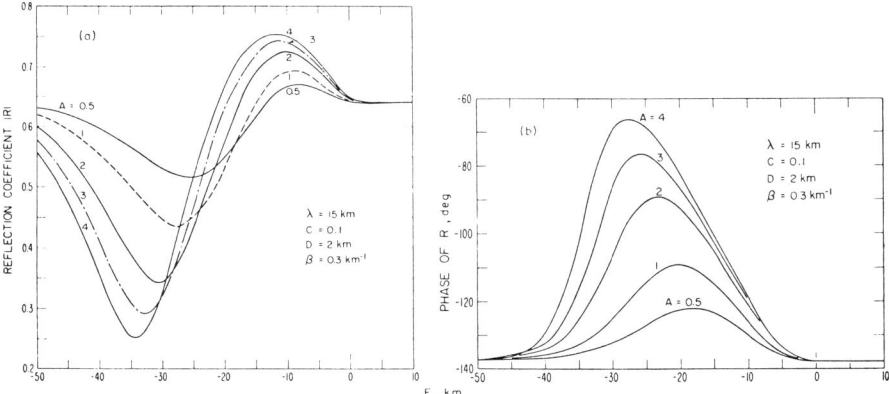

Figs. 13a and b. The reflection coefficient, as a function of F, for various amplitudes A of the perturbation.

whereas, at greater heights, it enhances the reflection. At the steeper angles of incidence, the situation becomes more complicated. It is probable that this results from interference between multiple reflected rays between the upper side of the "bump" and the exponentially varying layer. Such an interference phenomenon becomes more pronounced at steeper incidence because the vertical component of the wavelength is becoming comparable with typical values of F.

The phase of R is shown in figure 9b for the same conditions as in figure 9a. Again, it is apparent that, when the Gaussian "bump" is such that F is near 0 or above, the phase of R attains its undisturbed value. When the angle of incidence is highly oblique the phase undergoes an increase (i.e., decrease of lag) as the "bump" comes down to lower heights. Sufficiently far below the reference level, the phase of R returns to its undisturbed value. It is well to note that as C becomes small (i.e., approaching grazing incidence), the phase of R is approaching $-180°$.

For highly oblique incidence the influence of the "bump" is to lower the effective height of reflection for the whole range of F. However, a very interesting phenomenon occurs at steeper incidence. As can be seen in figure 9b, when $C=0.2$ the phase undergoes a rather rapid change as F varies from about -2 km to -27 km. As C is increased further there is an apparent discontinuity when the phase changes by 360°. Such a change of 2π radians is quite permissible since the ordinate is arbitrary to within any integral number of 2π radians. Thus, the phase curves for $C=0.3$ and 0.4 could have been drawn in the range below $-160°$.

The curves in figure 9b, even if they show nothing else, demonstrate that phase shifts in reflection phenomena may have some unusual cycle ambiguities.

The influence of the width of the Gaussian perturbation or "bump" is shown in figure 10a for the amplitude $|R|$ and in figure 10b for the phase of R. Here $A=2$, $\lambda=15$ km, and $C=0.2$. The amplitude curves show that, when D is increased, the overall influence of the layer becomes somewhat greater. There is some tendency for the thinner layers (i.e., smaller D) to be more effective at greater heights. The corresponding phase curves show that the thicker layers always produce a larger phase change. Furthermore, as D exceeds 2 km, a point is reached where the "360° jump" takes place.

The curves in figures 11a and b are for the same conditions as figures 10a and b except that now $C=0.1$, corresponding to nearer grazing incidence. The amplitude curves have a very similar shape. The phase curves are also similar except that the "360° jump" is no longer present.

The wavelength dependence of the reflection coefficient is shown in figures 12a and b. For these $C=0.1$, $A=2$, and $D=2$. The wavelengths chosen (10, 15, 20, 25, 30 km) correspond to frequencies of 30, 20, 15, 12, and 10 kHz. Qualitatively, the curves have a very similar shape. There is some tendency for the shorter wavelengths to be accompanied by more pronounced changes. In all cases the "bump" acts as an absorber at low heights while it enhances the reflection at greater heights.

Finally, in figures 13a and b, the influence of A, the relative magnitude of the anomalous electron density, is shown. As expected, the individual curves are similar in shape with the larger values of A corresponding to an increased change over the undisturbed values. It is important to note that the phase anomaly is almost directly proportional to A.

PART III EXPONENTIAL MODEL WITH HYPERBOLIC TRANSITION

Introduction

In Parts I and II oblique reflection of (VLF) radio waves from continuously stratified ionized media was considered. In (I), the profile of the effective conductivity was taken to be exponential in form. It was pointed out that an isotropic exponential model is a fairly good representation of the D layer of the ionosphere under quiet DAYTIME conditions. In (II), an idealized perturbed exponential model was considered. The perturbation consists of a localized increase of electron density which itself has a Gaussian profile. The reflection coefficient was shown to be influenced by the vertical location of this Gaussian perturbation. In the present Part III the exponential profile model introduced in (I), is modified by allowing the electron density to be increased for all heights above a specified level. Such a modification to the quiescent ionosphere could result from ionizing radiations associated with a solar flare [Jean and Crary, 1962; Pierce, 1963; Chilton et al., 1963].

Reflection of VLF Radio Waves

For the detailed numerical results given here, it is assumed that the earth's magnetic field may be neglected. This is justified for highly oblique incidence at VLF provided that attention is restricted to effects which result from ionization in lowest daytime ionosphere [Johler, 1962, Galejs 1972].

Description of the Profile

The notation follows that used in (I) as closely as possible. Thus, the undisturbed profile, as a function of height z, is defined by the conductivity parameter $1/L(z)$ where

$$1/L(z) = (1/L)\exp(\beta z), \tag{38}$$

Here L is a constant, β is a gradient parameter, and z is the height above the reference level. Under the isotropic assumption, it is known that Wait, 1962]

$$L = \frac{\omega(\nu + i\omega)}{\omega_0^2}, \tag{39}$$

in terms of the angular frequency ω, collision frequency ν, and the plasma frequency ω_0. Furthermore, at VLF, $\nu \gg \omega$ which leads to

$$L \simeq \omega/\omega_r, \text{ where } \omega_r = \omega_0^2/\nu. \tag{40}$$

In general, the conductivity parameter is proportional to $N(z)/\nu(z)$ where $N(z)$ and $\nu(z)$ are the electron density and collision frequency regarded as a function of height. The constant β, in the exponent, is a measure of the sharpness of the gradient. As in (II), it will be assumed that $\beta = 0.3$ km^{-1} typifies an undisturbed D-layer profile.

We shall now turn our attention to the modification of the exponential profile. As in (II), it is assumed that the collision frequency profile is unchanged whereas the ionization is to be increased by an amount $\Delta N(z)$ where

$$\Delta N(z) = \Delta N_0 \left[\tanh\left(\frac{z-F}{D}\right) + 1 \right], \tag{41}$$

and ΔN_0, F, and D are constants. It is evident that $\Delta N(z)$ is equal to ΔN_0 at $z = F$ which may be described as the "lower edge" of the modified layer. When z is somewhat less than F it is seen that $\Delta N(z)$ is zero whereas when z is much greater than F, $\Delta N(z)$ becomes equal to $2\Delta N_0$. It is clear that the vertical location of the "lower edge" is governed by the value of F. The rapidity of the transition from 0 to $2\Delta N_0$ is controlled by the magnitude of D; the quantity $2D$ could be described as the "transition thickness."

Following the discussion given in (I), the collision frequency profile is taken to be

$$\nu(z) = \nu_0 \exp(-\beta z/2), \tag{42}$$

where $\beta = 0.30$ km^{-1}. The resulting conductivity perturbation has the form

$$\frac{\Delta N(z)}{\nu(z)} = \frac{\Delta N_0}{\nu_0} \exp(\beta z/2) \left[1 + \tanh\left(\frac{z-F}{D}\right) \right]. \tag{43}$$

The complete conductivity profile is now given by

$$\frac{1}{L(z)} = \frac{1}{L(0)} \left\{ \exp(\beta z) + A \exp\left(\frac{\beta}{2} z\right) \left[1 + \tanh\left(\frac{z-F}{D}\right) \right] \right\}, \tag{44}$$

where the right-hand side is proportional to the effective conductivity of the medium as a function of height

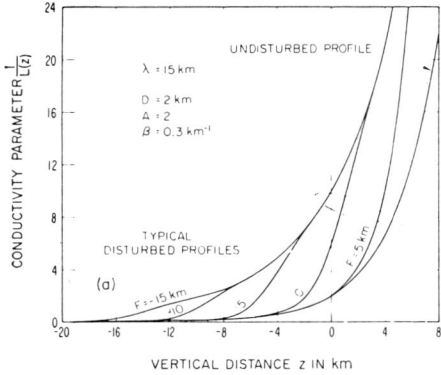
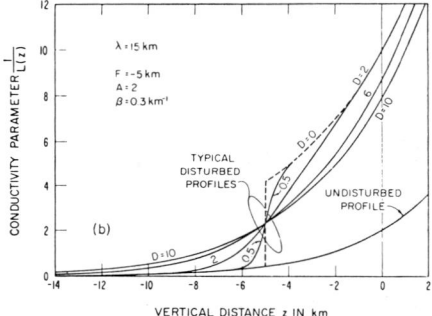
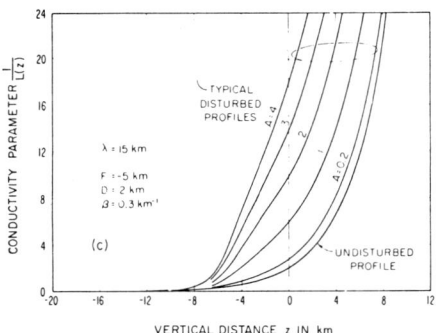

Figs. 14a–c. The profile of the conductivity of the model illustrating the influence of the various parameters.

Reflection of VLF Radio Waves

above (or below) the reference level at $z=0$. The coefficient A defines the strength or magnitude of the perturbation. The constant $L(0)$ is chosen to be equal to $7.5/\lambda$ where λ is the wavelength in kilometers.

To enable the reader to grasp the significance of the various possible forms of profiles, some typical examples are illustrated in figures 14a, b, and c. In figure 14a the vertical location of the "lower edge" of the perturbed layer is varied while other factors are kept constant. In 14b the abruptness of the transition at the "lower edge" is varied for fixed values of F and A. Finally, in 14c the strength of the perturbed layer is varied while keeping F and D constant.

3. Results of the Calculations

The method used in the reflection coefficient calculations is the same as that used in (I) and (II), so this aspect of the subject need not be discussed. As before, the amplitude $|R|$ and the phase of R for a vertically polarized wave are considered. The reflection coefficient is evaluated in the free-space region corresponding to $z \to -\infty$; however, it is to be remembered that the phase is referred to the level $z=0$.

In figure 15a the amplitude of the reflection coefficient is plotted as a function of F for $\lambda = 15$ km (i.e., $f = 20$ kH), $A = 2$, $D = 2$ km, and C values varying from 0.05 to 0.4. Small values of C correspond to angles near grazing; these are important for long distance propagation of VLF radio waves. It is evident that when F is somewhat above zero, $|R|$ assumes the value for an unperturbed exponential layer. As the "lower edge" of the perturbed layer moves down (i.e., F becoming negative), the tendency is for $|R|$ to increase somewhat. As the perturbed layer is moved farther down (i.e., becoming increasingly negative), $|R|$ tends to be reduced. The behavior illustrated here is not dissimilar to the corresponding case of a Gaussian "bump," described in (II), as it moves down to low heights. In both cases, the perturbation enhances the reflection at greater heights but degrades it at lower heights.

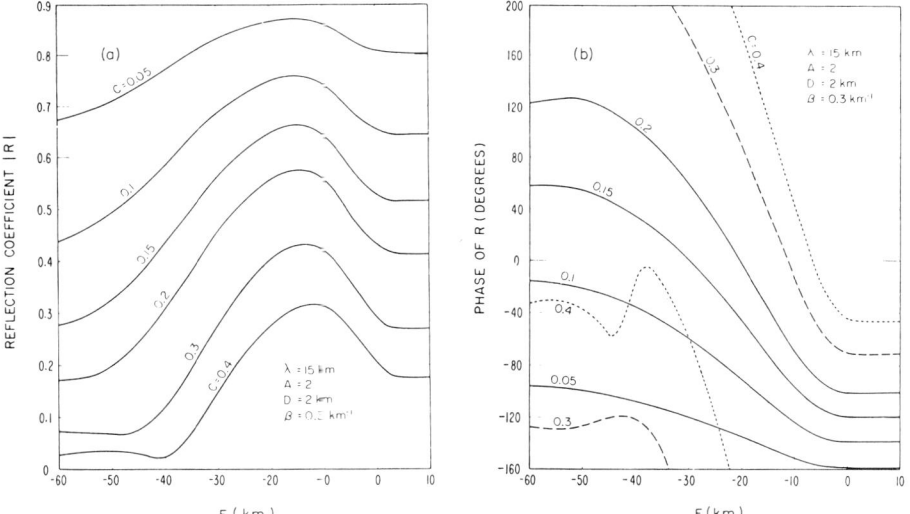

Figs. 15a and b. Amplitude and phase of the reflection coefficient for various values of C, the cosine of the angle of incidence. [F is the vertical distance, relative to the reference height, to the "lower edge" of the perturbed layer.]

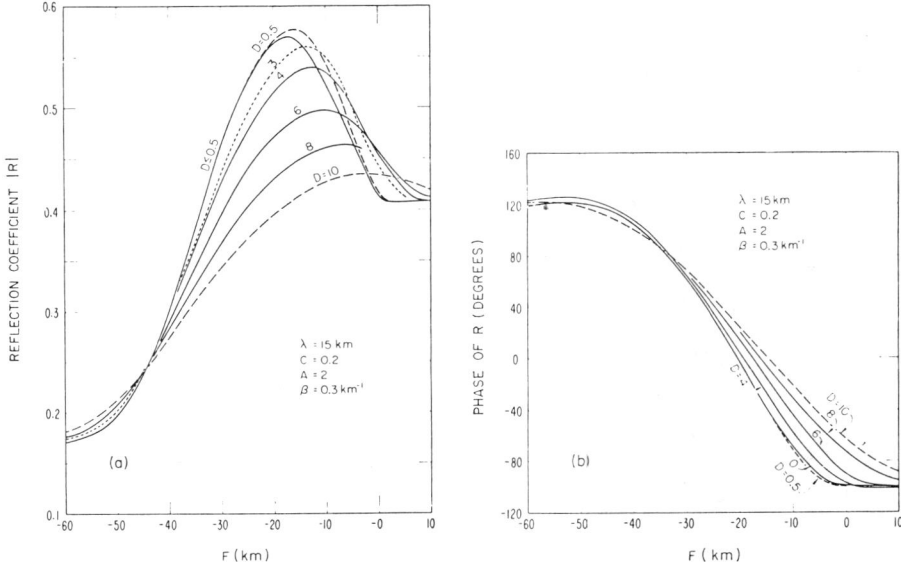

Figs. 16a and b. Amplitude and phase of the reflection coefficient for various values of D, the "half-thickness" of the layer transition for $C=0.2$.

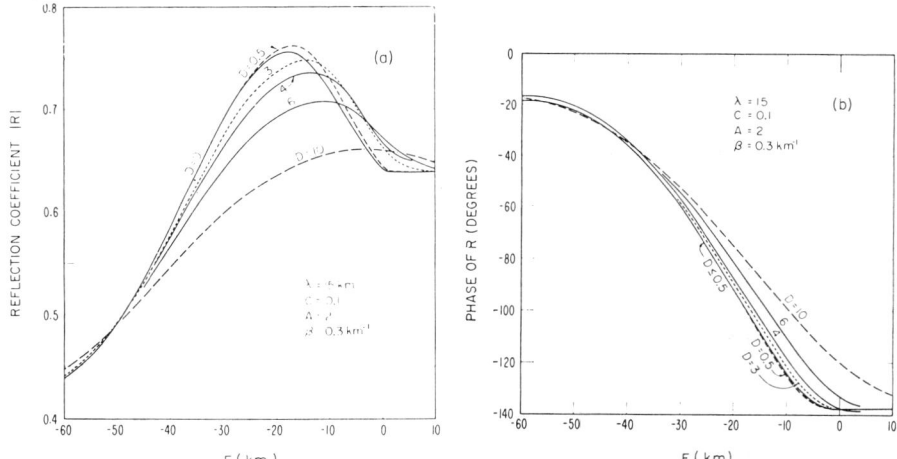

Figs. 17a and b. Amplitude and phase of the reflection coefficient for various values of D, the "half-thickness" of the layer transition for $C=0.1$.

Reflection of VLF Radio Waves

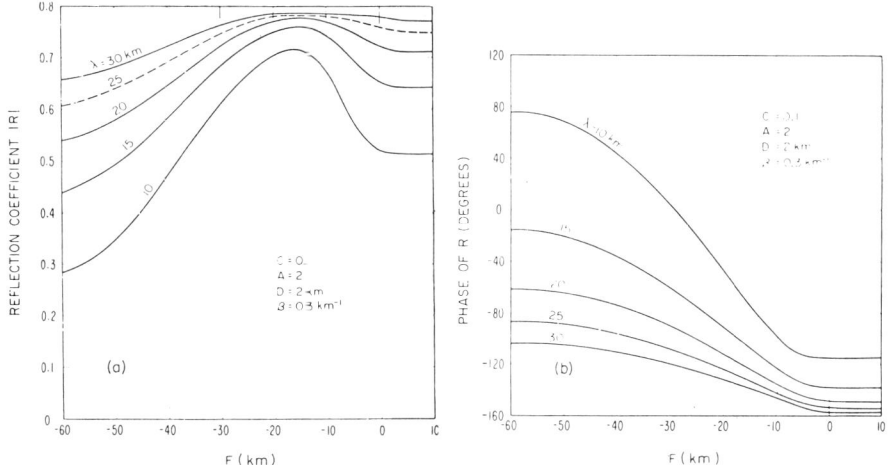

Figs. 18a and b. Amplitude and phase of the reflection coefficient for various values of λ, the wavelength.

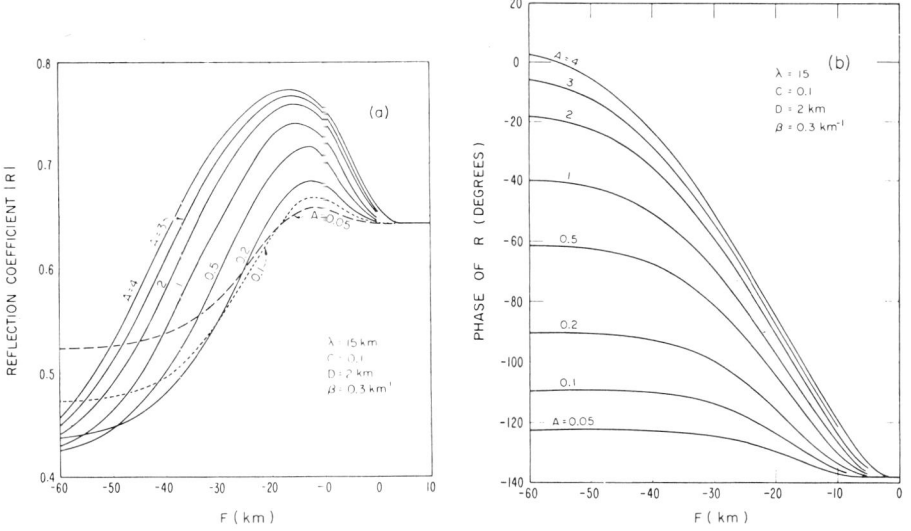

Figs. 19a and b. Amplitude and phase of the reflection coefficient for various values of A, the magnitude of the perturbation.

The phase of R is shown in figure 2b for the same conditions as in figure 2a. It is apparent that when the "lower edge" is somewhat above $F=0$, the phase attains its undisturbed value. As F becomes negative the phase increases and, for grazing angles, this increase is monotonic in nature. However, for the steeper angles of incidence there is some evidence of oscillations, which are presumably due to certain interference phenomena.

The dependence of the "transition thickness" on the reflection coefficient is shown in figures 13a, and 3b for $C=0.2$. These show, as one might expect, that a sharp transition is associated usually with an increased reflection coefficient. As the transition is stretched out, the tendency is for the maximum at $F \cong -15$ km to disappear. It is particularly interesting to note that the phase is not overly sensitive to changes of D.

The influence of varying D for a highly grazing situation is shown in figures 17a and b. These show behavior which is qualitatively the same as figures 16a and 16b.

The wavelength dependence of the reflection coefficient is shown in figures 18a and b for $\lambda = 10, 15, 20, 25$, and 30 km corresponding to frequencies of 30, 20, 15, 12, and 10 kHz. As expected, the curves have similar shapes to each other and the tendency is for the shorter wavelengths to be associated with more pronounced changes.

Finally, in figures 19a and b the influence of varying the strength A of the perturbed layer is indicated. Generally, as A is increased, the changes are more pronounced. It is interesting to note that at the very low heights the magnitude of the reflection coefficient is very sensitive to small changes of A. This behavior is similar to that found in the troposphere when considering the reflection of VHF radio waves from gradient changes of the refractive index [Wait, 1962].

References

BARRINGTON, R., LANDMARK, B., HOLT, O., and THRANE, E., (1962), Experimental studies of the ionospheric D-region, Report No. 44, Norwegian Defence Research Establishment, Kjeller, Norway.

BELROSE, J.S., (1963), The oblique reflection of low-frequency radio waves from the ionosphere, *AGARDograph*, 74, 149–165 (Pergamon Press).

BELROSE, J.S., (1963), Electron density measurements in the D region by the method of partial reflection, *Proc. of NRDE Conference, Electron Density Profiles*, Pergamon Press, Oxford, London.

BOOKER, H.G., and CRAIN, C.M., (1968), Simple methods for calculating L.F. and VLF reflection loss in the disturbed lower ionosphere, *Radio Science*, 3, No. 8, 775–781.

BOOKER, H.G., FEJER, J.G., and LEE, F.L., (1968), A theorem concerning reflection from a plane stratified medium, *Radio Science*, 3, No. 3, 207–212.

BOOKER, H.G., and LEFEUVRE, (1977), The relation between ionosphere profiles and E.L.F. propagation in the earth ionosphere transmission line, *Jour. Atmos. and Terr. Phys.*, 39, 1277–1292.

BUDDEN, K.G., (1961), *Radio Waves in the Ionosphere*, Cambridge University Press.

BURMAN, R., (1966), The reflection of radio waves from stratified perturbed ionospheric profiles, *Trans. IEEE*, AP-14, 661–662.

CHILTON, C.J., STEELE, F.K., and NORTON, R.B., (1963), VLF phase observations of solar flare-produced ionization in the D region of the ionosphere, *J. Geophys. Res.*

FIELD, E.C., and ENGEL, R.D., (1965), The detection of daytime nuclear bursts below 150 km by prompt VLF phase anomalies, *Proc. IEEE*, 2009–2019.

GALEJS, J., (1972), *Terrestrial Propagation of Long Electromagnetic Waves*, Pergamon Press, Oxford.

HEADING, J., (1972), Generalisation of Booker's Theorem, *Quart. Jour. Mech. and Appl. Math.*, 25, 207–224.

JEAN, A.G., and CRARY, J.H., (1962), VLF phase observations on the ionospheric effects of the solar flare of Sept. 28, 1969, *J. Geophys. Res.*, 67, 4903.

JOHLER, J.R., and HARPER, J.D., Jr., (1962), Reflection and transmission of radio waves at a continuously stratified plasma with arbitrary magnetic induction, *J. Res. NBS*, 66D, (Radio Prop.) 88–99.

KANE, J.A., (1962), Re-evaluation of ionospheric electron densities and collision frequencies derived from rocket measurements, Chapter 29 in *Radio Wave Absorption in the Ionosphere*, Pergamon Press, Oxford.

LEDINEGG, E., PAPOUSEK, E., and SCHNIZER, B., (1979), Fictitious surface impedance along the lower boundary of the ionosphere for an arbitrary geomagnetic field, *Archiv für Elect. und Übertragung Tech.*, 33, 278–284.

LEDINEGG, E., PAPOUSEK, E., and SCHNIZER, B., (1979), Electromagnetic wave propagation in a plasma with a static magnetic field of arbitrary direction, *Archiv für Elec. und Übertragung Tech.*, 33, 278–284.

PAPPERT, R.A., and MOLER, W.F., (1974), Propagation theory and calculations at extremely low frequencies, *IEEE Trans.*, COM-22, 438–451.

PIERCE, E.T., (1963), Stanford Research Institute, private communications.
WAIT, J.R., (1962), *Electromagnetic Waves in Stratified Media*, Pergamon Press, Oxford-New York, see Chap. IV.
WAIT, J.R., (1963), Influence of the lower atmosphere on propagation of VLF waves to great distances, *J. Res. NBS*, *67D* (Radio Prop.), No. 4, 375–381.
WAIT, J.R., and WALTERS, L.C., (1963a), Reflection of VLF radio waves from an inhomogeneous ionosphere. Part I. Exponentially varying isotropic model, *J. Res. NBS*, *67D* (Radio Prop.), No. 3, 361–367.
WAIT, J.R., and WALTERS, L.C., (1963b), Reflection of VLF radio waves from an inhomogeneous ionosphere. Part II. Perturbed exponential model, *J. Res. NBS*, *67D* (Radio Prop.), No. 5, 509–523.
WAIT, J.R., and WALTERS, L.C., (1963c), Reflection of VLF radio waves from an inhomogeneous ionosphere, Part III, Exponential model with hyperbolic transition, *J. Res. NBS.*, *67D*, (Radio Prop.), No. 6, 747–752.

19
Reflection from a Lossy Magnetoplasma Half-Space

Introduction

The lower ionosphere is primarily responsible for the propagation of VLF radio waves to great distances. In theoretical treatments of this problem it is often assumed that the lower edge of the ionosphere may be represented by a sharply bounded and homogeneous ionized medium. Actually, such a model was used by G. N. Watson [1919] over 40 years ago. Applications and refinements of such a model have been discussed frequently in the recent literature [e.g., Budden, 1962; Wait, 1962; Johler, 1963]. One such refinement is the inclusion of the earth's magnetic field in the analysis. This, of course, renders the medium anisotropic. If the vertical inhomogeneity (or horizontal stratification) of the ionosphere is also considered simultaneously, the situation becomes very complicated indeed. An extensive study of analytical methods to treat such problems has been carried out by K. G. Budden and his colleagues at Cambridge University. Much of this work is summarized in a monumental text [Budden, 1961] which will be the standard work on the subject for some time. Budden makes extensive use of "full wave" methods which may be described as a frontal assault on the differential equations satisfied by the field components in the medium.

In this note we shall consider a special case of a horizontally stratified and anisotropic ionosphere. Specifically, the earth's magnetic field is assumed to be purely transverse to the direction of propagation. Strictly speaking, this is applicable only to the situation when the path of propagation is along the magnetic equator. However, the characteristics in this special case prevail at other latitudes if the transverse component of the field is appreciable. At least this is borne out by a numerical study of the sharply bounded ionosphere for an arbitrary magnetic dip angle [Johler, 1961]. In any case, the resulting simplicity of the differential equations for the limiting case of a purely transverse magnetic field encourages one to consider this situation in more detail. In particular, it is desirable to investigate the influence of gradient of both the electron density and collision frequency. In much of the previous work on this subject the collision frequency has been assumed constant.

On examining the literature [e.g. Wait, 1962] it is found that both the electron density $N(z)$ and the collision frequency $\nu(z)$ vary approximately in an exponential manner with height z. For example, in the undisturbed daytime ionosphere we may assume that

$$N(z) = N_0 \exp(bz), \qquad (1)$$

and

$$\nu(z) = \nu_0 \exp(-az), \qquad (2)$$

where a and b are positive constants and z is some specified level in the ionosphere. From a study of the experimental data [Belrose, 1963], it appears that, if the reference level is 70 km above the earth's surface, $N_0 \sim 10^2$ electrons/cm^3 and $\nu_0 \sim 10^7$ sec^{-1}. The gradient parameters are then expected to be given approximately by $b \cong 0.15$ km^{-1} (± 0.1) and $a \sim 0.15$ km^{-1} (± 0.02). The quoted values of these constants must be considered tentative and certainly subject to change. Furthermore, it must be understood that significant departures from the exponential shape are to be expected under disturbed conditions.

---

*Reprinted, in part, from J.R. Wait and L.C. Walters, *Radio Science, J. Res. NBS*, 68D, 95–101, 1964.

Reflections from a Lossy Magnetoplasma Half-Space

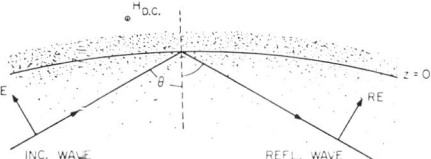

Fig. 1a. Illustrating oblique reflection of VLF radio waves from an inhomogeneous (horizontally stratified) ionosphere with a transverse d-c magnetic field.

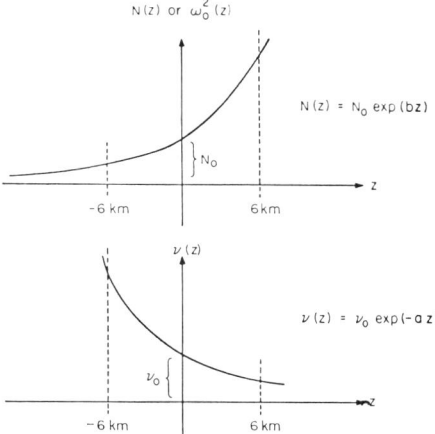

Fig. 1b. Sketch of the profiles of electron density $N(z)$ and the collision frequency $\nu(z)$.

Formulation

The situation is shown explicitly in figure 1a. A vertically polarized plane wave is incident at angle θ on to a horizontally stratified ionosphere. The z axis is taken to be positive in the upward direction. At the reference level $z=0$, the electron density and the collision frequency have values designated by N_0 and ν_0, respectively. The "scale height" which is equal to $1/b$, or $1/a$, as indicated in figure 1b is of the order of 6 km for both of these profiles. As mentioned above, these are typical of the daytime D layer for both the N and ν profiles.

The lower ionosphere, which is idealized here as a stratified ionized medium, may be regarded as an electron plasma. The (angular) electron plasma frequency ω_0 is thus given by

$$\omega_0^2 = 3.18 \times 10^9 \times N, \tag{3}$$

where N is the electron density in electrons per cubic centimeter and ω_0 has dimensions of radians per second. The dielectric properties of such a (cold) plasma may be described in terms of a tensor dielectric constant (ε). Thus, the displacement vector $\vec{D}$ and the electric field $\vec{E}$ are connected by

$$\vec{D} = (\varepsilon)\vec{E}. \tag{4}$$

Choosing the d-c magnetic field to be along the axial direction,[1] the tensor has the form

$$(\varepsilon) = \begin{pmatrix} \varepsilon' & -iq & 0 \\ iq & \varepsilon' & 0 \\ 0 & 0 & \varepsilon'' \end{pmatrix}, \tag{5}$$

where harmonic time dependence according to $\exp(+i\omega t)$ is assumed. The coefficients ε', ε'', and q are given by [Wait, 1962]

$$\frac{\varepsilon'}{\varepsilon_0} = 1 - \frac{i(\nu + i\omega)\omega_0^2/\omega}{\omega_T^2 + (\nu + i\omega)^2}, \tag{6}$$

$$\frac{q}{\varepsilon_0} = -\frac{\omega_T \omega_0^2/\omega}{\omega_T^2 + (\nu + i\omega)^2}, \tag{7}$$

$$\frac{\varepsilon''}{\varepsilon_0} = 1 - \frac{i\omega_0^2}{(\nu + i\omega)\omega}, \tag{8}$$

where $\varepsilon_0 = 8.85 \times 10^{-12}$ is the dielectric constant of free space and ω_T is the (angular) gyro frequency.
Maxwell's equations, which are applicable to this situation, are

$$i(\varepsilon)\omega \vec{E} = \text{curl } \vec{H}, \tag{9}$$

$$-i\mu\omega \vec{H} = \text{curl } \vec{E}, \tag{10}$$

where μ is the magnetic permeability of the plasma. The first equation above may be written

$$i\omega \vec{E} = (\varepsilon^{-1}) \text{ curl } \vec{H}, \tag{11}$$

where

$$\varepsilon_0(\varepsilon^{-1}) = \begin{pmatrix} M & -iK & 0 \\ iK & M & 0 \\ 0 & 0 & \varepsilon_0/\varepsilon'' \end{pmatrix}, \tag{12}$$

with

$$M = \frac{\varepsilon'\varepsilon_0}{(\varepsilon')^2 - q^2}, \tag{13}$$

and

$$K = \frac{-q\varepsilon_0}{(\varepsilon')^2 - q^2}. \tag{14}$$

For convenience in what follows, we now choose the gyro axis to be in the x direction (i.e., the d-c magnetic field is taken to be parallel to the x direction). As a result $\partial/\partial x = 0$ since the incident wave is specified to be

[1] In what follows, the x axis is taken to be in the axial direction. Thus equ. (4) is given by

$$\begin{pmatrix} D_y \\ D_z \\ D_x \end{pmatrix} = (\varepsilon) \begin{pmatrix} E_y \\ E_z \\ E_x \end{pmatrix}$$

Reflections from a Lossy Magnetoplasma Half-Space

transverse to the d-c magnetic field. It is now a straightforward matter to show that

$$\left(\frac{\partial^2}{\partial y^2} + \frac{\partial^2}{\partial z^2} + \frac{k^2}{M\hat{M}}\right)H_x = 0, \tag{15}$$

$$\left(\frac{\partial^2}{\partial y^2} + \frac{\partial^2}{\partial z^2} + \frac{k^2\varepsilon''}{\hat{M}\varepsilon_0}\right)E_x = 0, \tag{16}$$

where $k=(\varepsilon_0\mu_0)^{1/2}\omega=2\pi/\lambda$ and $M=\mu_0/\mu$. Here λ is the free space wavelength and μ_0 is the magnetic permeability of free space.

From Maxwell's equations, the other field components may be obtained from H_x and E_x. For example,

$$i\varepsilon_0\omega E_y = M\frac{\partial H_x}{\partial z} + iK\frac{\partial H_x}{\partial y}, \tag{17}$$

$$i\varepsilon_0\omega E_z = iK\frac{\partial H_x}{\partial z} - M\frac{\partial H_x}{\partial y}, \tag{18}$$

$$-i\mu_0\omega H_y = \hat{M}\frac{\partial E_x}{\partial z}, \tag{19}$$

$$-i\mu_0\omega H_z = -\hat{M}\frac{\partial E_x}{\partial y}. \tag{20}$$

It is immediately evident that the general problem splits into two parts. The total field may be regarded as the superposition of two partial fields; one is characterized by $\vec{E}=(E_x,0,0)$ and the other by $\vec{H}=(H_x,0,0)$. These may be called TE (transverse electric) and TM (transverse magnetic) waves, respectively. For a vertically polarized incident wave, the first partial field (i.e., the TE waves) is not excited, so further attention is restricted to the TM waves.[2]

Preliminary Problem

As a simple preliminary problem we shall consider the ionosphere to be sharply bounded and homogeneous. Thus, for $z>0$, the dielectric properties are to be characterized by the tensor (ε) which does not vary with z. The region $z<0$ is free space.

The incident wave is of the form

$$H_x^{\text{inc}} = h_0 e^{-ikCz} e^{-ikSy}, \tag{21}$$

where h_0 is a constant and where $C=(1-S^2)^{1/2}=\cos\theta$ in terms of the angle of incidence θ. Then the reflected wave must be of the form

$$H_x^{\text{refl}} = h_0 e^{+ikCz} e^{-ikSy} R, \tag{22}$$

where R, by definition, is the reflection coefficient. For the region $z>0$,

$$H_x = f(z)\exp(-ikSy), \tag{23}$$

where, by virtue of (15), $f(z)$ satisfies

$$\left[\frac{\partial^2}{\partial z^2} + k^2\left(\frac{1}{M\hat{M}} - S^2\right)\right]f(z) = 0. \tag{24}$$

---

[2] Actually, the TE waves in this case are not influenced by the d-c magnetic field and the medium is effectively isotropic.

Therefore,

$$H_x = h_0 T \exp\left[-ik\left[\frac{1}{M\hat{M}} - S^2\right]^{1/2} z\right] \exp(-ikSy), \qquad (25)$$

where T, by definition, is a transmission coefficient. The unknown functions R and T are found by applying the boundary conditions which require the continuity of E_y and H_x at $z=0$. This readily leads to

$$T = \frac{2C}{C+\Delta}, \qquad (26)$$

and

$$R = \frac{C-\Delta}{C+\Delta}, \qquad (27)$$

where

$$\Delta = M\left[\frac{1}{M\hat{M}} - S^2\right]^{1/2} + iKS. \qquad (28)$$

If $\hat{M} = \mu_0/\mu = 1$, which is the usual case,

$$\Delta = \frac{\left[C^2 + \frac{1+iL}{iL-L^2-\gamma^2}\right]^{1/2}(iL-L^2-\gamma^2) - i\gamma S}{(1+iL)^2 - \gamma^2}, \qquad (29)$$

where

$$L = \frac{(\nu + i\omega)\omega}{\omega_0^2} \quad \text{and} \quad \gamma = \frac{\omega_T \omega}{\omega_0^2}.$$

The latter result was given by Barber and Crombie [1959] who employed a somewhat more involved derivation.

The General Problem

We shall now return to our originally stated problem. The continuous profiles of $N(z)$ and $\nu(z)$ are replaced by a very large, but finite, number of steps. In other words, the inhomogeneous medium is replaced by a stack of thin homogeneous layers. For purposes of discussion, there shall be P such layers while a typical layer is the pth layer. Thus, p ranges from 1 to P through integral values. Somewhere at a sufficiently negative value of z, the medium may be regarded as free space. This level is denoted $z = -z_0$.

Within the pth layer, the fields may be regarded as a superposition of upgoing and downgoing waves. These are characterized by the functions $\exp(-i\beta_p kz)$ and $\exp(+i\beta_p kz)$, respectively, where

$$\beta_p = \left[\frac{1}{M_p \hat{M}_p} - S^2\right]^{1/2}. \qquad (30)$$

The characteristic impedances associated with these wave types are

$$K_p^+ = -\frac{E_y^+}{H_x^+}, \qquad (31)$$

and

$$K_p^- = \frac{E_y^-}{H_x^-}, \tag{32}$$

where the $+$ sign denotes upgoing and the $-$ sign denotes downgoing waves. From (17) and (18), it is seen that

$$K_p^+ = \eta_0 (M_p \beta_p + iK_p S), \tag{33}$$

and

$$K_p^- = \eta_0 (M_p \beta_p - iK_p S), \tag{34}$$

where

$$\eta_0 = (\mu_0/\varepsilon_0)^{1/2} = 120\pi \text{ ohms}.$$

The problem may now be solved by an application of nonuniform transmission line theory [Schelkunoff, 1943; Wait, 1962]. Thus, the reflection coefficient which is referred to the lower edge of the bottom slab is given by

$$R_0 = \frac{K_0^+ - Z_1}{K_0^- + Z_1} = \frac{C - \Delta}{C + \Delta}, \tag{35}$$

where $\Delta = Z_1/\eta_0$ and where Z_1 is the input impedance at the bottom of layer number 1. Now, Z_1 may be expressed in terms of Z_2 which, in turn, may be expressed in terms of Z_3. The process is continued until the topmost layer is reached where $Z_p = K_p^+$ is assumed known. The details of this derivation are given elsewhere [Wait, 1962].

The required number of layers is best determined by studying the stability of the solution as the number is increased. Because of the relatively long wavelength involved and because of the finite losses in the medium, the solution converges nicely as the number of layers is increased. For the cases discussed here, the step size was never greater than 0.1 km and this appeared to be smaller than necessary.

Presentation of Results

The final results of the numerical calculations are presented in such a fashion that the phase of the reflection coefficient R is *referred to the level* $z=0$. Thus, by definition,

$$R = [R_0 \exp(i2kCz_0)]_{z_0 \to \infty}. \tag{36}$$

Physically, this means that the observer at $z=-z_0$ is sufficiently far below the ionosphere that the medium may be regarded as free space. In practice z_0 is chosen to be large enough that the phase of R does not vary with further changes in z_0. For the case here, z_0 was of the order of 40 km. This particular normalization of the phase has been used on previous occasions [Wait and Walters, 1963].

Following the usage in previous papers [Wait and Walters, 1963], the quantity $\omega_r = \omega_0^2(0)/\nu(0)$ is specified. In particular, $\omega/\omega_r = 1/2$ at 15 kc/s, or $\omega_r = 6\pi \times 10^4 \text{ sec}^{-1}$. The "effective" conductivity σ_e of the medium at this level $z=0$, is then given by $\sigma_e = \varepsilon_0 \omega_r \sim 1.7 \times 10^{-6}$ mhos/meter. With exponential-type profiles the fixing of the parameter of ω_r is not an essential restriction. It is a simple matter to shift the reference level from $z=0$ to any other value if desired.

The parameters of the problem are thus λ, C, b, a, and ω_T/ν_0. In order to display the relative influence of these quantities, it is desirable to plot the amplitude and phase of R as a function of ω_T/ν_0 from -3 to $+3$ for a range of values of λ, C, b, and a. It should be noted that λ is in km, C is dimensionless, while b and a

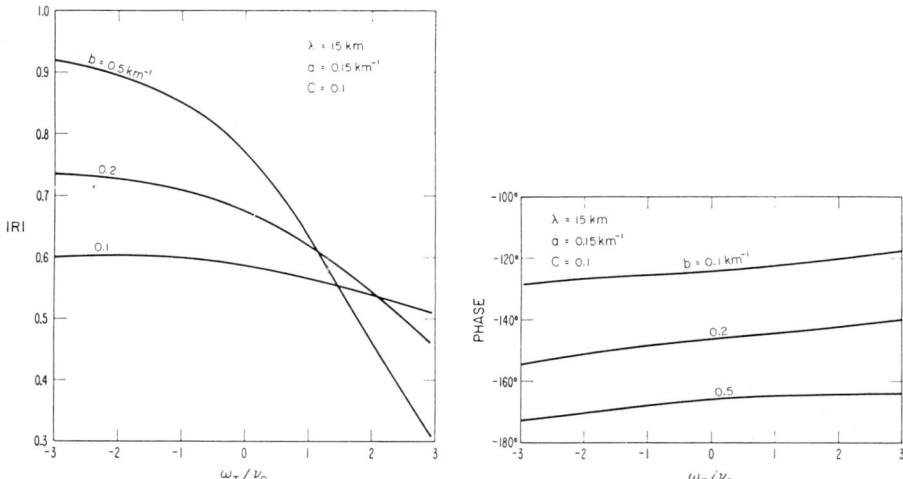

Figs. 2a and 2b. Reflection coefficient curves illustrating the dependence on electron density profile.

have dimensions of km^{-1}. Consequently, the scale length in the present problem is the kilometer. By changing this scale, the results may also have significance at higher frequencies.

In figures 2a and 2b the magnitude of the reflection coefficient $|R|$ and the phase of R are plotted as a function of ω_T/ν_0. Negative values of the abscissa correspond to propagation from west to east along the magnetic equator. The cosine angle of incidence is fixed at 0.1. Thus, the angle is highly oblique, being only 5.7° from grazing. For long-distance propagation of VLF radio waves, such highly oblique conditions prevail.[3] For the curves in figures 2a and 2b, the collision profile is chosen so that the collision parameter $a=0.15$ km^{-1} and the wavelength $\lambda=15$ km correspond to a frequency of 20 kc/s. For these curves, the electron density parameter b takes the values 0.1, 0.2, and 0.5 km^{-1}. It is evident that for $\omega_T/\nu_0 \cong 0$, the steep gradient of electron density is associated with maximum amplitudes of reflection. However, when ω_T/ν_0 is finite, this may no longer be the case. In fact, the asymmetry of the curves about $\omega_T/\nu_0 = 0$ is a measure of nonreciprocity in the reflection process. As indicated, the reflection coefficient for propagation from west to east is greater than for propagation from east to west. There is also some nonreciprocity in the phase curves but it is not great.

In figures 3a and 3b, a set of curves show the influence of varying the collision frequency parameter while keeping the electron density parameter fixed at $b=0.15$ km^{-1}. For these curves, as before, $C=0.1$ and $\lambda=15$ km. It is evident that the steeper gradient of the collision frequency corresponds to larger reflection coefficients.

In figures 4a and 4b, $|R|$ and the phase of R are shown as a function of ω_T/ν_0 for various frequencies in the range from 6 to 60 kc/s. For these curves, $C=0.1$, $B=0.15$, and $a=0.15$. The tendency is for the reflection coefficient to be diminished at the higher frequencies. In this case, the medium is acting like a good absorber rather than a reflector.

In figures 5a and 5b, $|R|$ and the phase of R are plotted for different values of the angle of incidence. For these curves, $\lambda=15$ km, $b=0.15$ km^{-1}, and $a=0.15$ km^{-1}. In general, it may be seen that the reflection coefficient is diminished for the steeper angle of incidence. It is rather interesting to note that the asymmetry (or nonreciprocity) in the phase curves is more pronounced at the steeper angles.

[3] In fact, the attenuation of the dominant mode in the earth-ionosphere waveguide at VLF is approximately proportional to $1-|R|$ for highly oblique incidence [Wait, 1962].

Reflections from a Lossy Magnetoplasma Half-Space

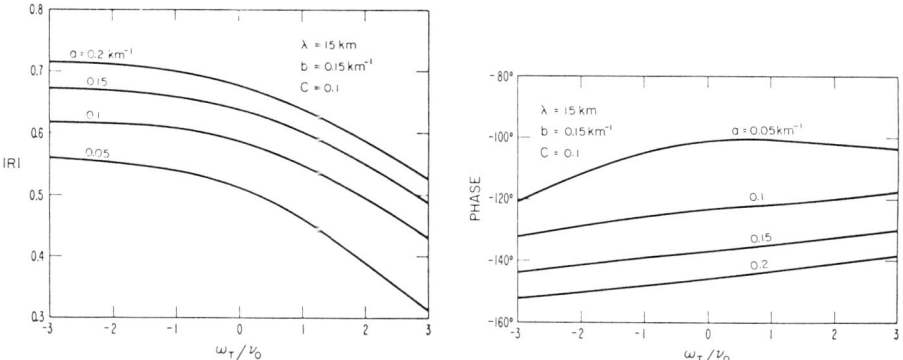

Figs. 3a and 3b. Reflection coefficient curves illustrating the dependence on collision frequency profile.

Finally, in figures 6a and 6b, $|R|$ and the phase of R are plotted for various values of the collision frequency parameter a for $\lambda=15$ km and $C=0.1$. These curves differ from figures 3a and 3b in that here the parameter $\beta=b+a$ is fixed rather than just a. In other words, the profile of N/ν as a function of z is fixed while the gradient of ν is changed. In the isotropic case, where $\omega_T=0$, it is interesting to note that R is determined only by the gradient of N/ν. However, for a finite gyrofrequency, the situation is changed significantly. In general, the nonreciprocity is accentuated when a is diminished. For example, if ν were assumed to be a constant, the dependence of the gyrofrequency is much greater than for a collision frequency which varies with height. In much of the earlier work [e.g., Budden, 1955] on full wave solutions in ionospheric radio waves, it is often assumed that ν can be regarded as a constant. Clearly, such an assumption may lead to very misleading results.

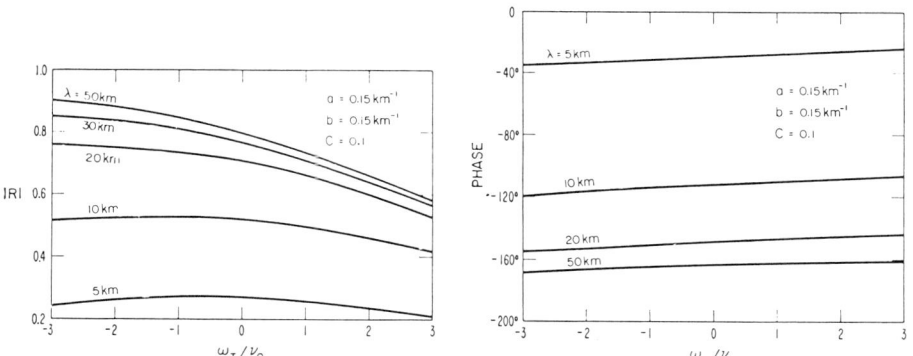

Figs. 4a and 4b. Reflection coefficient curves illustrating the dependence on wavelength.

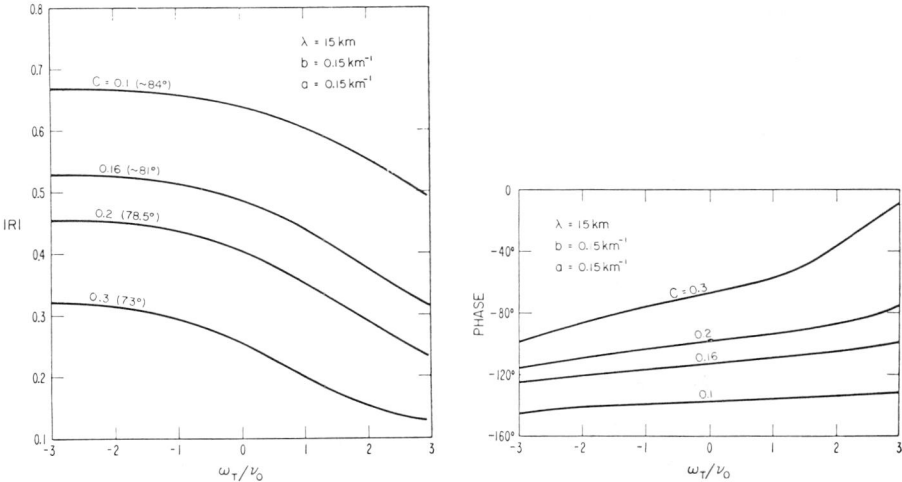

Figs. 5a and 5b. Reflection coefficient curves illustrating the dependence on angle of incidence.

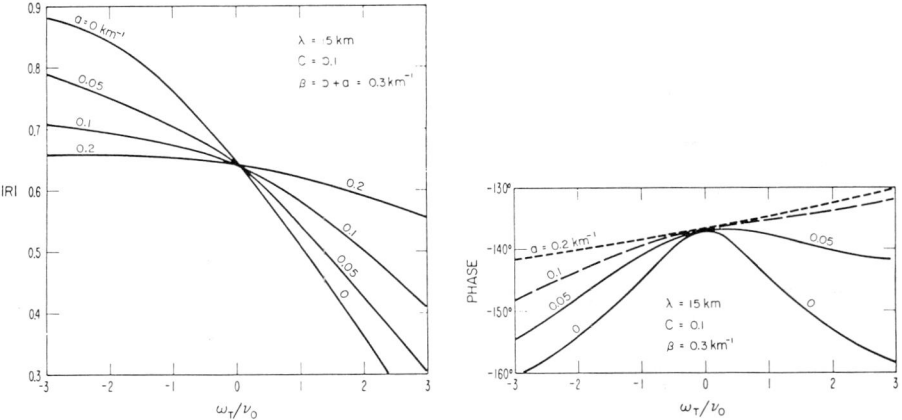

Figs. 6a and 6b. Reflection coefficient curves illustrating the dependence on collision frequency profile when the N/ν profile is fixed.

Discussion and Concluding Remarks

The numerical results given here should provide some insight into the nature of reflection from an inhomogeneous ionized medium. The nature of the dependence on N, ν, and ω_T is quite complicated. Nevertheless, it appears that the sharper gradients of electron density are usually associated with higher reflection coefficients. The dependence on the collision frequency profile is not so clear-cut.

In nearly every case it may be seen that the presence of the transverse magnetic field is to cause the reflection coefficient $|R|$ to be nonreciprocal. Furthermore, for a wide range of the parameters, $|R|$ is greater for west-to-east propagation than for east-to-west propagation. This is in accord with experimental data of Round et al. [1925] who observed that, for propagation over distances of the order of 6000 km, signals from VLF transmitters to the west are received more strongly than from those to the east. This observation has also been confirmed by Crombie [1958] in a series of field strength measurements in New Zealand and by Taylor [1960], who analyzed the waveforms of atmospherics. For some inexplicable reason, Budden [1955] deduces from a full wave solution that the directional dependence is just opposite to this. Although his model is not the same as the one considered here, it is difficult for this writer to accept the validity of his results in this regard. However, it is possible that, because of the complexity of the various phenomena, a reversed trend may emerge for certain special conditions, particularly for the nighttime ionosphere [Rhoads et al., 1963]. It is also worth mentioning that Budden [1955] has some qualms concerning the accuracy of his numerical data at small values of C.

References

BARBER, N.F., and CROMBIE, D.D., (1959), VLF reflections from the ionosphere in the presence of a transverse magnetic field, *J. Atmos. and Terrest Phys.*, *16*, 37.

BELROSE, J.S., (1963), Present knowledge of the lowest ionosphere, a chapter in *Radio Wave Propagation* (ed. by W. T. Blackband), Pergamon Press, Oxford.

BUDDEN, K.G., (1955), The solution of differential equations governing the reflexion of long radio waves from the ionosphere II, *Phil. Trans. Roy. Soc.* (London), Series A, *248*, 45–72.

BUDDEN, K.G., (1961), *Radio Waves in the Ionosphere*, (Cambridge University Press).

BUDDEN, K.G., (1962), *The Waveguide Mode Theory of Wave Propagation*, Prentice-Hall, Englewood Cliffs, N.J.

BURMAN, R., (1966), Transverse propagation of electromagnetic waves in a cylindrically stratified axially magnetized plasma, *Quart. Jour. Mech. and Appl. Math.*, *19*, Pt. II, 247–258.

CROMBIE, D.D., (1958), Differences between east-west and west-east propagation of VLF signals over long distances, *J. Atmos. and Terrest. Phys.*, *12*, 110–117.

GALEJS, J., (1961), ELF waves in the presence of exponential profiles, *IRE Trans. Ant. Prop.*, *AP-9*, No. 6, 554–562.

GALEJS, J., (1972), *Terrestrial Propagation of Long Electromagnetic Waves*, Pergamon Press, Oxford.

JOHLER, J.R., (1961), Magneto-ionic propagation phenomena in low- and very-low-radiofrequency waves reflected by the ionosphere, *J. Res. NBS*, *65D*, (Radio Prop.), 53–61.

JOHLER, J.R., (1963), Radio wave reflections at a continuously stratified plasma with collisions proportional to energy and arbitrary magnetic induction, *Proc. International Conference on the Ionosphere*, 436–445, Chapman & Hall Ltd., London.

RHOADS, F.J., GARNER, W.E., and LOGERSON, J.E., (1963), Some experimental evidence of direction effects on VLF propagation (private communication).

ROUND, H.J.T., ECKERSLEY, T.L., TREMELLEN, K., and LUNNON, F.C., (1935), Report on measurements made on signal strength at great distances during 1922 and 1923 by an expedition sent to Australia, *J.I.E.E.*, *63*, 933–1011.

SCHELKUNOFF, S.A., (1943), *Electromagnetic Waves*, Van Nostrand, New York.

TAYLOR, W.L., (1960), VLF attenuation for east-west and west-east daytime propagation using atmospherics, *J. Geophys. Res.*, *65*, No. 7, 1933–1938.

WAIT, J.R., (1962), *Electromagnetic Waves in Stratified Media*, Pergamon Press, Oxford, and Macmillan, New York.

WAIT, J.R., and WALTERS, L.C., (1963), Reflection of VLF radio waves from an inhomogeneous ionosphere, Parts I, II, III, *J. Res. NBS*, *67D* (Radio Prop.) Nos. 3, 5, and 6.

WATSON, G.N., (1919), The transmission of electric waves round the earth, *Proc. Roy. Soc.*, *95*, 546.

20
EM Propagation in the Earth-Ionosphere Waveguide

Introduction

There is now a vast literature on the subject of radio propagation of long electromagnetic waves in the waveguide formed by the earth's surface and the lowest ionosphere. The attention is focused most heavily on the VLF (very low frequency) band, which corresponds to frequencies from 3 to 30 kHz, the principal reason being that the attenuation of the waves is very low and the phase stability of the waves is very high. Thus, these waves find numerous applications in global communications and in long-range navigation systems.

An excellent and comprehensive review of the engineering capabilities of VLF waves has been published by Watt(1967). Also, a general survey of VLF propagation phenomena has appeared in a monograph by Al'pert and Fligel(1970). In addition, special issues on VLF and ELF propagation have been edited by Crombie (1967) and Wait(1974) respectively. As indicated in these publications, the mode theory of terrestrial propagation in the earth-ionosphere waveguide has been used extensively as a model to predict and interpret the fields. In this present survey, we give a concise outline of the relevant VLF propagation theory.

We begin by considering ground wave propagation over an airless spherical earth without any ionosphere. This serves as an excellent mechanism to introduce the relevant asymptotic approximations. The generalization to include waves reflected from the ionosphere is then treated in a simple, direct fashion. By suitably summing the "rays", a modal representation for the total field is obtained without becoming embroiled in a mass of mathematical detail. The ends, if not the means, are justified by a comparison with other theoretical treatments that are mathematically "rigorous" (e.g. Makarov, Novikov and Rybachek(1970), Wait (1970), Krasnushkin (1962).

Extensive use is made of impedance concepts in formulating the propagation equations. Apart from its convenience, this approach is much broader in its range of validity than some believe.

The waveguide model is generalized to include such effects as coupling between TM and TE modes, earth-detached modes, and mode conversion phenomena. Also, in some detail, we consider the planar analog of the spherical earth-ionosphere waveguide, since the analogy is the basis of a microwave laboratory model. Although this survey is concerned primarily with theory, we also indicate some pertinent comparisons with experimental VLF data.

(Reprinted, in part in revised form, from: J.R. Wait, Advances in Electronics and Electron Phys., (ed. by L. Marton), 25, 145-209, Academic Press, 1968).

EM Propagation in the Earth-Ionosphere Waveguide

PROPAGATION ALONG A SPHERICAL SURFACE

A basic preliminary problem concerns the propagation of electromagnetic waves along a homogeneous spherical surface when the external medium is free space with intrinsic properties ε_0 and μ_0. The situation is illustrated in Fig. 1, where we have chosen a spherical coordinate system (r, θ, ϕ) centered at the sphere. The fields are excited by a radially oriented dipole located on the axis $\theta = 0$ which is external to the surface of the sphere at $r = a$. The fields are taken to vary everywhere as $\exp(i\omega t)$.

It is clear that, for the configuration chosen, the magnetic field has only a ϕ component H_ϕ. Thus, we may write, for $r > a$,

$$H_\phi = -i\varepsilon_0 \omega \, \partial U/\partial \theta, \tag{1}$$

where U is a scalar function that satisfies

$$(\nabla^2 + k^2)U = 0 \tag{2}$$

except at the source itself. Here, $k = (\varepsilon_0 \mu_0)^{1/2}\omega$ is the wave number for the external free-space region.

From Maxwell's equation,

$$i\varepsilon_0 \omega \mathbf{E} = \nabla \times \mathbf{H}, \tag{3}$$

we readily find that the components of the electric field are

$$E_r = (k^2 + \partial^2/\partial r^2)(rU), \tag{4}$$

$$E_\theta = (1/r)(\partial^2/\partial r \, \partial \theta)(rU), \tag{5}$$

and

$$E_\phi = 0.$$

In order to facilitate the solution and avoid extraneous details, we invoke an impedance boundary condition at the surface of the sphere. Thus, we say that

$$E_\theta = -ZH_\phi|_{r=a}, \tag{6}$$

where Z is by definition the surface impedance. This is a valid description of the sphere, provided the tangential fields vary slowly over the surface. Certainly, in the case of VLF radio waves over the earth's surface, a suitable value of Z can always be found (Wait, 1970).

Solutions of (2) which are finite everywhere except at $\theta = 0$ are of the form

$$z_\nu(kr)P_\nu(-\cos \theta),$$

where z_ν is a spherical Bessel function (7, p. 437) and P_ν is a Legendre function, both of order ν with arguments as indicated. In the present case, we choose z_ν to be the spherical Hankel function $h_\nu^{(2)}(kr)$, since asymptotically it behaves as $r^{-1}\exp(-ikr)$ as $r \to \infty$. Thus, it exhibits the proper outgoing wave character.

With the foregoing considerations, we are led to write the field as a sum of modes. Thus,

$$U = \sum_s A_s h_{\nu_s}^{(2)}(kr)P_{\nu_s}(-\cos \theta), \tag{7}$$

where the discrete values of ν (i.e., $\nu_1, \nu_2, \nu_3, \ldots$) are found by invoking (6). Explicitly, the modal condition is

$$[(d/dx)\log[xh_\nu^{(2)}(x)] - i\Delta]_{x=ka} = 0, \tag{8}$$

where
$$\Delta = Z/\eta_0, \qquad \eta_0 = (\mu_0/\varepsilon_0)^{1/2}.$$

The summation sign in (7) is to indicate that all the modes are included, and their excitation is characterized by the coefficient A_s.

Using a certain amount of hindsight, we now utilize the fact that, for $ka \gg 1$, the important modes are such that $|v - ka| \ll ka$. In this case, we may use the Nicholson-Lorenz type of representation *(Abramowitz and Stegun, 1964, p. 371)*, which is expressed here in the form *(Wait, 1970)*,

$$kr h_v^{(2)}(kr) \simeq i(ka/2)^{1/6} w_1(t - y), \qquad (9)$$

where
$$y = [2/(ka)]^{1/3} k(r - a), \qquad (ka/2)^{1/3} t = (v + \tfrac{1}{2}) - ka,$$

and $w_1(t - y)$ is an Airy function of argument $(t - y)$. In writing (9), we have assumed that $(r - a)/a \ll 1$ and $(ka/2)^{2/3} \gg 1$ (see Appendix A).

The Airy function used here may be written as a contour integral of the form

$$w_1(t) = \pi^{-1/2} \int_{\infty e^{i2\pi/3}}^{\infty} \exp(tz - z^3/3) \, dz, \qquad (10)$$

where the contour of integration in the z plane is conveniently taken along a straight line from $\infty \exp(i2\pi/3)$ to the origin and then out along the real axis to ∞. Later on, we also need the related Airy function $w_2(t)$, which has a similar definition:

$$w_2(t) = \pi^{-1/2} \int_{\infty e^{-i2\pi/3}}^{\infty} \exp(tz - z^3/3) \, dz. \qquad (11)$$

For real t, it is clear that $w_2(t)$ is the complex conjugate of t.

Using (9), the modal condition (8) is equivalent to

$$w_1'(t) - q w_1(t) = 0, \qquad (12)$$

where the prime indicates a derivative with respect to t, and $q = -i(ka/2)^{1/3} \Delta$. Solutions of (12) are designated t_s (where $s = 1, 2, 3, \ldots$). This equation has been studied extensively, and a summary is available *(Wait, 1964a)*.

For cases of most interest, the Legendre function may be approximated by

$$P_v(-\cos \theta) \cong \left(\frac{2}{\pi v \sin \theta}\right)^{1/2} \cos\left[\left(v + \frac{1}{2}\right)(\pi - \theta) - \frac{\pi}{4}\right], \qquad (13)$$

which is valid if $|v| \gg 1$ and θ is not near 0 or π.

With this bare outline of the essential approximations, we see that the representation (7), for the mode sum, may be put into a tractable form. In the case of propagation over a spherical homogeneous earth with conductivity σ and dielectric constant ε, we conveniently express the radial electric field in the form

$$E_r = \frac{e^{-ika\theta}}{a(\theta \sin \theta)^{1/2}} V_0, \qquad (14)$$

where V_0 is, by definition, the attenuation function. When $\theta \ll 1$, we see that (14) is equivalent to

EM Propagation in the Earth-Ionosphere Waveguide

$$E_r \simeq \frac{e^{-ikd}}{d} V_0, \tag{15}$$

where $d = a\theta$ is the great circle distance, as indicated in Fig. 2. In the present context, it is convenient to normalize the source so that, for a dipole located on the surface, the radiation field E_r at $r = a$ is $2\exp(-ikd)/d$ in the limit of a flat, perfectly conducting surface (i.e., a and σ_g both approaching infinity).

For the present problem, the attenuation function may be written as (8, 9)

$$V_0 \simeq 2(\pi x)^{1/2} e^{-i\pi/4} \sum_{s=1,2,3,\ldots} \frac{e^{-ixt_s}}{t_s - q^2} \frac{w_1(t_s - y)}{w_1(t_s)}, \tag{16}$$

where $x = (ka/2)^{1/3}\theta = (\kappa a/2)^{1/3} d/a$ and $y = (2/ka)^{1/3} k(r-a)$. As indicated previously, the roots t_s are solutions of (12), where, for a homogeneous earth, the appropriate normalized surface impedance is

$$\Delta = \left(\frac{i\varepsilon_0 \omega}{\sigma_g + i\varepsilon_g \omega}\right)^{1/2} \left[1 - \frac{i\varepsilon_0 \omega}{\sigma_g + i\varepsilon_g \omega}\right]^{1/2}. \tag{17}$$

Here, the conductivity and permittivity of the ground are designated by σ_g and ε_g, respectively.

If the source dipole were located at $r = b$ (where $b > a$), it is not difficult to generalize (16) to

$$V_0 \simeq 2(\pi x)^{1/2} e^{-i\pi/4} \sum_{s=1,2,3,\ldots}^{\infty} \frac{e^{-ixt_s}}{t_s - q^2} \frac{w_1(t_s - y) w_1(t_s - \hat{y})}{w_1(t_s) w_1(t_s)}, \tag{18}$$

where $\hat{y} = (2/ka)^{1/3}(b-c)$. However, because of our approximation (9), we require that $(b - a) \ll a$ in addition to $(r - a) \ll a$. In other words, all heights should be small compared with the radius of the earth.

The modal sums given by (16) and (18) are often described as the residue series for the ground wave, since they were obtained in essentially this form by van der Pol and Bremmer via a Watson transformation that involved contour deformation in the complex v plane. This approach to the problem is described completely in the book by Bremmer(1949). The equivalence with the final form of the van der Pol-Bremmer theory is not at all obvious until one notes that Airy functions are closely akin to Hankel functions of order one-third. For example,

$$w_1(t) = \exp(-2\pi i/3)(-\pi t/3)^{1/2} H^{(2)}_{1/3}[\tfrac{2}{3}(-t)^{3/2}], \tag{19}$$

which is an identity.

For application in what follows, we now express (16) as a contour integral in the form

$$V_0 = e^{i\pi/4} \left(\frac{x}{\pi}\right)^{1/2} \oint \frac{e^{-ixt} w_1(t-y)}{w_1'(t) - qw_1(t)} dt. \tag{20}$$

The contour is here chosen so that it encloses the pole singularities at $t = t_s$ in the complex plane. The equivalence between (20) and (16) is readily established by noting that $w_1'(t_s) = qw_1(t_s)$, and, for any t, we have $w_1''(t) = tw_1(t)$. Thus, it is easy to see that $-2\pi i$ times the sum of the residues of the integrand of (20) leads back to (16).

When the distance parameter x is small compared with unity, the modal series given by (16) is poorly convergent. For numerical work, it becomes more convenient and certainly more economical to develop an asymptotic series for V_0 whose leading term is the flat earth result. The method, which is straightforward but tedious, is to express the contour integral as an inverse Laplace transform where the transform variable s is to be identified with $-it$. Thus, we observe immediately that the behavior of V_0, for x tending to zero, is related to the integrand for s tending to ∞. The details of the method are described elsewhere *(Wait, 1956, Bremmer, 1958)*, and the ranges of validity have been scrutinized rather carefully*(Spies and Wait, 1966)*. The relevant expansion has the form

$$V_0 \simeq 2G\{F(p) - (\delta^3/2)[1 - i(\pi p)^{1/2} - (1 + 2p)F(p)]$$
$$+ \text{ terms in } \delta^6, \delta^9, \text{ etc.}\}, \qquad (21)$$

where

$$F(p) = 1 - i(\pi p)^{1/2} e^{-p} \operatorname{erfc}(ip^{1/2}), \qquad (22)$$

$$p = -ikd\,\Delta^2/2, \qquad \delta^3 = i(ka\,\Delta^3)^{-1}, \qquad d = a\theta,$$

$$G \simeq 1 + ik\,\Delta z, \qquad z = r - a.$$

In the limit of a flat earth, $\delta \to 0$, and the attenuation function is proportional to Sommerfeld's function $F(p)$, which involves the complementary error function $\operatorname{erfc}(ip^{1/2})$ of complex argument $ip^{1/2}$.

INTRODUCTION OF THE SKY WAVES

The contour integral representation for the attenuation function V_0 given by (20) is in a form that permits a straightforward extension to account for sky waves. Although this technique *(Wait, 1961a)* is somewhat heuristic, it does provide insight into the reflection phenomena. Furthermore, the final results are essentially the same as those obtained by a more complicated method that utilizes the Watson transformation *(e.g. Wait 1970, Chapter 6)*.

To understand the present approach, we note that $\exp(-ixt)w_1(t-y)$ has an outgoing wave character for increasing y (i.e., increasing height). Thus, as indicated in Fig. 3, a concentric inhomogeneity at $y = y_0$ will couple this to a downcoming wave characterized by the function $\exp(-ixt)w_2(t-y)$. This reflection or coupling process is complicated when the inhomogeneity is an ionospheric layer. However, for the moment, we shall assume that the level $y < y_0$ is free space and the tangential fields at $y = y_0$ satisfy an impedance boundary condition (i.e., $E_\theta = Z_i H_\phi$, where Z_i is the surface impedance). The combination of a single upgoing and a single downcoming wave is obviously of the form

$$w(t-y) = w_1(t-y) + A(t)w_2(t-y), \qquad (23)$$

where $A(t)$ is of the nature of a reflection coefficient. The impedance boundary condition requires that

$$[(d/dy)w(t-y) - q_i w(t-y)]|_{y=y_0} = 0, \qquad (24)$$

where $q_i = -i(ka/2)^{1/3}(Z_i/\eta_0)$. Insertion of (23) into (24) yields

$$A(t) = -\left[\frac{w_1'(t - y_0) + q_i w_1(t - y_0)}{w_2'(t - y_0) + q_i w_2(t - y_0)}\right]. \qquad (25)$$

We now apply the same reasoning to calculate the coupling at the earth's surface between a downcoming and an upgoing wave, as indicated in Fig. 4.

The relevant combination is

$$w(t-y) = w_2(t-y) + B(t)w_1(t-y), \quad (26)$$

which satisfies

$$[(d/dy)w(t-y) + qw(t-y)]|_{y=0} = 0 \quad (27)$$

by virtue of the impedance condition $E_\theta = -ZH_\phi$ at $r = a$. Then, on using (26) and (27), we find that

$$B(t) = -\left[\frac{w_2'(t) - qw_2(t)}{w_1'(t) - qw_1(t)}\right], \quad (28)$$

where $q = -i(ka/2)^{1/3}Z/\eta_\bullet$.

ACCOUNTING FOR THE MULTIPLE REFLECTIONS

Since we now, presumably, understand the reflection process at the earth's surface and the upper reflecting layer, it is not difficult to sum up all the multiple reflections. The situation is illustrated in Fig. 5, where the reflection layer is located at $r = a + h$ (or $y = y_0$), and the earth's surface is $r = a$ (i.e., $y = 0$). Thus, we are led to write the total field for a ground-based electric dipole in the form

$$E_r = \frac{e^{-ika}}{a(\theta \sin \theta)^{1/2}} V, \quad \text{where} \quad V = \sum_{j=0,1,2,\ldots} V_j, \quad (29)$$

$$V_j = e^{i\pi/4}\left(\frac{x}{\pi}\right)^{1/2} \oint \frac{e^{-ixt}w_1(t-y)}{w_1'(t) - qw_1(t)}[A(t) B(t)]^{j/2} dt \quad (30)$$

for $j = 0, 2, 4, \ldots$, and

$$V_j = e^{i\pi/4}\left(\frac{x}{\pi}\right)^{1/2} \oint \frac{e^{-ixt}w_2(t-y)}{w_1'(t) - qw_1(t)}[A(t)]^{(j+1)/2}[B(t)]^{(j-1)/2} dt \quad (31)$$

for $j = 1, 3, 5, \ldots$.

It is obvious that the term for $j = 0$ (i.e., V_0) is the ground wave, whereas the term for $j = 1$ is the "first hop" sky wave, since it involves one reflection at the upper reflecting level. It is thus convenient to describe V_j (for $j > 0$) as the jth hop sky wave. The asymptotic evaluation of the V_j integrals leads to geometrical-optical representations. We shall return to this question later on.

The next step is to invert the order of the summation and the integration. Thus, for a sum over N hop sky waves, we have

$$V = \sum_{j=0,1,2,\ldots N} V_j$$

$$= e^{i\pi/4}\left(\frac{x}{\pi}\right)^{1/2} \int e^{-ixt} \frac{[w_1(t-y) + A(t)w_2(t-y)]}{[w_1'(t) - qw_1(t)][1 - A(t)B(t)]} dt + R_N, \quad (32)$$

where

$$R_N = -e^{i\pi/4}\left(\frac{x}{\pi}\right)^{1/2} \int \frac{e^{-ixt}}{w_1'(t) - qw_1(t)}$$

$$\times \left[\frac{w_1(t-y) + A(t)w_2(t-y)}{1 - A(t)B(t)} - w_1(t-y)\right][A(t)B(t)]^{N/2} dt \quad (N \text{ even})$$

$$= -e^{i\pi/4}\left(\frac{x}{\pi}\right)^{1/2} \int \frac{e^{-ixt}[w_1(t-y) + A(t)w_2(t-y)][A(t)B(t)]^{(N+1)/2}}{[w_1'(t) - qw_1(t)][1 - A(t)B(t)]} dt$$

$$(N \text{ odd}). \quad (33)$$

In the case where the observer is on the earth's surface, Zabina (1969) writes (29) in the following related form

$$E_r = \frac{300(P)^{1/2} e^{-ika\theta}}{a(\theta \sin \theta)^{1/2}} \; V$$

where

$$V = V_o + \sum_{m=1}^{\infty} V_m$$

where

$$V_m = -e^{-i\pi/4} \left(\frac{x}{\pi}\right)^{1/2} \int \frac{[A(t)]^m [B(t)]^{m-1} e^{-ixt}}{[w_1'(t) - qw_1(t)]^2} \, dt$$

Here P is the power of the dipole expressed in kilowatts (e.g. the dipole current moment is fixed so that it would radiate P kilowatts over a flat perfectly conducting ground plane). Zabina and her colleagues (1969) have used this approach (Wait, 1961b) to obtain numerical results for the various sky waves. Although the ionospheric model is different, their calculations are similar to earlier results published by this writer and his former colleague, Alyce Conda (Wait and Conda, 1961).

In arriving at the forms just given, we have simply summed the geometric progressions that involve $A(t)B(t)$ as the common ratio. In most applications, we chose N to be sufficiently large that the remainder integral R_N may be neglected. Formally, we can argue that if N tends to infinity a contour can be selected such that the product $[A(t)B(t)]^{N/2}$ tends to zero for all values of t on this contour. Thus, we claim that the contour integral given by (32) is the complete field, which includes the combined effect of the ground wave and all the sky wave hops. However, as indicated below, there are some subtleties that are glossed over in this treatment. For the time being, we shall not attempt to justify the individual steps of the argument. The interested reader is referred to some important extensions of this "wave hop" theory by Barry et al (1969) and Jones (1968).

Assuming that the contour integral in (32) represents the total field, we now note that the contour may be chosen to enclose the poles of the integrand at $t = t_n$, where t_n are solution of

$$1 - A(t)B(t) = 0. \tag{34}$$

It is important to observe that the "ground wave" poles that are solutions of $w_1'(t) - qw_1(t)$ are no longer poles of the integrand of (32).

Following Cauchy's theorem, we then find that $-2\pi i$ times the sum of the residues at $t = t_n$ gives a series representation,

$$V = -2e^{-i\pi/4}(\pi x)^{1/2} \sum_{n=0,1,2,\ldots} e^{-ixt_n} \frac{[w_1(t_n - y) + A(t_n)w_2(t_n - y)]}{[w_1'(t_n) - qw_1(t_n)][(\partial/\partial t) A(t)B(t)]_{t=t_n}}. \tag{35}$$

As we shall see below, this is a waveguide mode representation. Carrying out differentiations indicated and using the fact that $A(t_n)B(t_n) = 1$, (35) is cast into the form

EM Propagation in the Earth-Ionosphere Waveguide

$$V = \frac{4(\pi x)^{1/2}}{y_0} e^{-i\pi/4} \sum_{n=0, 1, 2, \ldots} \Lambda_n e^{-ixt_n} G_n(y), \quad (36)$$

where the "excitation factor" Λ_n is given by

$$\Lambda_n = \frac{y_0}{2} \left\{ (t_n - q^2) - \frac{(t_n - y_0 - q_i^2)[w_2'(t_n) - qw_2(t_n)]^2}{[w_2'(t_n - y_0) + q_i w_2(t_n - y_0)]^2} \right\}^{-1}, \quad (37)$$

and $G_n(y)$ is a "height-gain function" given by

$$G_n(y) = \frac{\Phi(t_n, y)}{\Phi(t_n, 0)} \quad (38)$$

for $0 < y < y_0$, where $\Phi(t_n, y) = w_1(t_n - y) + A(t)w_2(t_n - y)$.

There is a simple generalization of (36) to account for a finite height of the source. If the vertical electric dipole is located at $r = b$ (where $a < b < a + h$), we have, in place of (36),

$$V = \frac{4(\pi x)^{1/2}}{y_0} e^{-i\pi/4} \sum_{n=0, 1, 2, \ldots} \Lambda_n e^{-ixt_n} G_n(y) G_n(\hat{y}), \quad (39)$$

where $\hat{y} = (2/ka)^{1/3} k \hat{z}$ where $\hat{z} = b - a$. Reciprocity considerations dictate that the same height-gain function applies for both source and observer locations.

Previously, we have shown that the modal representation (39) can be obtained by a two-dimensional treatment in cylindrical geometry without the explicit introduction of the sky wave hops *(Wait, 1964b)* As a by-product of the investigation, it was found that the excitation factor is related to the height-gain function by

$$\Lambda_n = \frac{y_0}{2 \int_0^{y_0} [G_n(y)]^2 \, dy}. \quad (40)$$

This result may be verified by employing the easily proved result

$$\int [\Phi(t_n, y)]^2 \, dy = -(t_n - y)[\Phi(t_n, y)]^2 + [\Phi'(t_n, y)]^2 \quad (41)$$

and making use of the Wronskian condition

$$w_1(t) v_2'(t) - w_1'(t) w_2(t) = -2i \quad (42)$$

and the differential equation

$$(d^2/dy^2)\Phi(t_n - y) - (t_n - y)\Phi(t_n - y) = 0. \quad (43)$$

The result given by (40) indicates that, when $G_n(y)$ increases with height (i.e., with y), Λ_n may be substantially reduced, corresponding to weak excitation of the mode.

A slightly more complicated form of (37) is given by Galejs *(1972)* who uses a more general form of the Airy function representations. From a practical standpoint, the two forms do not seem to differ. This exhaustive text on the subject by the late Janis Galejs is recommended as required reading for any serious student of VLF and ELF propagation theory.

The Flat Earth Limit

The function $B(t)$ defined by (28) was introduced as a ground reflection coefficient. In order to add insight into its behavior, we examine its form in the asymptotic limit where $(-t) \gg 1$.[1] Thus, we may use the asymptotic expansions

$$w_1(t) \simeq e^{-i\pi/4}(-t)^{-1/4} \exp[-i\tfrac{2}{3}(-t)^{3/2}] \tag{44}$$

and

$$w_2(t) \simeq e^{+i\pi/4}(-t)^{-1/4} \exp[+i\tfrac{2}{3}(-t)^{3/2}]. \tag{45}$$

It is then found that

$$B(t) \simeq \left[\frac{-i(-t)^{1/2} - q}{i(-t)^{1/2} - q}\right] e^{-i\pi/2} \exp[i\tfrac{4}{3}(-t)^{3/2}], \tag{46}$$

which, on using the substitution $(-t)^{1/2} = (ka/2)^{1/3} C$, may be written as

$$B(t) \simeq [(C - \Delta)/(C + \Delta)] e^{i\pi/2} \exp[i\tfrac{4}{3}(ka/2)C^3]. \tag{47}$$

If C is identified as $\cos \theta$, we see immediately that $(C - \Delta)/(C + \Delta)$ is a Fresnel reflection coefficient for a vertically polarized wave incident at angle θ on a locally flat surface of normalized surface impedance Δ. The situation is indicated in Fig. 6.

The function $A(t)$ defined by (25) may be reduced to its asymptotic form when $(y_0 - t) \gg 1$. Then, on using the substitution $(y_0 - t)^{1/2} = (ka/2)^{1/3} C'$, we find that

$$A(t) \simeq [(C' - \Delta_i)/(C' + \Delta_i)] e^{-i\pi/2} \exp[-i\tfrac{4}{3}(ka/2)(C')^3], \tag{48}$$

where $C' = (C^2 + 2h/a)^{1/2}$.

When C' is identified with $\cos \theta'$, the factor $(C' - \Delta_i)/(C' + \Delta_i)$ is recognized immediately as the Fresnel reflection coefficient for a wave incident at angle θ' on a surface of normalized surface impedance Δ_i. The situation is indicated in Fig. 7.

Now, if simultaneously both $(-t)$ and $(y_0 - t) \gg 1$, the pole condition (34) is equivalent to

$$R_g R_i e^{-i\Phi} = 1 = e^{-i2\pi n}, \tag{49}$$

where

$$\Phi = (2ka/3)[(C')^3 - C^3],$$

where n is an integer and where

$$R_g = \frac{C - \Delta}{C + \Delta}, \quad R_i = \frac{C' - \Delta_i}{C' + \Delta_i}.$$

It is immediately evident that (49) has the proper form. A self-consistent mode corresponds to the case where there is a transverse resonance in the multiple reflection between the top and bottom walls of the waveguide. It is particularly revealing if we write Φ as a "phase integral." For example, it is a simple matter to show that

$$\Phi = 2k \int_0^h [C^2 + (2z/a)]^{1/2} \, dz, \tag{50}$$

which, in the limit $a \to \infty$, reduces to

$$\Phi = 2khC,$$

---

[1] More correctly, we should say $|t_n| \to \infty$, where $|\arg(-t_n)| < 2\pi/3$.

EM Propagation in the Earth-Ionosphere Waveguide

and, at the same time, $C' \simeq C$. In this flat earth limit, the angles of incidence and the angle of reflection are the same for both boundaries. The geometries for the curved and flat earth models are illustrated in Figs. 8 and 9, respectively.

Earth-Detached Modes

An interesting phenomenon occurs when the following condition holds:

$$\text{Re}(t_n) \gg 1 \quad \text{for} \quad |\arg t_n| < \pi/3.$$

Thus, we approximate the Airy functions by the first term in the relevant asymptotic expansion valid for large positive arguments. The mode equation now has the form

$$\frac{w_1'(t_n - y_0) + q_i w_1(t_n - y_0)}{w_2'(t_n - y_0) + q_i w_2(t_n - y_0)} e^{-2i\delta(t_n)} = e^{-2\pi i n}, \tag{51}$$

where

$$\delta(t_n) \simeq \frac{e^{-2x}}{2x} \frac{1 + qt_n^{-1/2}}{1 - qt_n^{-1/2}}, \quad x = \tfrac{2}{3} t_n^{3/2}. \tag{52}$$

We see here immediately that, if t_n is sufficiently large (in a positive sense), $\delta(t_n)$ may be ignored and the mode equation loses its dependence on the ground impedance parameter. Such modes are called earth-detached *(Wait, 1970)*. They are closely related to the wispering gallery modes investigated originally by Lord Rayleigh*(1910)* for acoustic waves and by Budden and Martin*(1972)* case of radio waves in the ionosphere. A discussion of some of the approximations involved has been given by Wait*(1967)*. The problem was also considered by Krasnushkin *(1962)*.

The earth-detached modes are important at the upper end of the VLF band (i.e., 20–30 kHz), where the phase velocity of the lowest order modes is less than c. For further discussion of their properties, the reader is referred to the references quoted previously.

Some Concrete Results

A crude but useful representation of the lower ionosphere is a sharply bounded cold isotropic plasma. For example, in the daytime under undisturbed conditions, the electron density at the level of reflection is of the order of 10^3 electrons/cm^3, and the effective collision frequency v is about 10^7 sec^{-1}. The appropriate surface impedance to be used in the mode equation is then

$$\Delta_i = \frac{Z_i}{\eta_0} = \left(1 + \frac{\omega_r}{i\omega}\right)^{-1} \left(\frac{\omega_r}{i\omega}\right)^{1/2},$$

where $\omega_r = \omega_0^2/v$, and where ω_0 is the electron plasma frequency. From extensive studies of data *(Wait, 1970)* it would appear that the value $\omega_r = 2 \times 10^5$ is quite representative of daytime conditions for the VLF band.

For a mode of order n, we define the attenuation rate A_n by

$$A_n = -\text{Im } kS_n \simeq \text{Im}(kC_n^2/2) \text{ nepers/unit length},$$

where $C_n^2 = (-t_n)(2/ka)^{2/3}$, and t_n are solutions of (34). For purposes of graphical plotting, it is desirable to express A_n in terms of decibels per 1000 km of path length. Choosing the sharply bounded ionosphere model described previously, curves of A_1 and A_2 are shown in Figs. 10 and 11, respectively, for a perfectly conducting ground. The height of reflection (i.e., the assumed location of the lower edge of the ionosphere) is allowed to vary from 60 to 100 km. It is evident that, for the first mode (which is the mode of least attenuation), the reflection height does not influence strongly the attenuation rate. However, for the second mode, the attenuation rate is greatly decreased as the reflection height is raised.

The strong influence of ground conductivity on the attenuation rate is indicated in Fig. 12. These curves show the expected change of the attenuation rate as the ground conductivity changes from infinity to 1 mmho/m. The latter is typical of moderately poorly conductive earthen materials.

The phase velocity v_n, relative to the velocity of light c, is related to S_n by

$$(c/v_n) = \operatorname{Re} S_n \simeq 1 - \operatorname{Re}(C_n^2/2).$$

Thus, for purposes of calculation, we use

$$(v_n/c) - 1 \simeq \operatorname{Re}(C_n^2/2).$$

This normalized phase velocity increment is shown in Figs. 13–15 for the first two modes, where we have adopted the same sharply bounded ionosphere model. These curves are reminiscent of the dispersion curves for propagation in a rectangular waveguide. However, it is important to note that, in the latter case, the ratio $v_n/c > 1$. In the present case, it may well be $v_n/c < 1$ for the earth-detached type of modes.

The excitation factor Λ_n, for the modes in this sharply bounded guide, may be calculated from (37). As an example, we show the magnitude and phase of Λ_1 for the first mode in Fig. 16. As indicated, the magnitude of Λ_1 diminishes significantly as frequency increases. This behavior is again consistent with the expected earth-detached character of the dominant mode at the upper end of the VLF band. It is also important to note that the ground conductivity plays a major role in determining the phase of the excitation factor. The close relationship of Λ_1 with the height-gain function $G_1(y)$ as defined by (38) is seen in Figs. 17 and 18, where it is clear that, for 30 kHz the large height-gain is associated with weak excitation. This is also consistent with (40).

To provide further insight into the behavior of the height-gain functions and the excitation factors, we show results for the first two modes in Figs. 19 and 20 for lossless conditions. Here, we choose $\sigma_g = \infty$ or $q = 0$ and imagine the upper boundary of the waveguide to be a perfect magnetic conductor, corresponding to letting $\Delta_i = \infty$ or $q_i = \infty$. Now, both Λ_n and G_n are real functions.

The Planar Analog

It is both practically and conceptionally important to examine analogs to the curved earth-ionosphere waveguide. To this end, we consider the TE (transverse electric) modes, which will propagate in the straight rectangular waveguide shown in Fig. 21. The broad dimension of the cross section is $2h$, and the width is taken to be small compared with the free-space wavelength. The interior volume of the waveguide is now loaded with an inhomogeneous dielectric whose refractive index $N(z)$ varies with z but not with either x' or y'.

For the configuration illustrated in Fig. 21, it is evident that the electric field $E_{y'}$ satisfies the wave equation

$$[\nabla^2 + k^2 N^2(z)]E_{y'} = 0, \qquad (53)$$

EM Propagation in the Earth-Ionosphere Waveguide

where $\nabla^2 = \partial^2/\partial x'^2 + \partial^2/\partial z^2$. If we now seek solutions that have the form $f(z)\exp(-ikSx')$, it is evident that

$$(\partial^2/\partial z^2)f(z) + k^2[\chi(z) + C^2]f(z) = 0, \tag{54}$$

where $\chi(z) = N^2(z) - 1$ is the dielectric susceptibility. Now, since we are looking for a transverse variation that is to be analogous to the height-gain function in the earth-ionosphere waveguide, it is suggested that we choose $\chi(z)$ to have a linear dependence on z in order that the solutions are Airy functions. For example, we let $\chi(z) = 2|z|/a$ for $|z| \leq h$, so that the susceptibility varies from 0 at the centerline to a maximum value of $2h/a$ at the walls. Then, for the range $0 < z < h$, Eq. (54) becomes

$$\frac{\partial^2 f}{\partial z^2} + k^2\left[C^2 + \frac{2z}{a}\right]f = 0. \tag{55}$$

By a simple change of variables, (55) is equivalent to

$$(d^2/dy^2)\Phi(t, y) + (y - t)\Phi(t, y) = 0, \tag{56}$$

where

$$\Phi(t, y) = \text{const} \times f(z),$$

$$(-t) = (ka/2)^{2/3}C^2,$$

$$y = (2/ka)^{1/3}kz.$$

Solutions of (56) are linear combinations of the Airy functions $w_1(t - y)$ and $w_2(t - y)$. For $0 < y < y_0$, we choose

$$\Phi(t, y) = w_1(t - y) + A(t)w_2(t - y), \tag{57}$$

where $A(t)$ is to be determined, and $y_0 = (2/ka)^{1/3}kh$. For the range $0 > z > -h$, we impose the condition that $E_{y'}$ will be even about the centerline. Thus, $\Phi(t, y) = \Phi(t, -y)$, and, for the range $0 > y > -y_0$,

$$\Phi(t, y) = w_1(t + y) + A(t)w_2(t + y). \tag{58}$$

It might be mentioned in passing that solutions that are odd about the centerline also exist, but we shall not need these, since the excitation is to be symmetrical.

We now impose the boundary condition on the narrow walls of the guide. These are taken to be of the form

$$H_{x'} = -Y_i E_{y'}|_{z=h}, \qquad H_{x'} = Y_i E_{y'}|_{z=-h}, \tag{59}$$

where Y_i is the surface admittance. Since $i\mu_0 \omega H_{x'} = \partial E_{y'}/\partial z$, we can write these in the form

$$\left[\frac{1}{i\mu_0 \omega}\frac{\partial E_{y'}}{\partial z} \pm Y_i E_{y'}\right]\bigg|_{z=\pm h} = 0, \tag{60}$$

which is equivalent to

$$\left[\frac{\partial \Phi}{\partial y} \mp q_i \Phi\right]\bigg|_{y=\pm y_0} = 0, \tag{61}$$

where

$$q_i = -i\eta_0 Y_i(ka/2)^{1/3}.$$

When (61) is applied to (57) and (58), we obtain

$$A(t) = -\frac{w_1'(t - y_0) + q_i w_1(t - y_0)}{w_2'(t - y_0) + q_i w_2(t - y_0)}, \qquad (62)$$

which, not surprisingly, is the same form as (25). In order to obtain the modes for the waveguide, we merely equate the $H_{x'}$ fields on both sides of the centerline. This is equivalent to imposing the condition

$$\left.\frac{\partial \Phi}{\partial y}\right]_{y=0^+} = \left.\frac{\partial \Phi}{\partial y}\right]_{y=0^-} \qquad (63)$$

Utilizing (57) and (58), this leads immediately to the condition that

$$A(t)B_0(t) - 1 = 0, \qquad (64)$$

where

$$B_0(t) = -w_2'(t)/w_1'(t). \qquad (65)$$

We see that (64) has the same form as the mode equation (34) for the earth-ionosphere waveguide; however, $B_0(t)$ is only the same as $B(t)$ if $q = 0$ in the latter. Thus, the TE modes in the inhomogeneously loaded waveguide are equivalent to the TM modes in the terrestrial wave when the earth's surface may be regarded as a perfect conductor. This equivalence was the basis of a suggestion *(Wait, 1962a)* to simulate VLF ionospheric propagation in a laboratory model *(Bahar and Wait, 1964)*.

Although in many cases of practical interest in VLF radio propagation the earth may be regarded as a perfect conductor (e.g., for propagation over sea water), there are numerous instances where the attenuation resulting from the finite ground conductivity is dominant. Therefore, it is desirable to extend our planar analog to allow for absorption along the centerline of the guide. The idea is to locate a thin conductive film of thickness d at the center of the guide, as indicated in Fig. 22. The conductivity and dielectric constant of this uniform film are denoted σ_f and ε_f, respectively. It is argued that, by a suitable selection of the parameters σ_f and ε_f, the influence of finite ground conductivity in the terrestrial waveguide can be simulated.

In order to facilitate the analysis, we shall use the following approximate boundary condition for the conductive film:

$$(\sigma_f + i\varepsilon_f \omega) dE_{y'}|_{x'=0} = H_{x'}|_{x'=d/2} - H_{x'}|_{x'=-d/2}. \qquad (66)$$

In essence, this states that the discontinuity of the tangential magnetic field across the film is equal to the transverse current in the film. At the same time, we require that $E_{y'}$ be continuous across the film. Boundary conditions of this type are valid when the thickness of the film or sheet is small compared with a skin depth in the conductor. In other words, we require that $|\gamma_f d| \ll 1$, where $\gamma_f = [i\mu_0 \omega(\sigma_f + i\varepsilon_f \omega)]^{1/2}$ is the propagation constant of the conductive material. The idealization of such a thin sheet has been utilized in electromagnetic induction studies of the earth's crust *(e.g., Wait, 1951, 1953; Rititake, 1966)*. (It is also understood that $|\gamma_f| \gg k$ and $h \gg d$.)

Evidently, the boundary condition (66) may be written as

$$-2q\Phi(t, 0) = \Phi'(t, 0^+) - \Phi'(t, 0^-), \qquad (67)$$

where

$$\Phi'(t, 0^\pm) = (d/dy)\Phi(t_n, y)|_{y=\pm \Delta y} \quad \text{as} \quad \Delta y \to 0,$$

and

$$q = -i(ka/2)^{1/3} \eta_0 (\sigma_f + i\varepsilon_f \omega) d/2.$$

EM Propagation in the Earth-Ionosphere Waveguide

In addition, the continuity of the tangential electric field means that

$$\Phi(t, 0) = \Phi(t, 0^+) = \Phi(t, 0^-). \tag{68}$$

Application of (67) and (68) to the functional forms (57) and (58) leads to the self-consistent condition that

$$A(t)B(t) - 1 = 0, \tag{69}$$

where $A(t)$ is given by (62), and

$$B(t) = -\left[\frac{w_2'(t) - qw_2(t)}{w_1'(t) - qw_1(t)}\right]. \tag{70}$$

We see now that (69) is fully analogous to (34). In both cases, the roots $t = t_n$ give the waveguide modes of the system. Furthermore, we observe, from the respective definitions of q, that the normalized ground impedance Z_g/η_0 is now replaced by the normalized admittance of the film, namely, $\eta_0(\sigma_f + i\varepsilon_f \omega)d/2$.

We see now that (69) may be cast into the approximate phase integral form

$$R_g F_i \exp(-i\Phi) = 1 = \exp(-2\pi i n), \tag{71}$$

where

$$R_g = (C - \Delta)(C + \Delta), \qquad R_i = (C' - \Delta_i)/(C' - \Delta_i),$$

$$\Phi = 2k \int_0^h [C^2 + \chi(z)]^{1/2} \, dz, \qquad \chi(z) = 2z/a.$$

In complete analogy with the earth-ionosphere waveguide, $(-t)^{1/2} = (ka/2)^{1/3}C$ and $(y_0 - t)^{1/2} = (ka/2)^{1/3}C'$. However, in this planar waveguide model, $\Delta_i = \eta_0 Y_i$, where Y_i is the surface admittance of the narrow walls and $\Delta = \eta_0(\sigma_f + i\varepsilon_f \omega)d/2$ is the effective admittance of the central conductive film. It is understood, of course, that the phase integral form given by (71) is only valid if both $(y_0 - t)$ and $(-t) \gg 1$. Actually, these are both satisfied if $(ka/2)^{1/3} \text{Re } C > 2$ or 3.

Within the limits of the phase integral approximation, the ray picture is depicted in Fig. 21, which shows that the angles $\theta = \arccos C$ and $\theta' = \arccos C'$ are preserved in this planar model. Therefore, it appears to be rather convincing that all essential features of the earth-ionosphere waveguide can be simulated in an inhomogeneous loaded planar device.

In order to provide further insight into the analogies between the spherical earth-ionosphere waveguide and the planar analog, we shall say something about the excitation of the modes. In the planar model, we imagine that the sources are two symmetrically placed current line sources, as indicated in Fig. 23. We again assume that all variations perpendicular to the broad walls are zero. Thus, we are interested in the solution for a two-dimensional configuration. As indicated previously, the broad walls are to have perfect conductivity, but the two narrow walls exhibit a surface admittance Y_i, and the central conductive film has thickness d and properties σ_f and ε_f.

Now, the transverse field $E_{y'}$ can be written as a sum of modes as follows:

$$E_{y'} = \sum_n \Phi(t_n, y) \exp(-ikS_n|x'|)A_n, \tag{72}$$

where $\Phi(t_n, y)$ is given by (57) and (58), and t_n are solutions of (69). The summation is to extend over all modes, and the coefficient A_n is as yet undefined. Anticipating the nature of the final results, it is convenient to rewrite (72) in the equivalent form

$$E_{y'} = \sum_n b_n G_n(y) \exp(-ikS_n|x'|), \tag{73}$$

where $G_n(y) = \Phi(t_n, y)/\Phi(t_n, 0)$ is the height-gain function, which has the property that $G_n(y) = G_n(-y)$ for symmetrical excitation. In (73), b_n is a new dimensionless coefficient proportional to the preceding A_n.

The initial or source condition on the fields is now taken to be

$$H_z|_{x'=0^+} - H_z|_{x'=0^-} = -I[\delta(z-\hat{z}) + \delta(z+\hat{z})], \tag{74}$$

which is simply a consequence of Ampere's law. The delta or impulse functions are appropriate, since we assume filamental currents I at $z = \pm \hat{z}$. In terms of the "natural" y coordinate, we note that

$$\delta(z \pm \hat{z}) = k(2/ka)^{1/3} \delta(y \pm \hat{y}), \tag{75}$$

where $\hat{y} = (2/ka)^{1/3} k\hat{z}$.

The vertical magnetic field is obtained from (73) and may be written in the form

$$H_z = \pm(\eta_0)^{-1} \sum_n S_n b_n G_n(y) \exp(-ikS_n|x'|), \tag{76}$$

where the plus sign is used for $x' > 0$ and the minus sign for $x' < 0$.

On combining (74)–(76), the required condition on b_n is

$$(\eta_0)^{-1} \sum_n S_n b_n G_n(y) = -(I/2)k(2/ka)^{1/3}\delta(y - \hat{y}), \tag{77}$$

which applies over the range $0 < y < y_0$. In the usual manner, we find b_n by multiplying both sides of (77) by $G_m(y)$ and integrating over y from 0 to y_0. Since

$$\int_0^{y_0} G_n(y) G_m(y)\, dz = \begin{cases} 0 & \text{for } m \neq n \\ y_0/(2\Lambda_n) & \text{for } m = n, \end{cases} \tag{78}$$

we obtain

$$b_n = -\frac{I\eta_0}{h\, S_n} \Lambda_n G_n(\hat{y}). \tag{79}$$

The vertical magnetic field resulting from the double-line source excitation is thus given by

$$H_z = \mp(I/h) \sum_n \Lambda_n G_n(\hat{y}) G_n(y) \exp(-ikS_n|x'|), \tag{80}$$

where, as indicated before, the summation is over all modes. The excitation coefficient Λ_n, used here, has the same form as that given by (37). We should remember, of course, that the q and q_i parameters have a different meaning.

It is a simple exercise to show that, in the limit of the unloaded guide (i.e., $a \to \infty$) and for the absence of a conductive strip, Λ_n tends to unity. In this limit, we also find that the height-gain function is given by

$$G_n(y) = \cos(kC_n z) \tag{81}$$

for $-h > z > h$, where C_n are symmetrical solutions of the "planar" mode equation

$$R_i \exp(-i2khC) = \exp(-i2\pi n), \tag{82}$$

where $R_i = (C - \Delta_i)/(C + \Delta_i)$ and $\Delta_i = \eta_0 Y_i$.

We might mention that there are some subtle points that arise if one tries to scrutinize the fine distinctions between the full spherical terrestrial waveguide and the planar model. First of all, we should remember that, in approximating the spherical wave functions by the Airy function form given by (9), we require among other things that $|v - kr| \ll (ka/2)^{2/3}$. In terms of the complex cosine of the ground reflection coefficient, this means that $|C_n^2| \ll 1$ for all important modes. Fortunately, in all cases of prime interest, this is a valid approximation in the actual earth-ionosphere waveguide. The phase factor $\exp(-ikS_n|x'|)$ may be replaced by $\exp(-ik|x'|)\exp(+ikC_n^2|x'|/2)$, where, of course, we require that, for finite wall losses, $+\operatorname{Im} C_n^2 > 0$. In fact, the attenuation rate is $+k \operatorname{Im} C_n^2/2$ in nepers per meter along the guide centerline of the planar model. The corresponding phase velocity v_n, for a mode of order n, is obviously given by

$$v_n = (\omega/k)(\operatorname{Re} S_n)^{-1} \simeq (\omega/k)(1 - \operatorname{Re} C_n^2/2)^{-1} \quad \text{m/sec} \tag{83}$$

(i.e., we always use rationalized mks units).

We should like to mention that the earth-detached phenomenon in the terrestrial waveguide has an interesting analogy in the planar model. To illustrate, we consider the mode equation (69) under the condition that $|t_n| \gg 0$ with $|\arg t_n| < \pi/3$ and $|y - t_n| \gg 0$ with $|\arg y - t_n| < \pi/3$. Using the leading term of the appropriate asymptotic expansion, we then find that

$$\frac{C_n' - \Delta_i}{C_n' + \Delta_i} \exp\left[i\left(\frac{\pi}{2} - \frac{2ka}{3}(C_n')^3\right)\right] = \exp[-i2(\pi n - \delta_n)], \tag{84}$$

where

$$\delta_n = \frac{\exp[-\frac{4}{3}t_n^{3/2}]}{2(1 - qt_n^{-1/2})}(1 + qt_n^{-1/2}), \tag{85}$$

where $t_n = -(ka/2)^{2/3}C_n^2$ and $q = -i(ka/2)^{1/3}\Delta$, and we have used the relation

$$(y_0 - t)^{1/2} = (ka/2)^{1/3}C_n'. \tag{86}$$

The phase integral form of (84) may be written as

$$2k \int_{z_{0n}}^{h} [C_n^2 + \chi(z)]^{1/2} dz + \Phi(\theta_n') - [(\pi/2) - 2\delta_n] = 2\pi n, \tag{87}$$

where, as usual, the susceptibility $\chi(z) = 2z/a$, and z_{0n} is the "turning point" defined by $\chi(z_{0n}) + C_n^2 = 0$ or simply $z_{0n} = -aC_n^2/2$. Here, Φ is a "complex phase" defined by

$$\exp[-i\Phi(\theta_n')] = (C_n' - \Delta_i)/(C_n' + \Delta_i). \tag{88}$$

The ray picture that is suggested by (87) is indicated in Fig. 24, where the distance from the waveguide centerline to the caustic surface is z_{0n}. Actually, in the case of losses, this quantity is complex. However, if, for convenience of understanding, we imagine the narrow walls to be purely reactive (i.e., $\Delta_i = i|\Delta_i|$) and the centerline to be a purely dielectric film (i.e., $\Delta = i|\Delta|$ or $q = |q|$), we now see that Φ is a measure of the phase lag for a reflection of the ray from the waveguide walls at the points I. On the other hand, there is actually a $(\pi/2 - 2\delta_n)$ phase lead as the ray skims the point P. The $\pi/2$ factor is what one would expect for the turning points of the ray system (i.e., at the caustic surfaces). However, the additional correction $2\delta_n$ is a result of the finite value of z_{0n}. For example, if we set

$$e_n = (1 - v_n/c),$$

where v_n is the phase velocity of the mode and $c = \omega/k$, we readily find that

$$\delta_n \simeq \frac{\exp[-2^{5/2}(ka/3)e_n^{3/2}]}{2[1 - |\Delta|2^{-1/2}e_n^{-1/2}]}[1 + |\Delta|2^{-1/2}e_n^{-1/2}], \tag{89}$$

which vanishes exponentially when the phase velocity v_n is somewhat less than c. This corresponds to the earth-detached phenomenon discussed before [e.g., as exemplified by (51)].

In (29) and (30), we indicated that the field in the earth-ionosphere wave-waveguide could be expressed as a sum of multiply reflected waves. Each of these was denoted V_j, where j is the order of the "hop." Within the geometrical-optical approximation, we may write, for ground-based terminals,

$$V_j = R_i^j R_g^{j-1}(1 + R_g)^2 \alpha_j e^{+i\Omega_j}, \tag{90}$$

where $R_i = (C_j' - \Delta_i)/(C_j' + \Delta_i)$ is the ionosphere reflection coefficient for the jth hop, and where $R_g = (C_j - \Delta)/(C_j + \Delta)$ is the ground reflection coefficient for the jth hop. Here, C_j and C_j' are the cosine of the complex angles of incidence at the ground and ionosphere, respectively. The convergence coefficient is defined by $\alpha_j = [1 + x(2jp)^{-1}]^{1/2}$, where $p^2 = -t = (ka/2)^{2/3}C_j^2$, with $p \gg 1$. Finally, the phase factor in (90) is

$$\Omega_j = -xp^2 + \tfrac{4}{3}j(y_0 + p^2)^{3/2} - \tfrac{4}{3}jp^3,$$

where $x = (ka/2)^{1/3}\theta$, $\theta = d/a$, and $y_0 = (2/ka)^{1/3}kh$. The situation is illustrated in Fig. 25, where only the first three hops are shown. The number of arrows indicate the order of the hop.

Actually, (90) is a stationary phase approximation of the integral representation for the various rays. Thus, it is only valid when t is sufficiently negative for the various stationary phase points. In essence, this means that (90) is only valid if $(ka/2)^{1/3} \operatorname{Re} C_j > 2$ or 3. [Because $C_j' \cong (C_j^2 + 2h/a)^{1/2}$, the condition that $(ka/2)^{1/3} \operatorname{Re} C_j' > 2$ or 3 is automatically satisfied if the foregoing inequality holds.] Now, it is obvious that this condition will never hold for $j = 0$ (i.e., the ground wave). Thus, we should always use (20) or some variant of it.

The theory for the ray treatment in the loaded planar waveguide is essentially the same as in the curved, unloaded waveguide. Again, we choose the susceptibility $\chi(z)$ in the guide of width $2h$ to be $2|z|/a$ for the range $-h < z < h$. The corresponding picture of the first three rays is illustrated in Fig. 26. As indicated, the great circle distance TP in the curved model is now identified with the linear distance TP in the planar model. In the latter case, the rays are curved, but the angles of incidence are all preserved. The remarks made before about wall loading and the resistive strip still apply.

The geometrical-hop picture obviously breaks down when the ground reflection angles become near grazing. As the range increases, it is evident that difficulty is first encountered when C_1 (i.e., $j = 1$) tends to zero. In this caustic region, $C_1' = (C_1^2 + 2h/a)^{1/2} \simeq (2h/a)^{1/2}$, whence $d \simeq 2(2ah)^{1/2} \simeq 2000$ km. The first-hop ray is shown in Figs. 27 and 28 for this situation. Of course, one must realize that the physical meaning of the "ray" becomes clouded when θ tends toward 90°. Nevertheless, the geometry is valid.

To calculate the field in the region where the ground ray is grazing, or near grazing, we must return to the integral representations (29) and (30), which are uniformly valid for d greater than or less than the critical range $2(2ah)^{1/2}$. To facilitate understanding, we assume that, although θ is near

$90°$, the angle θ' is sufficiently different from $90°$ that $(ka/2)^{1/3}C' > 2$ or 3. As usual, $C = \cos\theta$ and $C' = \cos\theta'$, where the subscript $j = 1$ is dropped for the present discussion.

Under the circumstances indicated, we are able to write the first hop contribution *(Wait, 1961a)* in the form

$$V_1 \simeq 2(x/\pi)^{1/2} e^{i\pi/4} e^{-i(4/3)y_0} R_i I_1, \tag{91}$$

where

$$R_i \simeq \frac{(2h/a)^{1/2} - \Delta_i}{(2h/a)^{1/2} + \Delta_i}, \tag{92}$$

and

$$I_1 = \int_0^\infty \left\{ \frac{e^{-iXt}}{[w_1'(t) - qw_1(t)]^2} + \frac{e^{-i\pi/3} e^{-\sqrt{3}Xt/2} e^{iXt/2}}{[w_2'(t) - qe^{-i2\pi/3} w_2(t)]^2} \right\} dt, \tag{93}$$

where $X = x - 2y_0^{1/2} = (ka/2)^{1/3}[(d - 2(2ah)^{1/2})/a]$.
Some rather obvious simplifications have been made here, which are a result that $|d - 2(2ah)^{1/2}| \ll d$, for the region of interest.

To interpret the integral I_1, we look at its asymptotic approximation, valid for $p = -X/2 \gg 1$, whence

$$I_1 \sim \frac{1}{2} \left(\frac{\pi}{2p} \right)^{1/2} e^{-i\pi/4} \left(\frac{2p}{p + iq} \right)^2 e^{-i(2/3)p^3}. \tag{94}$$

Actually, this is consistent with the geometrical-optical formula given by (90) when d is near $2(2ch)^{1/2}$. It is evident that I_1 would become infinite at $p = 0$, which, of course, is not permitted in view of the restriction $p \gg 1$. This infinity at $p = 0$ also appears in the convergence coefficient α_j when the jth hop arrives tangentially at the earth's surface. However, as indicated, geometrical optics is not valid near these caustic points.

To illustrate the range of validity of the asymptotic approximation, we show in Fig. 29 an Argand plot of I_1 where the results based on (93) and (94) are intercompared. For X sufficiently negative, the two forms become indistinguishable. However, the results for $-X$ less than about 2.5 differ appreciably, and, when X is near zero or when it becomes positive, the asymptotic form (corresponding to geometrical optics) is not valid.

Actually, when $d > 2(2ah)^{1/2}$, the observer at P is in the "shadow" of the source at T, as indicated in Figs. 30 and 31 for the two configurations. In this case, it may be shown *(Wait, 1961a)* that V_1 may be written as a sum of modes that correspond to the residues of the poles of the integrand of (30) when $j = 1$. This has the form

$$V_1 \cong \sum_{s=1, 2, 3, \ldots} A_s \exp[-i(x - 2y_0^{1/2})t_s],$$

where A_s is a coefficient and where t_s are the roots of the equation $w_1'(t) - qw_1(t) = 0$. Thus, we see that the once-reflected sky wave, at sufficiently large distances, may be represented by a sum of "ground wave" modes. The attenuation of these is determined by the magnitude of the distance $TA + BP$ in Figs. 30 and 31 which is proportional to $(x - 2y_0^{1/2})$. Similar remarks apply to the multiple-hop sky waves, but from a practical standpoint the latter are not particularly significant.

Some applications and extensions of the diffraction theory of wave-hop propagation have been carried out by Berry *(1964)* and by Berry and Chrisman *((1965a)* 'The latter authors have also presented extensive numerical calculations for the relevant integrals: *(Berry and Christmas, 1965b)* Their work is characterized by the thoroughness in which the problem is formulated. As a result, they are able to clarify where our own and more approximate methods are valid. Fortunately, the introduction of our Airy function representations is justified in the regions where they are meant to be used.

EXTENSION TO AN INHOMOGENEOUS IONOSPHERE

In the preceding discussions, we have characterized the ionosphere by a surface impedance at some height where we expect the waves to be reflected. In the case of a sharply bounded ionosphere, this approach makes a certain amount of sense, and the resulting procedures are straightforward. However, under realistic conditions, both the electron density $N(z)$ and the collision frequency $v(z)$ will vary with height z (above some reference level $z = 0$) in a complicated fashion. The form of these functions is influenced by many factors, and we shall not attempt to discuss these here. However, a study of a large body of data has indicated that exponential variations of both N and v are a fair description, at least for the daytime ionosphere *(Wait and Spies, 1964)* Thus we choose

$$\omega_r(z) = \omega_r(z_0) \exp \beta(z - z_0),$$

where $\omega_r(z_0) = \omega_0^2/v$ in terms of the plasma frequency and collision frequency at the reference height denoted $z = z_0$. For normalization purposes, $\omega_r(0) = 2.5 \times 10^5$ sec^{-1}. Here, β is the parameter that describes the rate of change of the effective conductivity parameter $\omega_r(z)$ as a function of height.

Calculations of the reflection coefficient R_v for a vertically polarized wave from an exponential layer of the type described have been carried out by Wait and Walters *(1963)* The situation is illustrated in Fig. 32, where we have a plane wave incident from the free-space region below (i.e., $z \to -\infty$), which, in turn, is reflected as a plane wave. The reflection coefficient R_i is then defined, in terms of the incident electric field E_0 and the reflected electric field E at $z = -\infty$, by

$$E = E_0 R_i(C') \exp(-2ikC'z_0) \exp(2ikC'z),$$

where $C' = \cos \theta'$ in terms of the angle of incidence θ'. It is important to recognize that $R_i(C')$ is the reflection coefficient of the inhomogeneous medium referred to the level $z = z_0$. In other words, we may replace the diffusely bounded layer by a sharp boundary at $z = z_0$ whose reflection coefficient is $R_i(C')$.

With the normalization indicated, the amplitude and phase of the reflection coefficient R_v are shown in Fig. 33, for 20 kHz, as a function of the gradient parameter β for various angles of incidence. As indicated, the smaller values of β are associated with lower reflection coefficients. As β increases, $|R_v|$ rises to a broad maximum. Further increases of β show a slight diminishing of R_v, which ultimately recovers and rises eventually to unity as β becomes very large. Further insight into the nature of the reflection process is indicated in Fig. 34, where the magnitudes of both the vertically polarized and the horizontally polarized reflection coefficients are plotted together as a function of C' or $\cos \theta'$. As predicted by theory *(Wait, 1970)* the function $\log |R_h|$ exhibits a linear dependence with C' over the whole range. This linear dependence does not hold, in general, for $\log |R_v|$, but, nevertheless, it is a very good approximation for the small values of C' (e.g., $C' < 0.3$). The curvature in the curves for vertical polarization leads to a minimum or at least a bulge downward that is related to a Brewster absorption phenomenon.

EM Propagation in the Earth-Ionosphere Waveguide

A convenient description of the reflection process is to write

$$R_i(C') = -\exp[(\alpha_1 + i\alpha_2)C'],$$

where the coefficient $(\alpha_1 + i\alpha_2)$ is nearly independent of C' for angles of incidence which are sufficiently great (i.e., near grazing). As an example, we show $-\alpha_1$ and α_2 as a function of the wavelength λ_0 in Fig. 35. Here, we choose $C' = 0.16$, which is appropriate for the modes of lowest attenuation. We now argue that, although the reflection coefficient calculations were carried out for only real angles of incidence (i.e., real C'), the exponential formula for R_i, given previously, may be used to continue it analytically into the complex C' plane. This step, although not essential, does permit a great saving in the amount of calculations required for determining the modal characteristics. The relevant modal equation for this case is thus written as

$$R_g(C)R_i(C')F \exp(-i\Phi) = \exp(-2\pi i n), \qquad (95)$$

where

$$R_g(C) = (C - \Delta)/(C + \Delta), \qquad R_i(C') = -\exp(\alpha C'),$$

$$\alpha = \alpha_1 + i\alpha_2, \qquad \Phi = 2k \int_0^h (C^2 + 2z/a)^{1/2} \, dz,$$

and

$$F = \frac{\left[\dfrac{w_2'(t) - qw_2(t)}{w_1'(t) - qw_1(t)}\right]}{i \exp\left[i\dfrac{4}{3}(-t)^{3/2}\right]\left[\dfrac{-i(-t)^{1/2} - q}{i(-t)^{1/2} - q}\right]}. \qquad (96)$$

In this case, we choose a the effective reflection height of the waveguide to be the same as z_0 the reference height, where $\omega_r = 2.5 \times 10^5$. The selection of h involves a certain arbitrariness in that, for a curved waveguide model, the final results will not be independent on how we define the reference level. Actually, earth curvature should be included in the wave calculations for the coefficient α, but, obviously, this would complicate our procedures. Fortunately, as adjudged by the recent work of Galejs *(1964)* and Pappert et al *(1967)* the method we used to define the reference height does not lead to significant discrepancies.

Extension to Anisotropic Ionosphere

By virtue of the earth's main magnetic field, the ionosphere is rendered anisotropic. Therefore, the assumption of a single scalar reflection coefficient for each type of incident polarization is not valid in general. The explanation for the anisotropy is found in magnetoionic theory, and a vast amount of literature is available. An excellent treatise on the subject is one by Budden *(1961)*.

In connection with the earth-ionosphere waveguide, the most easily understood effect is the change of polarization at reflection when the ionized medium is anisotropic. For example, if a vertically polarized wave, with electric field E_{0v}, is incident at an angle θ', the resultant electric field of the reflected wave is $E_{0v \parallel} R_\parallel$ in the plane of incidence, and $E_{0v \parallel} R_\perp$ perpendicular to this plane. Similarly, if the incident wave is horizontally polarized with electric field E_{0h}, the resultant electric field of the reflected wave is $E_{0h \perp} R_\parallel$ in the plane of incidence, and $E_{0h} \perp R_\perp$ perpendicular to the plane of incidence. The situation is illustrated in Fig. 36, where the plane of incidence is in the plane of the paper.

The reflection process just described can be written compactly in matrix form. Thus, the reflected wave, written as a column matrix, is given by

$$\begin{bmatrix} E_v \\ E_h \end{bmatrix} = \begin{bmatrix} {}_\parallel R_\parallel & {}_\perp R_\parallel \\ {}_\parallel R_\perp & {}_\perp R_\perp \end{bmatrix} \begin{bmatrix} E_{0v} \\ E_{0h} \end{bmatrix}, \quad (97a)$$

where now the reflection process is described by a 2 × 2 matrix that operates on the column matrix of the incident wave. In passing, we note that the reflection coefficient for an isotropic medium is described by a diagonal matrix. For example, the reflection coefficient matrix for a locally flat ground is simply

$$\begin{bmatrix} R_v & 0 \\ 0 & R_h \end{bmatrix},$$

where R_v and R_h are the Fresnel reflection coefficients for vertical and horizontal polarization, respectively.

When the upper boundary of the earth-ionosphere waveguide is an anisotropic reflector of the type described previously, the transverse magnetic (TM) and the transverse electric (TE) modes are obviously coupled. The mode equation in this case can be written in the physically meaningful form (Wait, 1970).

$$(1 - {}_\parallel R_\parallel R_v F_v e^{-i\Phi})(1 - {}_\perp R_\perp R_h F_h e^{-i\Phi})$$
$$- {}_\parallel R_\perp {}_\perp R_\parallel R_v F_v R_n F_h e^{-2i\Phi} = 0, \quad (97b)$$

where, as usual, $\Phi = 2k \int_0^h (C^2 + 2z/a)^{1/2} dz$, where h is the height of the reflector above the ground and where $R_v = (C - \Delta_v)/(C + \Delta_v)$, and $R_h = (C - \Delta_h)/(C + \Delta_h)$. Here, F_v is the earth curvature function given by (96), where $q = q_v = -i(ka/2)^{1/3} \Delta_v$, whereas F_h has the same form, but, instead, we use $q = q_h \cong -i(ka/2)^{1/3} \Delta_h$. It is evident that, under isotropic conditions where ${}_\parallel R_\perp = {}_\perp R_\parallel = 0$, (97b) breaks into two decoupled TM and TE modal equations; they are, respectively,

$$1 - {}_\parallel R_\parallel R_v F_v e^{-i\Phi} = 0,$$
$$1 - {}_\perp R_\perp R_g F_h e^{-i\Phi} = 0,$$

which are analogous to (95). The effect of the ionospheric anisotropy is to couple the TM and TE mode types. Also, it is understood that the dc magnetic field will change ${}_\parallel R_\parallel$, and, in some cases, nonreciprocal effects will be in evidence.

In the formal development of the theory (Wait, 1963) for propagation in the curved earth-ionosphere waveguide the modes are still characterized by an angular dependence of the form $\exp(-ika\theta - ixt_s)$, where $x = (ka/2)^{1/3}\theta$, and t_s are eigenvalues. Now however, in place of (34), we find that t_s are roots of

$$\det\{[A(t)][B(t)] - [1]\} = 0, \quad (98)$$

where

$$[A(t)] = \begin{bmatrix} {}_\parallel R_\parallel & {}_\perp R_\parallel \\ {}_\parallel R_\perp & {}_\perp R_\perp \end{bmatrix} \exp[-i\tfrac{4}{3}(y_0 - t)^{3/2} - i\pi/2)], \quad (99)$$

and
$$[B(t)] = \begin{bmatrix} -\dfrac{w_2'(t) - q_v w_2(t)}{w_1'(t) - q_v w_1(t)} & 0 \\ 0 & -\dfrac{w_2'(t) - q_h w_2(t)}{w_1'(t) - q_h w_1(t)} \end{bmatrix}, \quad (100)$$

and where
$$[1] = \begin{bmatrix} 1 & 0 \\ 0 & 1 \end{bmatrix}$$

is the identity matrix. An extensive review of this theory *(Wait, 1963)* is given in the monograph by Al'pert and Fligel *(1970)*.

As indicated in (98), the relevant mode equation is obtained by taking the determinant of the resultant square matrix. On noting the usual identity $(-t)^{1/2} = (ka/2)^{1/3}C$, it may be seen that (98) is actually the same as (97b).

To compute the field for a vertical electric dipole located on the surface of the earth, we find it again convenient to employ matrix notation. Thus, the radial components of the fields at $(r, 0)$ within the earth-ionosphere waveguide are obtained from

$$\begin{bmatrix} E_r \\ \eta H_r \end{bmatrix} = \frac{e^{-ika\theta}}{(a \sin \theta)^{1/2}} \begin{bmatrix} V_e \\ V_h \end{bmatrix}, \quad (101)$$

where V_e and V_h are the elements of the "attenuation function" column matrix. In analogy to (35),

$$\begin{bmatrix} V_e \\ V_h \end{bmatrix} = -2(\pi x)^{1/2} e^{-i\pi/4} \sum_{s=0,1,2,\ldots} \frac{e^{-ixt_s}}{w_1'(t_s) - qw_1(t_s)} [\Omega_s'(t_s)]^{-1}$$
$$\times [w_1(t_s - y) + [A(t_s)]w_2(t_s - y)] \begin{bmatrix} 1 \\ 0 \end{bmatrix}, \quad (102)$$

where
$$[\Omega(t)] = [A(t)][B(t)] - [1], \quad (103)$$

$$[\Omega'(t)] = \frac{(\partial/\partial t)\{\det[\Omega(t)]\}}{\det[\Omega(t)]} [\Omega(t)]. \quad (104)$$

As mentioned before, t_s are solutions of the mode equation
$$\det[\Omega(t)] = 0. \quad (105)$$

Calculations based on (102) require a rather complicated sequence of operations. Fortunately there is considerable simplification in one important situation. When the propagation is purely transverse to the dc magnetic field and the dip angle is zero, there is no cross-coupling between the modes. Some extensive numerical results for this case are available in a published technical note *(Wait and Spies, 1964)*. In those cases, it was assumed that both the electron density $N(z)$ and the collision frequency varied with height in an exponential manner. For example, we assumed that

$$N(z) = N(h) \exp[b(z - h)],$$
$$v(z) = v(h) \exp[a(h - z)],$$

where h is the reference height. The latter was chosen again, where
$$v_r = \omega_0^2(h)/v(h) = 2.5 \times 10^5.$$

Finally, the strength of the transverse dc magnetic field was indicated by the parameter Ω, which was defined by

$$\Omega = \omega_T/v(h),$$

where ω_T was the gyrofrequency. For propagation from west to east, Ω was negative, whereas Ω was positive for propagation from east to west. Under isotropic conditions, $\Omega = 0$.

An example of the available calculations *(Wait and Spies, 1964)* is indicated in Fig. 37, where the attenuation rate of the dominant mode is shown as a function of frequency, for a perfectly conducting ground. In this case we choose $\beta = b + a = 0.3$ km^{-1} and $a = 0.15$ km^{-1}. The reference height was allowed to assume four different values, and so the results are indicative of the variation to be expected as a function of the time of day.

Actually, Galejs *(1967)* has presented calculations of the attenuation rates and phase velocities for an ionospheric model which incorporate the D-region profiles proposed by Deeks *(1966)*. The latter are believed to be very realistic. Also, the interested reader is well advised to study the extensive work of R.A. Pappert *(1967)* and his colleagues who have demonstrated the effect of the earth's magnetic field on the mode characteristics. They have emphasized the importance of the transverse electric type modes that can be excited electric dipoles in the terrestrial waveguide bounded by an anisotropic ionosphere.

COMPARISON WITH SOME EXPERIMENTAL DATA

It is of interest to compare some of the calculated curves with appropriate experimental data as reported in the literature. There are two distinct sources of such data. These are recordings of field strengths of distant VLF transmitters and the observations of the waveforms of atmospherics which originate in lightning discharges. Comprehensive surveys of the various experimental methods are given in papers by Watt and Croghan *(1964)* and by Horner *(1964)*

The phase velocity is a rather crucial characteristic in the theory of the propagation of VLF radio waves. The U.S. Navy Electronics Laboratory (N.E.L.) has obtained some valuable experimental data for phase velocity in the frequency range from 9.2 to 15.2 kHz. The results *(Tibbals, 1960; Pierce and Nath, 1961)* are shown in Fig. 38 for both daytime and nighttime paths predominantly over the sea. The vertical bars encompass the range of several independent measurements for the frequencies indicated. These results were obtained by employing an ingenious technique that combined the results observed from widely spaced transmitting stations operating in sequence. As a consequence, the data for east-to-west and west-to-east paths are averaged in a sense.

In a further study (during July 1963), Steele and Chilton *(1964)* conducted an experiment to measure phase velocity by utilizing frequency-stabilized signals radiated in sequence from transmitters NPG and NBA at frequencies of 18 kHz. They measured the phase in Colorado, Alaska, Hawaii, and Argentina. From a combined analysis, they deduced the average phase velocities for night and day as indicated in Fig. 38.

Theoretical curves are shown in Fig. 38 which correspond to the calculations for an exponential isotropic ionosphere with $\beta = 0.5$, $\sigma_g = \infty$, $n = 1$, and heights $h = 70$ and 90 km. Also shown is a similar pair of calculated curves for a homogeneous and sharply bounded isotropic ionosphere characterized by $\omega_r = 2 \times 10^5$ sec^{-1}. It is evident that the agreement between the experimental points and the calculated curves is quite good. Here, there is not much to choose between the two theoretical models.

A comparison between theory and experiment, showing the effect of finite ground conductivity, is shown in Fig. 39. The two indicated experimental points, for propagation over sea and land, at 10.2 kHz are quoted from the work of Pierce and Nath (1961) The difference between these, expressed as a ratio to c, corresponds to about 5×10^{-4}. Corresponding theoretical curves for earth conductivities of ∞, 5, and 1 mmho/m are also shown. As somewhat of a coincidence, the $\sigma_g = \infty$ curve passes through the experimental point for sea water, and the $\sigma_g = 5$ curve passes through the point for land. Thus, the theoretical prediction that the finite ground conductivity slows the wave down is confirmed experimentally.

The excitation factor Λ is a rather elusive quantity that is not always understood. Fortunately, Watt and Croghan (1964) in a noble effort, have taken a vast amount of experimental daytime data for VLF propagation over sea-water paths and extrapolated it back to zero distance in such a manner that experimentally deduced values of $|\Lambda|$ may be estimated. In some cases, they required a knowledge of effective radiated power of the transmitting antenna. The vertical bars indicated in Fig. 40 show rather crudely the range of the data points which they deduced for mode 1. As indicated, it seems to fall on a theoretical curve calculated for $h = 80$ km and $\beta = 0.5$. It would have been more satisfying to see closer agreement with the $h = 70$ km curve, which is several decibels higher. However, two factors might contribute to this apparent discrepancy. In the first place, the assumed radiated powers might be lower than claimed. Secondly, the effect of conversion of energy from mode 1 to higher modes would also tend to reduce the apparent excitation efficiency.

Experimental data on attenuation rates relating to nighttime propagation over sea are indicated in Fig. 41. The vertical bars indicate the range of data points quoted by Taylor and Lange (1958), who analysed the waveforms of atmospherics observed simultaneously at several stations. The dashed curve at the bottom of Fig. 41 is quoted directly from Watt and Croghan (1964) who deduced it mainly from the early data of Round et al. (1925). Calculated curves are shown for mode 1 and h = 90km for exponential ionospheres with both $\beta = 0.3$ and 0.5 km^{-1}. It is apparent that the $\beta = 0.5$ curve is certainly more representative for nighttime propagation as suggested. The theoretical curve for mode 2 and $\beta = 0.5$ is also shown, which indicates that modal interference will be significant at frequencies above 20 kHz. The spread of the experimental data points attests to this fact. Also, for comparison, the curve for an ionosphere with $\beta = 1$ is shown in Fig. 41. It appears to be quite near the experimental curve attributed to Watt and Croghan (1964). Finally a calculated attenuation curve for a homogenous, sharply bounded, and isotropic ionosphere (with $h = 90$ km and $\omega_r = 2 \times 10^5$) is shown on Fig. 41 which is in only fair agreement with the experimental data.

Measured daytime attenuation rates (*Watt and Croughan*) for propagation over sea are shown in Fig. 42 by a curve that is the average for north-to-south paths at temperate latitudes. This is seen to agree reasonably well with a calculated curve for an exponential ionosphere with $h = 70$ km and $\beta = 0.3$ km^{-1}. The calculated curve for $\beta = 0.5$ is decidedly too low, which is also the case for the homogeneous, sharply bounded model for $h = 70$ km.

The measured dependence of attenuation rate on direction of propagation is indicated in Fig. 43 for daytime propagation over sea. The data attributed to Watt and Croghan (1964) are based on the analysis of field strength for paths at temperate latitudes which are predominantly west to east, north to south, or east to west. Again, some of their data are taken from the classic paper by Round et al (1964). Corresponding attenuation data deduced from atmospheric waveforms by Taylor (1960 a and b) are also shown in Fig. 43. These data are also for paths at temperate latitudes over sea water in daytime.

Calculated attenuation curves for mode 1 with $\beta = 0.3$ km^{-1}, $a = 0.15$ km^{-1}, $\sigma_g = \infty$, and $h = 70$ km are shown in Fig. 43 for comparison with the experimental curves. As indicated, the values of Ω given by $-1, 0,$ and $+1$ seem to be quite appropriate. Any other choice of the magnitude of Ω would not give the right amount of nonreciprocity.

Swanson (1964) has communicated his experimental results to the author. For propagation at 10.2 kHz over the sea in daytime and in summer, he finds attenuation rates as follows: 2.5 dB/1000 km for west to east; 3.4 dB/1000 km for north to south or south to north; and 4.1 dB/1000 km for east to west. These may be compared with the respective values in Fig. 43 of 2.65, 3.33, and 4.28 dB/1000 km. The dip angle for these measurements was of the order of 55°, whereas the calculations are for a purely horizontal magnetic field. The good agreement is attributed to the fortuitous choice of the magnitude of the parameter Ω. Swanson also found that, for the same conditions, at night the attenuation rate at 10.2 kHz was 1.6 dB/1000 km for north-south propagation, whereas Fig. 41 indicates a calculated value of 1.5 dB/1000 km when $\beta = 0.5$ km^{-1}, $n = 1$, $\Omega = 0$, and $h = 90$ km. Furthermore, he quotes the average nonreciprocal variation at night of ± 0.5 dB/1000 km, which is certainly consistent with the calculated variation indicated in Fig. 37. At the same time, Swanson comments that the N.E.L. data in Fig. 38, for 10.2 kHz, now indicate that $(v/c) - 1 \simeq 0.34 \times 10^{-2}$ for day and $\simeq 0.06 \times 10^{-2}$ for night, for propagation over sea water in the temperate latitudes during the summer.

The interference pattern between the first and second modes has been observed recently by Rhoades and Garner (1967), who showed some very convincing comparisons with the mode calculations for an exponential ionosphere. Other investigators (Lynn, 1967; Ries, 1967; Steele and Crombie, 1967; Bickel, 1967; Burgess and Jones, 1967) have also found that the waveguide calculations (Wait and Spies, 1964) for an exponential ionosphere are consistent with much of their experimental data on both amplitude and phase.

PROPAGATION IN THE EARTH-IONOSPHERE WAVEGUIDE WITH SLOWLY VARYING CHANGES IN HEIGHT

In the previous discussion, we have assumed that the boundaries of the earth-ionosphere waveguide are uniform in the sense that their properties do not change along the path between transmitter and receiver. However, under actual conditions, the effective surface impedance and the reflecting heights will not, in general, be constant. For example, the strength and direction of the terrestrial dc magnetic field will vary along the path except in the unlikely situation that propagation is around the magnetic equator. Fortunately, however, this is the type of lateral variation which is sufficiently slow that the waveguide is locally uniform. In other words, the height-gain function for a propagating mode is determined only by the local height of the guide and its effective surface impedance. Thus, under such conditions we can neglect the effects of mode conversion.

In order to illustrate the nature of the slowly varying nonuniform waveguide, we consider the two-dimensional model shown in Fig. 44. The parameters are the same as those used in the uniform model, but now we permit them to be slowly varying functions of x', which is the normalized distance from the source to a variable point on the path. As indicated, the reflection height $y_0(x')$ varies from $y_0(0)$ to $y_0(x)$ over the range $0 < x' < x$. The corresponding surface impedance parameters for the upper and lower boundary are $q_i(x')$ and $7(x')$, respectively. (For simplicity, we are assuming that the upper boundary is characterized by a scalar surface impedance.)

The magnetic field $H(x, y)$ at the receiving location is related to the reference field $H_0(x, y)$ in the usual manner:

$$H(x, y) = H_0(x, y)\tilde{V},$$

where $\tilde{V}$ is an attenuation function. Some consideration (Wait, 1964b) shows that

$$\tilde{V} \simeq \frac{4(\pi x)^{1/2}}{[y_0(0)y_0(x)]^{1/2}} e^{-i\pi/4} \sum_{n=1,2,3,\ldots} \exp\left[-i\int_0^x t_n(x')\,dx'\right]$$
$$\times \tilde{\Lambda}_n G_n(\hat{C}, \hat{y})G_n(x, y), \tag{106}$$

where $t_n(x')$ is a solution of the "local" mode equation

$$A[t_n(x')]B[t_n(x')] = \exp(-2\pi i n), \tag{107}$$

and

$$\tilde{\Lambda}_n = \frac{[y_0(0)y_0(x)]^{1/2}}{2\left\{\int_0^{y_0(0)}[G_n(0, y)]^2\,dy \cdot \int_0^{y_0(x)}[G_n(x, y)]^2\,dy\right\}^{1/2}} \tag{108}$$

is the effective excitation factor. The functions A and B occurring here have the same form as they do in the fully uniform guide. Also, of course, the height-gain functions have the same form.

The key property of (106) is the total field at the receiving point which involves an integration of the complex phase function along the path. Thus, the properties of the whole intervening path influence the total field. As expected, the series formula (106), the corresponding mode equation (107), and the excitation factor (108) all reduce to the appropriate forms for the fully uniform waveguide [e.g., compare with (34), (39), and (40)].

Another example of a nonuniform but slowly varying waveguide is produced when the effective reflection height is depressed over a large circular region. For example, if an intense ionizing source such as a nuclear explosion occurs at a height of 1000 km, it can be expected that a large bowl-shaped depression of the lower ionosphere will occur. Its horizontal extent may be of the order of several thousand kilometers; thus, it is sufficiently slowly varying to treat it as a lens-like structure such that the horizontal phase paths can be calculated on the basis of negligible mode conversion. Several attempts at treating the problem from this viewpoint have appeared recently in the literature *(Wait, 1964c; Wait, 1964d, Crombie, 1964a)*.

Mode Conversion Effects

For a proper understanding of propagation in the nonuniform earth-ionosphere waveguide, one should consider the effects of mode conversion. For example, in the case of propagation across a sunrise or sunset line, there is now ample evidence that the modes are coupled *(Steele and Crombie, 1967, Crombie, 1964b)*. Thus, for this example, a single mode incident on the boundary region will produce two or more modes that continue to propagate toward the receiving terminal.

There has been a great deal of work done on microwave propagation in metallic waveguides with variable cross section   *(Solymar, 1959, Unger, 1965)*. Unfortumately these analyses cannot be directly applied to the earth-ionosphere waveguide because of certain essential features of the latter which are not considered when treating conventional waveguides. For example, the earth's curvature requires that the characteristic functions are not of a simple trigonometric type, but, instead, they involve Airy functions. Also, the boundaries of the guide change both in height and surface properties. In a series of papers, Wait and Bahar have addressed themselves to various approaches to the general problem of the nonuniform earth-ionosphere waveguide *(Bahar and Wait, 1964, 1965, Bahar, 1966, Wait and Bahar, 1966)*. The use of planar analog devices seems particularly promising in obtaining information that is difficult to get by any other means.

Here, we shall present a simplified version of the theory which already has shown some promise in predicting observed effects in the actual earth-ionosphere waveguide.

We consider the two-dimensional model illustrated in Fig. 45. As indicated, the waveguide is permitted to have a sudden change of height at a (normalized) distance x_1 from the source at point A. We imagine this to be an idealized representation of the day/night transition.

We designate the daytime height as h_1 and the nighttime height as h_2, and the corresponding surface impedances are Z_1 and Z_2; thus, $y_1 = (2/ka)^{1/3} kh_1$ and $y_2 = (2/ka)^{1/3} kh_2$ are the respective height parameters, and $q_1 = -i(ka/2)^{1/3} Z_1/\eta_0$ and $q_2 = -i(ka/2)^{1/3} Z_2/\eta_0$ are the respective impedance parameters for the upper boundary.

The method of solution consists of writing model expansions for the fields in the two-waveguide sections *(Wait, 1962b)*. Relationships between the coefficients are then obtained by matching the tangential field components across the aperture plane at $x = x_1$. In general, this leads to an infinite set of equations with an infinite number of unknowns. However, there is a great simplification permitted when the reflection at the junction of the waveguide may be neglected. Certainly, from a practical standpoint in the earth-ionosphere waveguide, this is well justified. In any case, this will be assumed in what follows.

For $x < x_1$, the field is given by

$$E(x, y) = \sum_m E_m^{(1)}(x, y), \quad (109)$$

where

$$E_m^{(1)}(x, y) = \frac{E_0(x)}{y_1} \Lambda_m^{(1)} G_m^{(1)}(y) G_m^{(1)}(\hat{y}) \exp(-ixt_m^{(1)}) \quad (110)$$

in terms of the height-gain function $G_m^{(1)}(y)$ and excitation factor $\Lambda_m^{(1)}$ for region (1). Here, $E_0(x)$ is a suitable reference field. The coefficient $t_m^{(1)}$ satisfies an equation of the type given by (34) where the appropriate values of $A(t)$ and $B(t)$ are used. As usual, we have the condition that

$$\Lambda_m^{(1)} = (y_1/2)\left[\int_0^{y_1} [G_m^{(1)}(y)]^2 \, dy\right]^{-1}. \quad (111)$$

Using a straightforward process, we now find that, in the region $x > x_1$, the field has the form

$$E(x, y) = \sum_m E_m^{(2)}(x, y), \quad (112)$$

where

$$E_m^{(2)}(x, y) = E_0(x)(y_1 y_2)^{-1/2} \sum_n [\Lambda_m^{(1)}]^{1/2} G_m^{(1)}(\hat{y})[\Lambda_n^{(2)}]^{1/2}$$

$$\times G_n^{(2)}(y) \hat{S}_{n,m} \exp(-ix_1 t_m^{(1)}) \exp[-i(x - x_1)t_n^{(2)}], \quad (113)$$

and where, as indicated, an incident mode of order m in region (1) "scatters" into n modes in region (2). In region (2), the appropriate height-gain and excitation factors are $G_n^{(2)}(y)$ and $\Lambda_n^{(2)}$, respectively, for a mode of order n. The coefficients t_n again satisfy (34) with the appropriate forms for $A(t)$ and $B(t)$. The all-important scattering coefficient $\hat{S}_{n,m}$ determines how much coupling exists between an incident mode of order m and the transmitted mode of order n.

In order to derive an expression for the scattering coefficient, we utilize the orthogonality property

$$\int_0^{y_2} G_n^{(2)}(y) G_{n'}^{(2)}(y) \, dy = 0 \quad \text{for} \quad n' \neq n. \quad (114)$$

On equating (113) and (110) at the aperture plane, we find that

$$(y_1)^{-1/2}[\Lambda_m^{(1)}]^{1/2} G_m^{(1)}(y) = \sum_n (y_2)^{-1/2}[\Lambda_n^{(2)}]^{1/2} G_n^{(2)}(y) \hat{S}_{n,m}. \quad (115)$$

On multiplying both sides by $G_{n'}^{(2)}(y)$ and integrating with respect to y from 0 to y_2, we find, on using (114), that

$$\hat{S}_{n,m} = \left[\frac{y_2 \Lambda_m^{(1)}}{y_1 \Lambda_n^{(2)}}\right]^{1/2} S_{n,m}, \quad (116)$$

where

$$S_{n,m} = \frac{\int_0^{y_1} G_m^{(1)}(y) G_n^{(2)}(y)\, dy}{\int_0^{y_2} [G_n^{(2)}(y)]^2\, dy}, \qquad (117)$$

and where we have assumed that $G_m^{(1)}(y) = 0$ for $y_1 < y < y_2$. First of all, we observe that the denominator of $S_{n,m}$ is $y_2/[2\Lambda_n^{(2)}]$. To evaluate the numerator, we note that

$$\frac{d^2 G_m^{(1)}(y)}{dy^2} = (t_m^{(1)} - y) G_m^{(1)}(y), \qquad (118)$$

$$\frac{d^2 G_n^{(2)}(y)}{dy^2} = (t_n^{(2)} - y) G_n^{(2)}(y). \qquad (119)$$

Thus,

$$G_n^{(2)} \frac{d^2 G_m^{(1)}}{dy^2} - G_m^{(1)} \frac{d^2 G_n^{(2)}}{dy^2} = [t_m^{(1)} - t_n^{(2)}] G_m^{(1)} G_n^{(2)}, \qquad (120)$$

whence, on integration from 0 to y_1, we get

$$G_n^{(2)} \frac{d}{dy} G_m^{(1)}\bigg|_{y_1} - G_m^{(1)} \frac{d}{dy} G_n^{(2)}\bigg|_{y_1} = [t_m^{(1)} - t_n^{(2)}] \int_0^{y_1} G_m^{(1)} G_n^{(2)}\, dy, \qquad (121)$$

where we have used the fact that

$$\frac{dG_m^{(1)}}{dy}\bigg|_{y=0} = \frac{dG_n^{(1)}}{dy}\bigg|_{y=0} = -q. \qquad (122)$$

The latter is a consequence of the surface impedance boundary condition on the homogeneous lower boundary. Now, at the upper boundaries of the waveguide, we have

$$\left[\frac{dG_m^{(1)}}{dy} - q_1 G_m^{(1)}\right]\bigg|_{y=y_1} = 0,$$

$$\left[\frac{dG_n^{(2)}}{dy} - q_2 G_n^{(2)}\right]\bigg|_{y=y_2} = 0. \qquad (123)$$

Using these, (121) simplifies to

$$\int_0^{y_1} G_m^{(1)}(y) G_n^{(2)}(y)\, dy = \frac{G_m^{(1)}(y_1)}{t_m^{(1)} - t_n^{(2)}} [q_1 G_n^{(2)}(y_1) - G_n^{(2)\prime}(y_1)], \qquad (124)$$

which is an explicit expression for the relevant integral in (117). Thus, we find that

$$S_{n,m} = \frac{2}{y_2} \Lambda_n^{(2)} \frac{G_m^{(1)}(y_1)}{t_m^{(1)} - t_n^{(2)}} [q_1 G_n^{(2)}(y_1) - G_n^{(2)\prime}(y_1)], \qquad (125)$$

which may be used for calculation of the coefficient $\hat{S}_{n,m}$ via (116). Some examples are given in Appendix B.

Some further simplification and insight is obtained if we utilize the following easily proved expansions:

$$G_n^{(2)}(y_1) = G_n^{(2)}(y_2)$$

$$\times \left[1 - q_2(y_2 - y_1) + \frac{\hat{t}_n^{(2)}(y_2 - y_1)}{2} + \frac{(1 - \hat{t}_n^{(2)} q_2)}{6}(y_2 - y_1)^3 + \cdots\right]$$

$$(126)$$

where $\hat{t}_n^{(2)} = t_n^{(2)} - y_1$

and

$$G_n^{(2)\prime}(y_1) = G_n^{(2)}(y_2)$$
$$\times [q_2 - \hat{\imath}_n^{(2)}(y_2 - y_1) - \tfrac{1}{2}(1 - \hat{\imath}_n^{(2)}q_2)(y_2 - y_1)^2 - \cdots]. \quad (127)$$

Combining these with (25), we see that

$$S_{n,m} = \frac{2}{y_2} \Lambda_n^{(2)} \frac{G_m^{(1)}(y_1) G_n^{(2)}(y_2)}{t_m^{(1)} - t_n^{(2)}}$$

$$\times \left[(q_1 - q_2) - (q_1 q_2 - \hat{\imath}_n^{(2)})(y_2 - y_1) \right.$$

$$\left. + \left(\frac{q_1 \hat{\imath}_n^{(2)}}{2} + \frac{1 - \hat{\imath}_n^{(2)} q_2}{2} \right)(y_2 - y_1)^2 + \cdots \right]. \quad (128)$$

The form of the expansion is rather revealing. The first term within the brackets, namely, $q_1 - q_2$, is proportional to the surface impedance contrast $Z_1 - Z_2$, whereas the second term involving $(y_2 - y_1)$ is proportional to the height change $h_2 - h_1$. The third and succeeding terms involve both the impedance *and* the height changes. Presumably, for sufficiently small changes, only the first two terms need be retained.

Using the internal definition of $S_{n,m}$ given by (117), Rugg *(1967)* has carried out some calculations at 21.4 kHz for a day-to-night propagation path. He assumed a daytime surface impedance Z_1 corresponding to an exponential isotropic ionosphere with $\beta = 0.3$ km^{-1} and reflection height $h = 70$ km. The nighttime surface impedance Z_2 corresponds to an exponential isotropic ionosphere with $\beta = 0.5$ km^{-1}, with a reflection height $h = 88$ km. Finally, the ground conductivity σ_g was assumed to be 5 mmho/m throughout the path. Table I is applicable to this special situation.

The fact that the magnitude of the coefficient $S_{2,1}$ is 0.68 is a striking indication that a day-night transition (i.e., a sunrise) will launch a strong second-order mode into the nighttime portion of the earth-ionosphere wave-guide.

A common type of phase and amplitude variation for the path NSS to Denver is shown in Fig. 46. Three-hour periods for both the sunrise and the sunset periods are shown The corresponding calculations are shown when it is indicated that a total phase change of 230° occurs over the sunrise. The rather striking comparison between the calculated and the experimental curves suggests that, at least for the sunrise period, the assumption of an abrupt change from a nighttime to a daytime waveguide is justified. Other examples that support this contention are also given by Rugg *(1967)*.

Mode Conversion in Cascaded Transition Sections

It is evident that, under some circumstances, the abrupt change from one uniform waveguide to another uniform waveguide is too idealized. For example, the sunset transition is sufficiently slow that a more refined model is needed. One possibility is to employ a number of cascaded sections. This is illustrated in Fig. 47, where we have chosen a three-step transition. The parameters have their usual meaning.

For a mode of order n incident from A, the field received at B will have the following form:

$$E_m(x, y) = \frac{E_0(x)}{y_1} \Lambda_m^{(1)} \exp(-ix_1 t_m^{(1)}) \sum_n S_{n,m}^{(1)} \exp[-i(x_2 - x_1)t_n^{(2)}]$$
$$\times \sum_p S_{p,n}^{(2)} \exp[-i(x_3 - x_2)t_p^{(3)}] \sum_q S_{q,p}^{(3)} \exp[-i(x - x_3)t_q^{(4)}],$$
(129)

where $E_0(x)$ is a suitable reference field. The coefficient $S_{n,m}^{(1)}$ is determined by the scattering from an incident mode of order m, in the first waveguide section, into a transmitted mode of order n, in the second section. It has the same form as (117). In a similar fashion, we then employ the coefficient $S_{p,n}^{(2)}$ as the scattering from a mode of order n in the second section to a mode of order p in the third section. Finally, $S_{q,p}^{(3)}$ describes the scattering into the uniform waveguide region on the right in Fig. 47.

Insofar as the field at B is concerned, it is possible to write (129) in the form

$$E_m(x, y) = \frac{E_0(x)}{y_1} \exp(-ix_0 t_m^{(1)}) \Lambda_m^{(1)} \sum_q S_{q,m}^{\text{eff}} \exp[-i(x - x_0)t_q^{(4)}],$$
(130)

where $S_{q,m}^{\text{eff}}$ is the effective conversion coefficient for the transition region being referred to in the distance x_0 as indicated in Fig. 47. An alternative form of (130) is

$$E_m(x, y) = E_0(x) \exp(-ix_0 t_m^{(1)})(y_1 y_4)^{-1/2} [\Lambda_m^{(1)}]^{1/2}$$
$$\times \sum_q [\Lambda_q^{(4)}]^{1/2} \hat{S}_{q,m}^{\text{eff}} \exp[-i(x - x_0)t_q^{(4)}], \quad (131)$$

where

$$\hat{S}_{q,m}^{\text{eff}} = \left[\frac{\Lambda_m^{(1)} y_4}{\Lambda_q^{(4)} y_1}\right]^{1/2} S_{q,m}^{\text{eff}}.$$

This form involving $\hat{S}_{q,m}^{\text{eff}}$ is physically more meaningful than (130), since the excitation factor $[\Lambda_m^{(1)}]^{1/2}$ for the source and the corresponding factor $[\Lambda_q^{(4)}]^{1/2}$ for the observer occur in a symmetrical fashion.

Some calculations based on (130) and (131) and further extensions of the theory appear elsewhere *(Wait and Spies, 1968)*. The extension to oblique incidence across the terminator has also been considered *(Wait, 1968)*. The elevated coast-line has also been treated using a similar approach *(Wait and Spies, 1970)*. Experimental techniques to study mode conversion effects have been developed by Mahmoud and Beal *(1971)*, Lynn *(1967)* and Kaiser *(1968)*.
This would appear to be a fruitful approach for further investigations involving mode conversion in the nonuniform earth-ionosphere waveguide.

In the foregoing discussion, we have assumed that the direction of propagation is from the daytime to the nighttime waveguide. If the appropriate scattering matrix for this case is designated $S_{n,m}^{DN}$, then, from considerations of reciprocity *(Bahar and Wait, 1964)*, it is possible to show that

$$y_2^{-1} \Lambda_m^{(2)} S_{n,m}^{ND} = y_1^{-1} \Lambda_n^{(1)} S_{m,n}^{DN}, \quad (132)$$

where $S_{n,m}^{ND}$ is the scattering matrix for propagation from night to day. An equivalent statement of (132) is simply

EM Propagation in the Earth-Ionosphere Waveguide

$$\hat{S}_{n,m}^{ND} = \hat{S}_{m,n}^{DN}. \tag{133}$$

This identity allows us to restrict our calculations only to the situation where the propagation is in the direction of increasing width of the waveguide.

Concluding Remarks

In spite of the numerous advances, the theory of VLF propagation in the earth-ionosphere waveguide is still in a relatively primitive state. For example, the usual practice of assuming that the ionosphere is a cold electron plasma is open to severe criticism. Although the finite temperature of the plasma does not have a noticeable effect on the waveguide characteristics, it will play an important role when the transmitting antenna is located within the ionosphere. Obviously, this is a subject that will have practical importance in the future. Also, we have not delved into the significance of the other ion species. When considering penetration of the waves well into the ionosphere, the resultant field strength will be critically dependent on the various particle constituents of the plasma. Again, this is a subject worthy of investigation.

The mechanism of launching "whistlers" from the earth-ionosphere waveguide has not been even mentioned in this article. This is a whole subject in itself, and we feel that it warrants a thorough and comprehensive wave treatment. Some of the complications are described briefly by Al'pert et al. *(1970)* in their recent monograph and by Wieder *(1967)*.

The relatively recent surge of VLF research in the Soviet Union is typified by the reviews by Makarov, Novikov and Orlov *(1970)* and by Orlov and Azarin *(1970)*. This work also has its controversial aspects; for example, Krashnushkin and Fedorov *(1973)* have claimed that Makarov made unjustified criticisms of their mode separability approach for treating a dipping terrestrial magnetic field. Also, the Krashnushkin school argue that the surface impedance techniques, that Makarov and company use, are invalid. This writer was greatly stimulated by discussions with both Dr. Makarov and Dr. Krashnushkin during a visit in 1971 to the Soviet Union.

Appendix A: Approximation of the Spherical Wave Functions

The spherical wave functions occurring in the formal theory for the earth-ionosphere waveguide can be represented in terms of Airy functions. This fact follows from the exact contour integral representation *(Sommerfeld, 1949)*

$$h_v^{(2)}(x) = [i/(2\pi x)^{1/2}] \int_C \exp[-x \sinh v + v + \tfrac{1}{2}]\, dv, \tag{A1}$$

where the contour is indicated in Fig. 48. We now change the variable to $Z = (x/2)^{1/3} v$ and expand the sinh v in the foregoing exponent. Thus, following Fock *(1965)*,

$$h_v^{(2)}(x) = i\left(\frac{\pi}{2x}\right)^{1/2} \cdot \left(\frac{2}{x}\right)^{1/3}$$

$$\times \int_{\infty \exp(i 2\pi/3)}^{\infty} \exp\left[tZ - \frac{Z^3}{3}\right]\left\{1 - \frac{1}{60}\left(\frac{2}{x}\right)^{2/3} Z^5 + \cdots\right\} dZ, \tag{A2}$$

where $t = (v + \tfrac{1}{2} - x)(2/x)^{1/3}$, and x is a large parameter. Here, t and Z are regarded as finite over the important range of the integration contour which runs from $\infty \exp(i 2\pi/3)$ along a straight line to the origin and then out along the real axis to infinity.

Using the integral definition of $w_1(t)$ given by (10), it follows that (A2) is equivalent to

$$h_v^{(2)}(x) = \frac{i}{(2x)^{1/2}} \left(\frac{2}{x}\right)^{1/3} \left[w_1(t) - \frac{1}{60}\left(\frac{2}{x}\right)^{2/3} w_1^5(t) + \cdots\right]. \quad (A3)$$

Now, since $w_1''(t) = tw_1(t)$, we see, from repeated differentiations, that $w_1^5(t) = t^2 w_1'(t) + 4t w_1(t)$. Thus, (A3) is rewritten as

$$xh_v^{(2)}(x) = i(x/2)^{1/6}$$
$$\times [w_1(t) - (\tfrac{1}{60})(2/x)^{2/3}[t^2 w_1'(t) + 4t w_1(t)] + \cdots]. \quad (A4)$$

If x is sufficiently large (i.e., $x^{2/3} \gg 1$ and t bounded), (A4) reduces to (9) in the main text.

We note, also, that

$$xh_v^{(1)}(x) = -i(x/2)^{1/6}$$
$$\times [w_2(t) - (\tfrac{1}{60})(2/x)^{2/3}[t^2 w_2'(t) + 4t w_2(t)] + \cdots]. \quad (A5)$$

The consequence of retaining only the leading Airy function terms in the spherical-earth mode equation has been examined by Rybachek (1970). She shows the resulting errors in the attenuation rates are insignificant while significant phase errors only occur at frequencies less than 10 kHz, and even here the differences are of the order of 1 part in 10^4.

APPENDIX B: SCATTERING COEFFICIENTS $\hat{S}_{n,m}$ FOR AN ABRUPT DAY-NIGHT TRANSITION

On combining (116) and (125), we obtain an explicit formula for the scattering coefficient $\hat{S}_{n,m}$ for the junction indicated in Fig. 45. Thus,

$$\hat{S}_{n,m} = 2T_n \left[\frac{\Lambda_m^{(1)} \Lambda_n^{(2)}}{y_1 y_2}\right]^{1/2} \frac{G_m^{(1)}(y_1)}{t_m^{(1)} - t_n^{(2)}},$$

where $T_n = q_1 G_n^{(2)}(y_1) - G_n^{(2)\prime}(y_1)$. To facilitate computation, we note th $G_n^{(2)}(y_1)$ is defined by

$$G_n^{(2)}(y_1) = \frac{w_1(t_n^{(2)} - y_1) + A(t_n^{(2)}) w_2(t_n^{(2)} - y_1)}{w_1(t_n^{(2)}) + A(t_n^{(2)}) w_2(t_n^{(2)} - y_1)},$$

and, thus,

$$T_n = \frac{[q_1 w_1(t_n^{(2)} - y_1) + w_1'(t_n^{(2)} - y_1)] + A(t_n^{(2)})[q_1 w_2(t_n^{(2)} - y_1) + w_2'(t_n^{(2)} - y_1)]}{w_1(t_n^{(2)}) + A(t_n^{(2)}) w_2(t_n^{(2)})}.$$

Selected numerical values of the scattering coefficient are given in Tables II–IV, where we have chosen $\sigma_g = \infty$ throughout.

An examination of the tabulated values reveals some interesting features. For example, on comparing the respective entries in Table II for 15 and 20 kHz, we see that mode conversion is stronger for the higher frequency if all other conditions are the same. The results in Tables III and IV demonstrate that the mode conversion produced by height changes is relatively more important than changes in the conductivity parameter. In fact, when the height is kept constant and the conductivity is changed for a factor of 4 to 1, the resulting amount of mode conversion is about the same as a height change of only 1 km.

EM Propagation in the Earth-Ionosphere Waveguide

TABLE I
$S_{n,m}$ (Day-to-Night Path)

n	$m = 1$	$m = 2$	$m = 3$
1	$0.436 - i0.031$	$-0.123 + i0.025$	$0.055 - i0.019$
2	$0.676 + i0.042$	$0.502 - i0.107$	$-0.153 + i0.047$
3	$-0.139 - i0.016$	$0.680 + i0.087$	$0.249 - i0.059$

TABLE II
Values of $\hat{S}_{n,m}$ in Complex Polar Form

$\omega_r^{(1)} = 2 \times 10^5$ $\omega_r^{(2)} = 2 \times 10^5$	$h_1 = 70$ km $h_2 = 90$ km	$f = 20$ kHz

n	$m = 1$	$m = 2$	$m = 3$
1	0.8493 (0.49°)	0.3180 (177.77°)	0.1934 (−4.17°)
2	0.5108 (−1.15°)	0.6973 (−0.64°)	0.2501 (179.18°)
3	0.1309 (176.50°)	0.6370 (1.27°)	0.4832 (−3.58°)

$\omega_r^{(1)} = 2 \times 10^5$ $\omega_r^{(2)} = 2 \times 10^5$	$h_1 = 70$ km $h_2 = 90$ km	$f = 15$ kHz

n	$m = 1$	$m = 2$	$m = 3$
1	0.9205 (0.08°)	0.2518 (175.46°)	—
2	0.3777 (−0.62°)	0.7693 (−1.89°)	—
3	0.0997 (179.52°)	0.5886 (3.61°)	—

$\omega_r^{(1)} = 2 \times 10^5$ $\omega_r^{(2)} = 4 \times 10^5$	$h_1 = 70$ km $h_2 = 90$ km	$f = 20$ kHz

n	$m = 1$	$m = 2$	$m = 3$
1	0.8693 (0.84°)	0.3063 (176.83°)	0.1846 (−5.24°)
2	0.4764 (−2.57°)	0.7326 (0.22°)	0.2524 (179.31°)
3	0.1316 (176.73°)	0.5997 (0.41°)	0.5388 (−1.75°)

$\omega_r^{(1)} = 2 \times 10^5$ $\omega_r^{(2)} = 4 \times 10^5$	$h_1 = 70$ km $h_2 = 70$ km	$f = 20$ kHz

n	$m = 1$	$m = 2$	$m = 3$
1	0.9994 (−0.03°)	0.0320 (17.22°)	0.0167 (−163.99°)
2	0.0306 (−165.19°)	0.9983 (−0.13°)	0.0506 (18.68°)
3	0.0163 (10.38)	0.0457 (−166.72°)	0.9941 (−0.35°)

TABLE III
Variation of $\hat{S}_{n,1}$ as a Function of $h_2 - h_1$

$h_2 - h_1$	1	5	10	15	20
$n=1$	0.9998 (−0.01°)	0.9913 (0.00°)	0.9628 (0.09°)	0.9149 (0.27°)	0.8493 (0.49°)
$n=2$	0.0217 (−1.37°)	0.1164 (−1.38°)	0.2467 (−1.38°)	0.3813 (−1.30°)	0.5108 (−1.15°)
$n=3$	0.0102 (−179.91°)	0.0516 (179.63°)	0.0964 (178.80°)	0.1245 (177.78°)	0.1309 (176.50°)

$\omega_r^{(2)} = \omega_r^{(1)} = 2 \times 10^5$, $f = 20$ kHz, $h_1 = 70$ km.

TABLE IV
Variation of $\hat{S}_{n,1}$ as a Function of $\omega_r^{(2)}/\omega_r^{(1)}$

$\omega_r^{(2)}/\omega_r^{(1)}$	$\frac{1}{4}$	$\frac{1}{2}$	1	2	4
$n=1$	0.9975 (0.01°)	0.9994 (−0.01°)	1.0000 (<0°)	0.9994 (−0.03°)	0.9974 (−0.23°)
$n=2$	0.0584 (−6.55°)	0.0278 (1.88°)	0	0.0306 (−165.19°)	0.0680 (−164.12°)
$n=3$	0.0295 (177.11°)	0.0144 (−176.87°)	0	0.0163 (10.38°)	0.0348 (5.73°)

$\omega_r^{(1)} = 2 \times 10^5$, $f = 20$ kHz, $h_1 = h_2 = 70$ km.

EM Propagation in the Earth-Ionosphere Waveguide

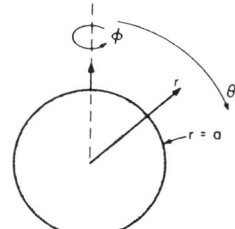

Fig. 1. The spherical earth model.

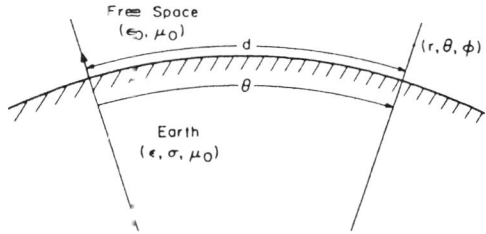

Fig. 2. A small section of the spherical earth.

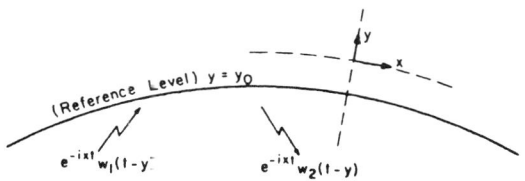

Fig. 3. Illustrating reflection of the upgoing wave type.

280   Wave Propagation Theory

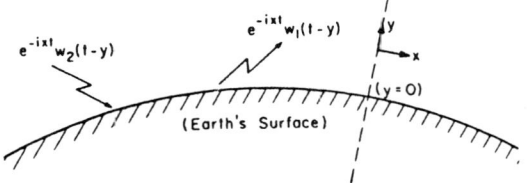

Fig. 4. Illustrating reflection of the downgoing wave type.

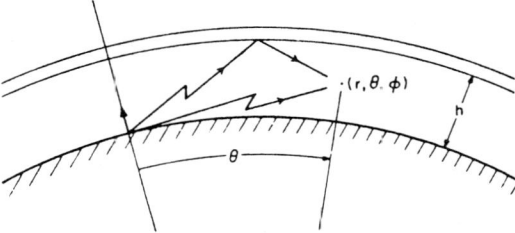

Fig. 5. Illustrating the multiple reflections.

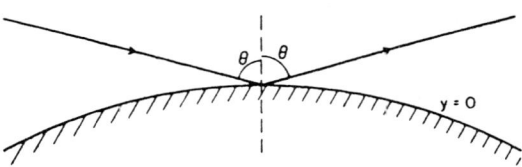

Fig. 6. Geometry for reflection from the ground.

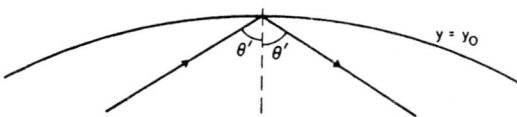

Fig. 7. Geometry for the reflected sky wave.

EM Propagation in the Earth-Ionosphere Waveguide

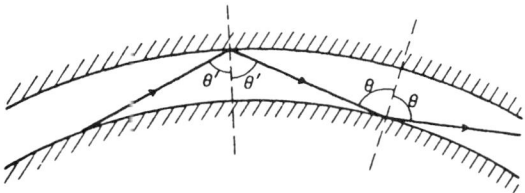

Fig. 8. Geometry for the spherical waveguide.

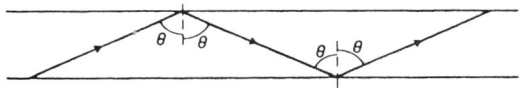

Fig. 9. Geometry for the planar waveguide.

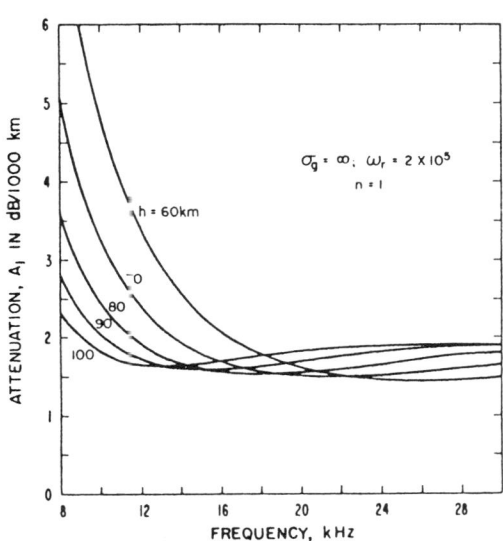

Fig. 10. Attenuation rate for mode 1 versus frequency for the sharpley bounded model.

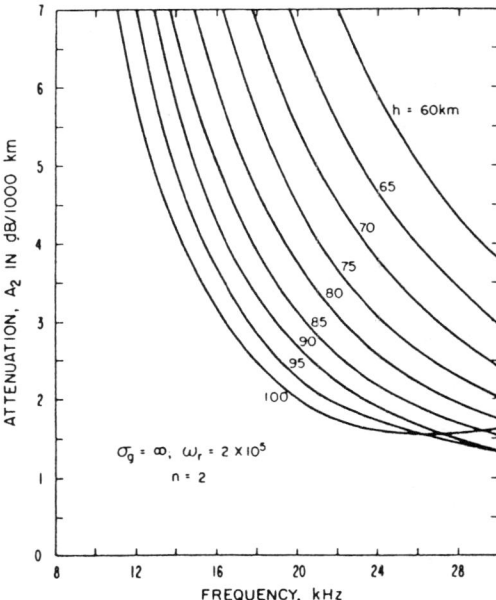

Fig. 11. Attenuation rate for mode 2 versus frequency for the sharply bounded model.

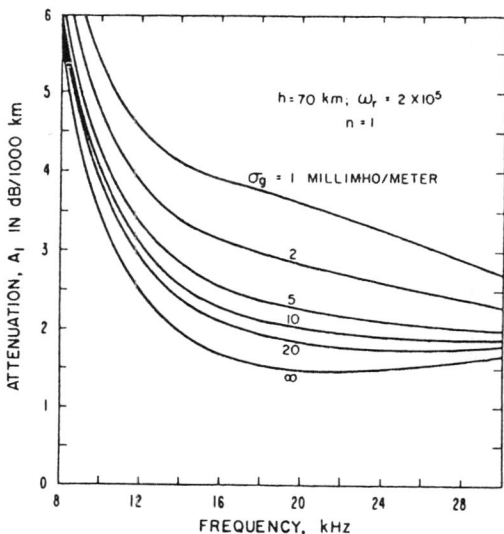

Fig. 12. Attenuation rate for mode 1 versus frequency for the sharply bounded model illustrating the effect of ground conductivity.

EM Propagation in the Earth-Ionosphere Waveguide

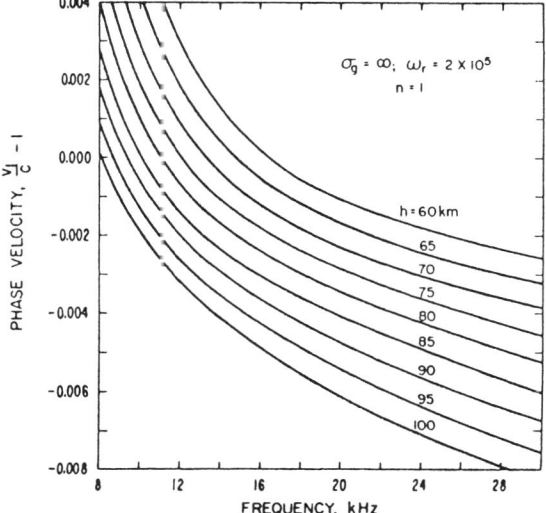

Fig. 13. Phase velocity versus frequency for mode 1 for sharply bounded model.

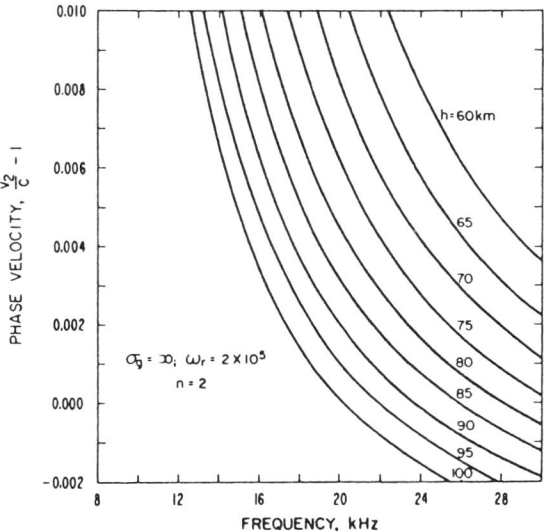

Fig. 14. Phase velocity versus frequency for mode 2 for sharply bounded model.

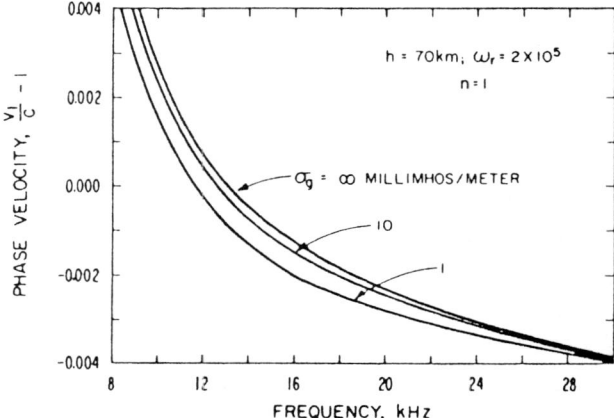

Fig. 15. Phase velocity versus frequency for mode 1 illustrating the effect of finite ground conductivity.

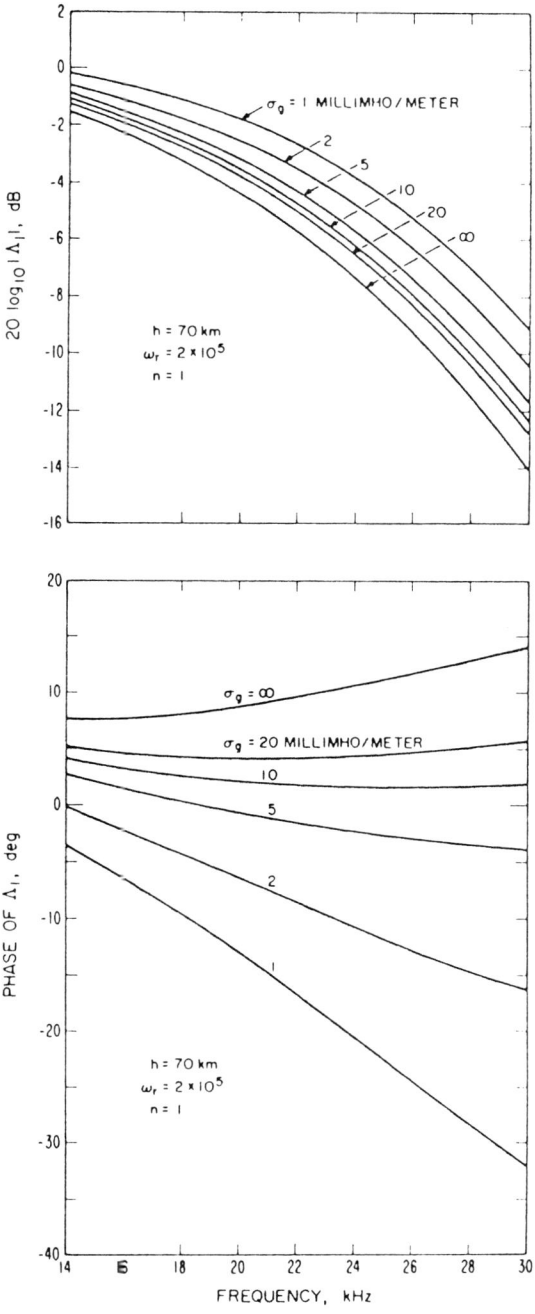

Fig. 16. Amplitude and phase of the excitation factor for mode 1 for sharply-bounded model.

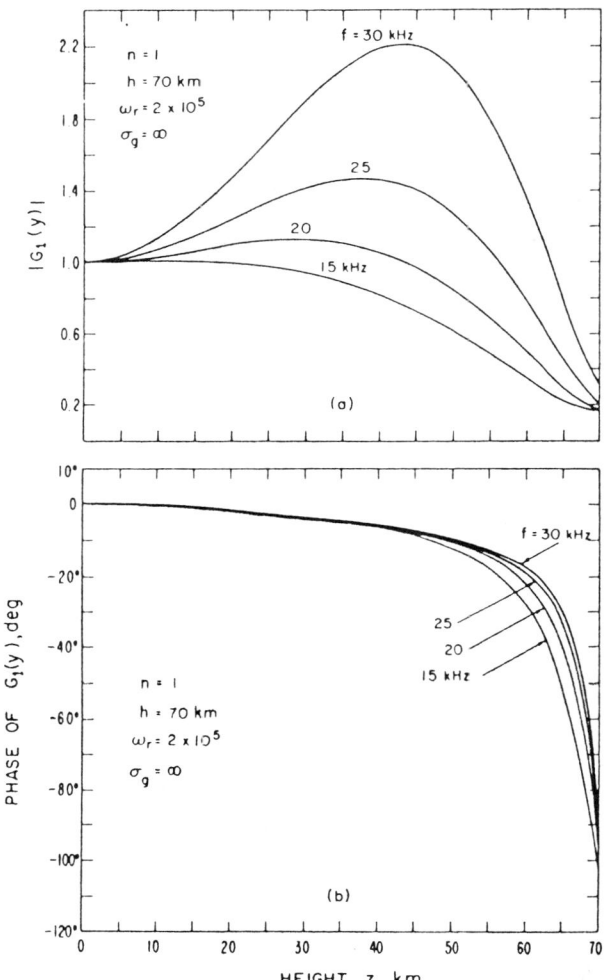

Fig. 17. Amplitude and phase of the height-gain function for mode 1 for sharply bounded model for σ_g infinite.

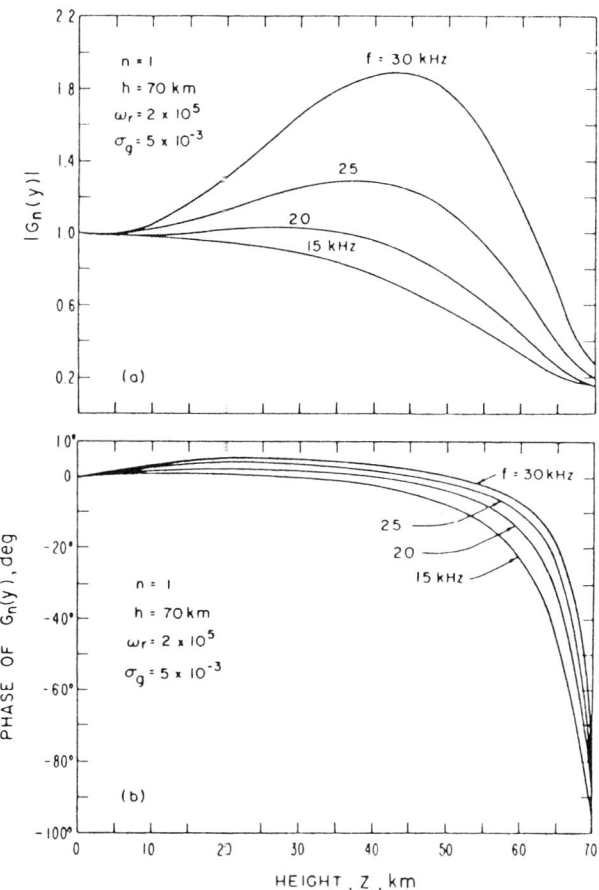

Fig. 18. Amplitude and phase of the height-grain function for mode 1 for sharply bounded model for σ_g finite.

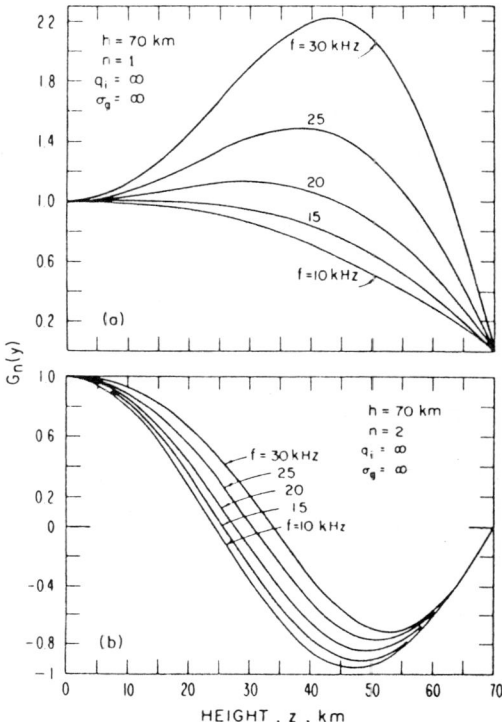

Fig. 19. Height-gain functions for perfectly reflecting boundaries.

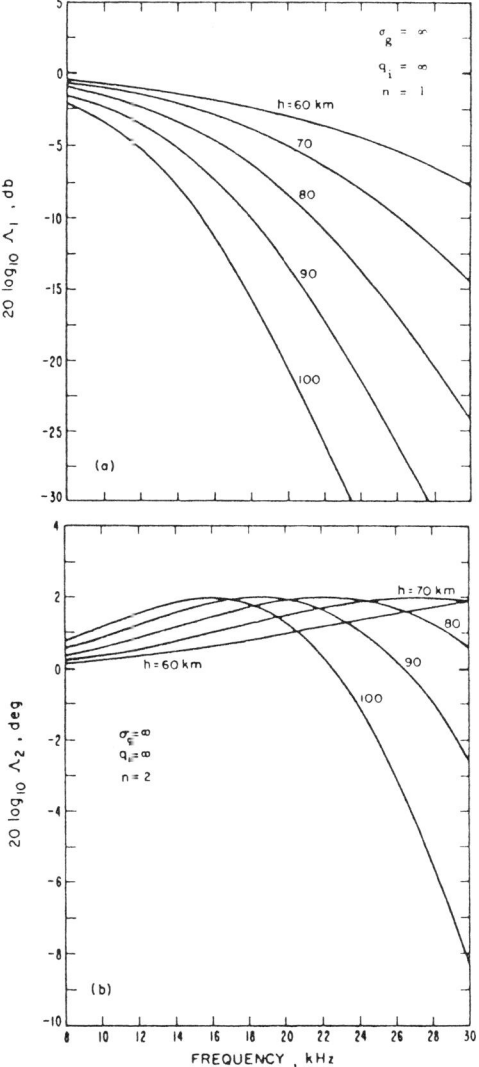

Fig. 20. Excitation factors for perfectly reflecting boundaries.

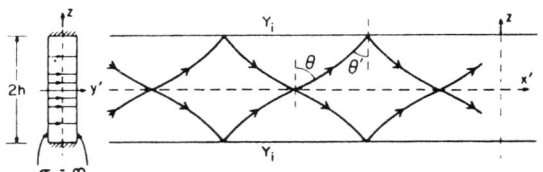

Fig. 21. The rays associated with a mode in the planar loaded waveguide.

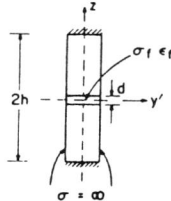

Fig. 22. Cross section of the planar waveguide showing centrally located conductive strip.

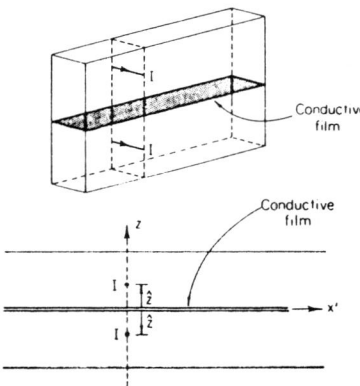

Fig. 23. Showing excitation of the planar waveguide by line source.

EM Propagation in the Earth-Ionosphere Waveguide

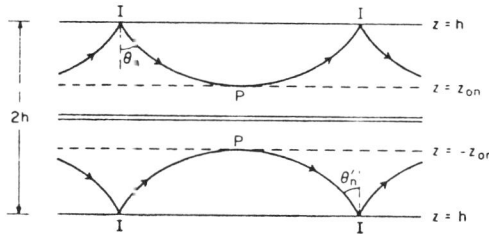

Fig. 24. Rays associated with an "earth-detached" type of mode in the planar waveguide model.

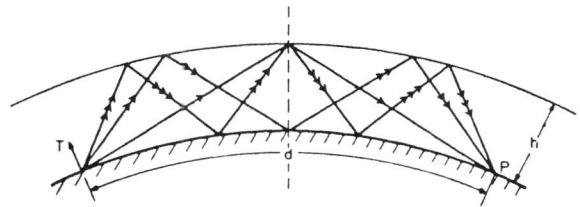

Fig. 25. Ray or hop picture for transmission from T to P (only first three hops are shown).

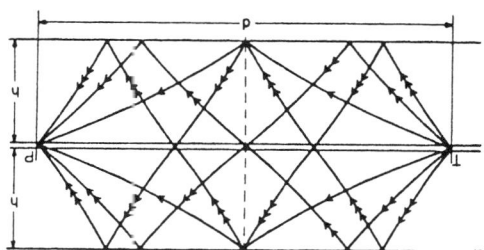

Fig. 26. The ray structure for three hops in the planar-loaded waveguide (compare with Fig. 25).

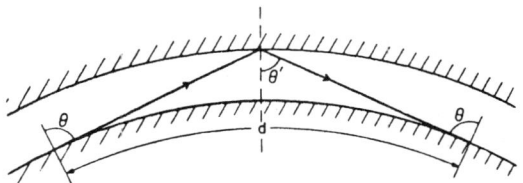

Fig. 27. The first hop ray in spherical geometry.

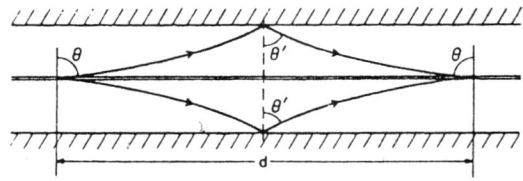

Fig. 28. The first hop ray in the planar analog.

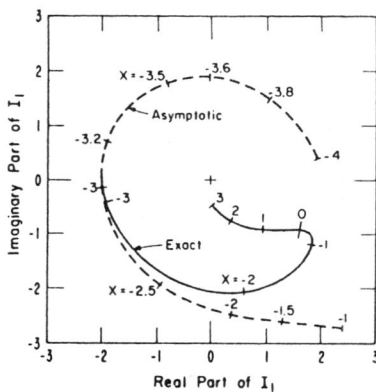

Fig. 29. Comparison between the integral formula for I_1 and its asymptotic approximation.

EM Propagation in the Earth-Ionosphere Waveguide

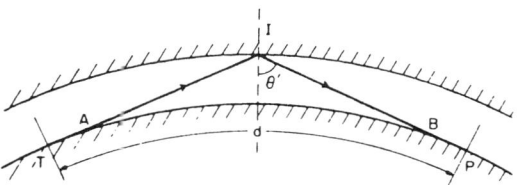

Fig. 30. The "one-hop" wave beyond the horizon.

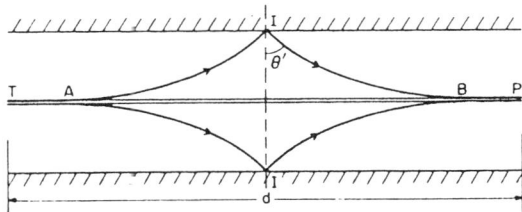

Fig. 31. The "one-hop" wave for shadow zone propagation in the planar waveguide.

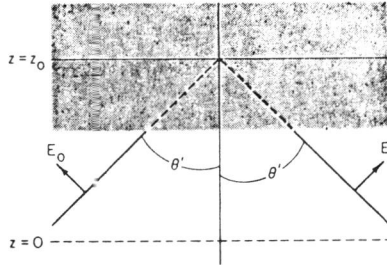

Fig. 32. Illustrating the nature of reflection from a diffuse boundary.

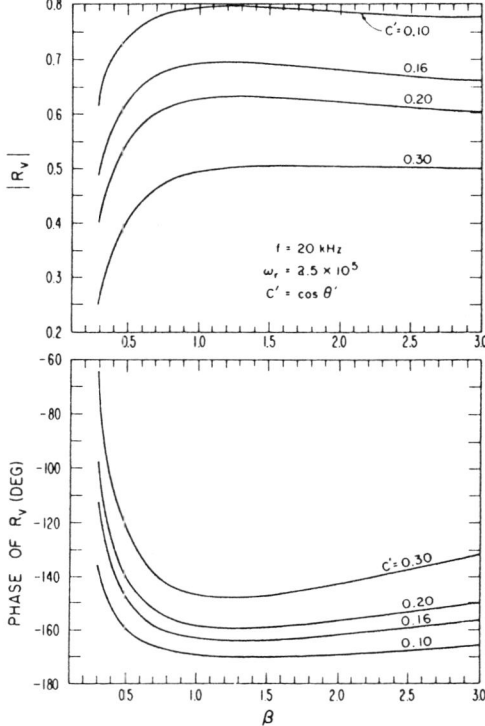

Fig. 33. The vertically polarized reflection coefficient for an exponential conductivity gradient.

EM Propagation in the Earth-Ionosphere Waveguide

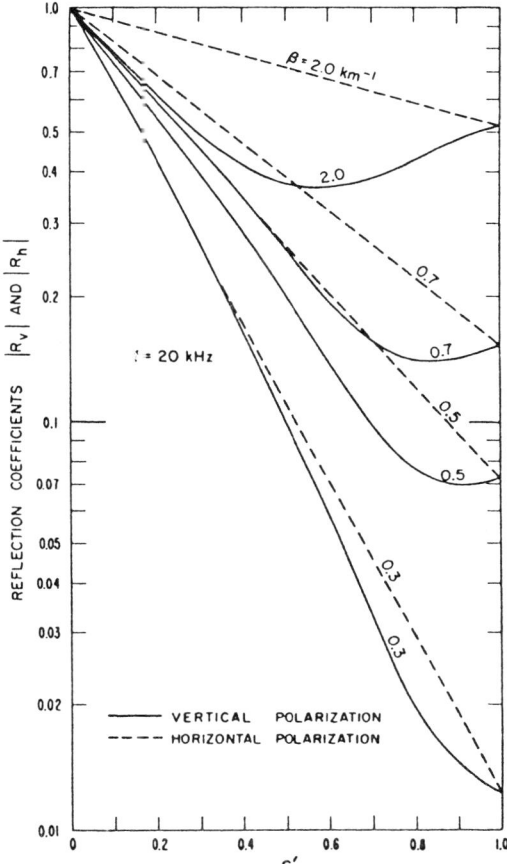

Fig. 34. The magnitude of the reflection coefficient for both vertical and horizontal polarization.

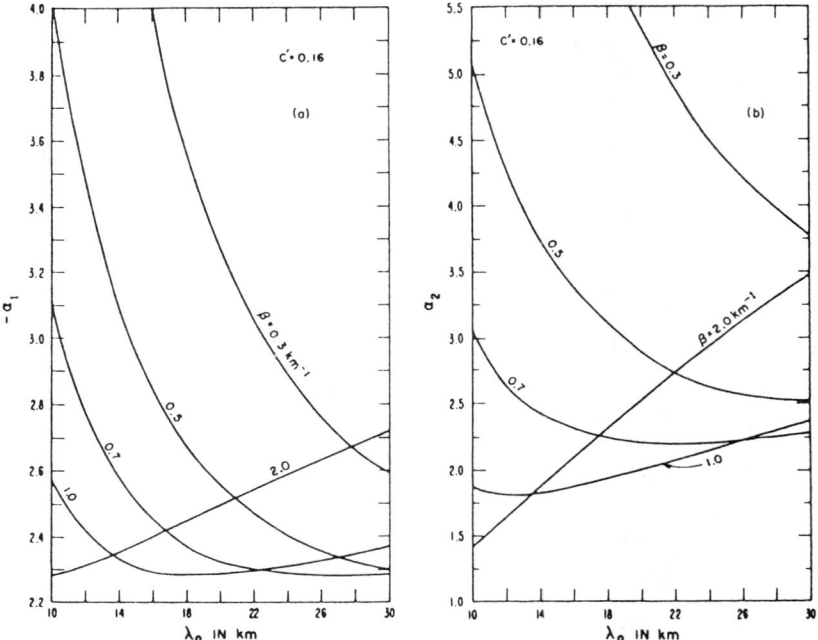

Fig. 35. The factor $\alpha_1 + i\alpha_2$ as a function of the wavelength λ_o.

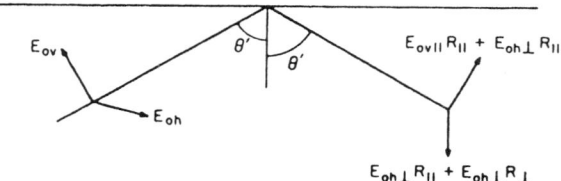

Fig. 36. Indicating the reflection of a plane wave from an anisotropic medium.

EM Propagation in the Earth-Ionosphere Waveguide 297

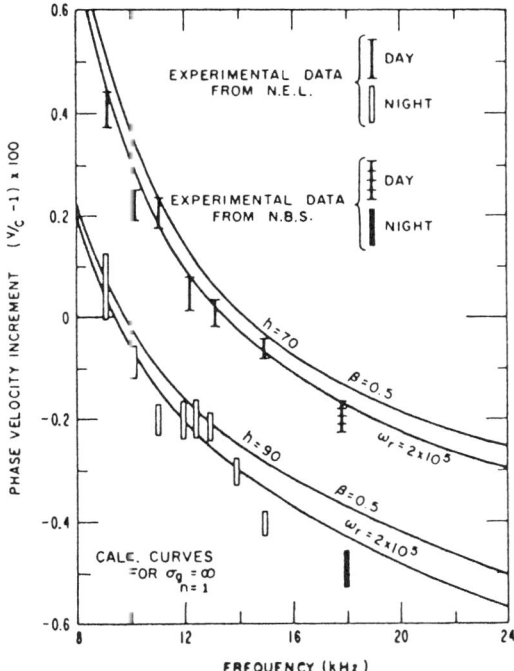

Fig. 37. Attenuation rates as a function of frequency for a transverse dc magnetic field.

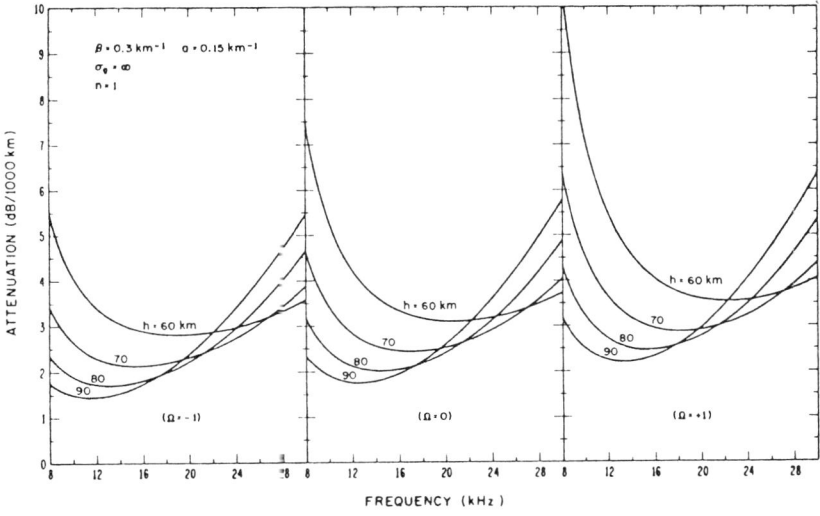

Fig. 38. Phase velocity versus frequency (theory and experiment).

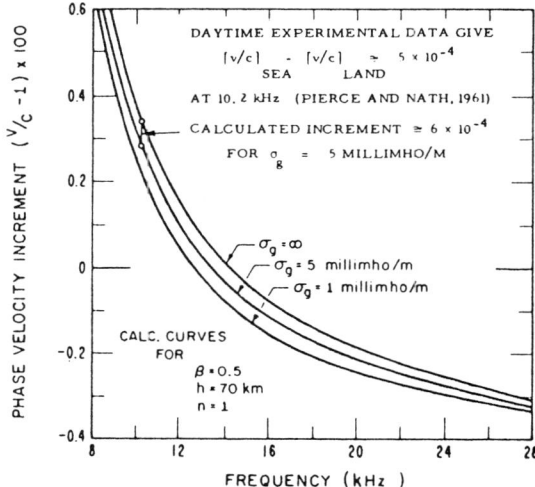

Fig. 39. Phase velocity versus frequency (theory and experiment).

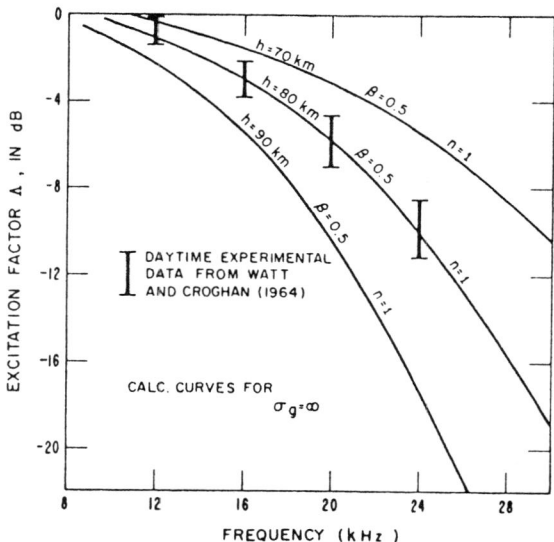

Fig. 40. Excitation factor versus frequency. (theory and experiment).

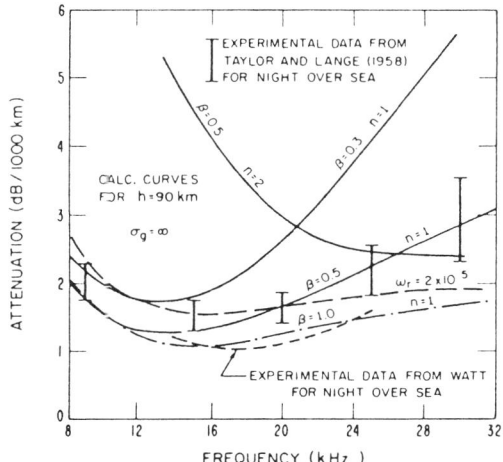

Fig. 41. Attenuation rate frequency (theory and experiment).

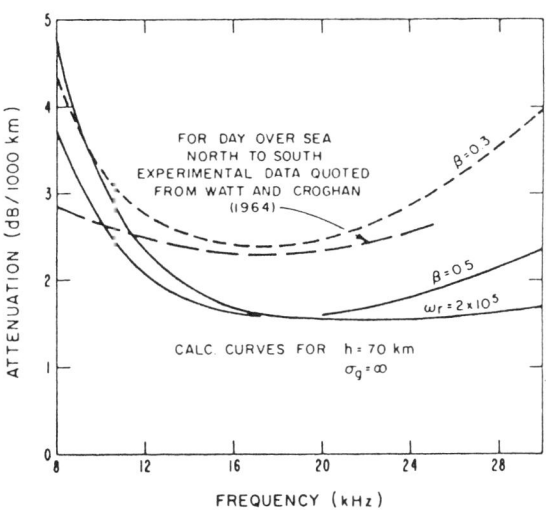

Fig. 42. Attenuation rate versus frequency (theory and experiment).

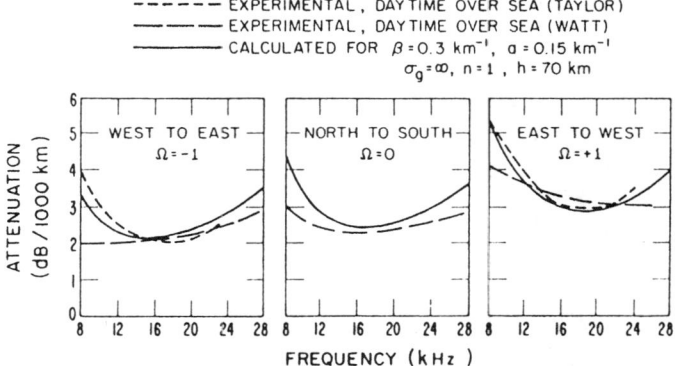

Fig. 43. Attenuation rate versus frequency (theory and experiment).

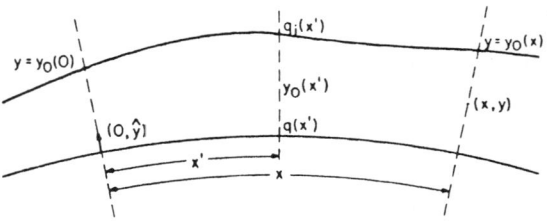

Fig. 44. Two-dimensional model of the nonuniform waveguide.

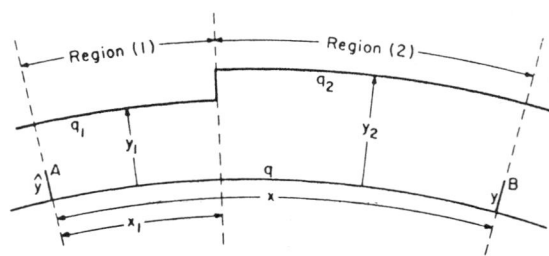

Fig. 45. Abrupt transition model for day-to-night propagation

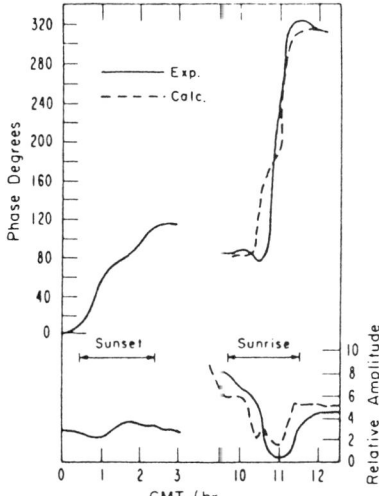

Fig. 46. Calculated and experimental phase and amplitude variations for NSS to Denver path at 21.4 kHz, April 22, 1964 [After Rugg, 1967].

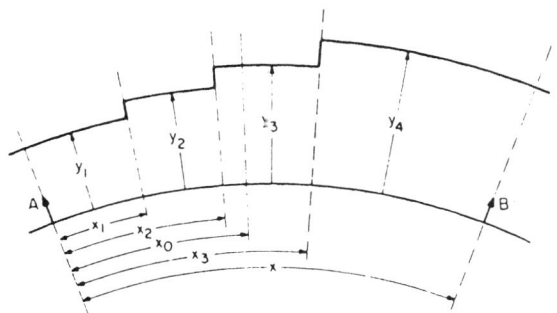

Fig. 47. Multistep model for day-to-night transition.

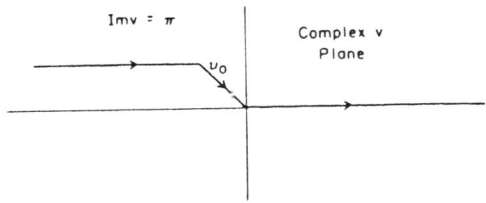

Fig. 48. Contour C in the complex v plane; here, $v_o = -(\pi/3^{1/2}) + i\pi$.

REFERENCES

ABRAMOWITZ, M., and STEGUN, I., eds., (1964), *Appl. Math. Ser. No. 55 Handbook Math. Functions*, U.S. National Bureau of Standards.

AL'PERT, YA. L., and FLIGEL, D.S., (1970), *Propagation of ELF and VLF Waves Near the Earth*, Consultants Bureau New York.

AL'PERT, YA. L., FLIGEL, D.S., KAPUSTINA, O.V., and ZABAVINA, I.N., (1972), On the amplitude and phase velocity of VLF and LF e.n. waves in the spherical earth-ionosphere waveguide, *Journ. Atmos. Terr. Physics*, $\underline{34}$, 877-892.

BAHAR, E., (1966), Propagation of VLF radio waves in a model earth-ionosphere waveguide of arbitrary height and finite surface impedance boundaries; Theory and experiment, *Radio Science*, $\underline{1}$, (New Ser.), No. 8, 925-938.

BAHAR, E., and WAIT, J.R., (1964), Microwave model techniques to study VLF radio propagation in the earth-ionosphere waveguide, in *Quasi-Optics*, J. Fox, ed., pp. 447-464, Polytech. Press of the Polytech. Inst. of Brooklyn, Brooklyn, New York.

BAHAR, E., and WAIT, J.R., (1965), Propagation in a model terrestrial wave-guide of nonuniform height: Theory and experiment, *J. Res. Nat'l. Bur. Std.*, $\underline{D69}$ No. 11, 1445-1463.

BERRY, L.A., (1964), Wave hop theory of long distance propagation of LF radio waves, *J. Res. Natl. Bur. Std.*, $\underline{D68}$, No. 12, 1275-1284

BERRY, L.A. and CHRISMAN, M.E., (1965a), The path integrals of LF/VLF wave hop theory, *J. Res. Natl. Bur. Std.*, $\underline{D69}$, No. 11, 1469-1480.

BERRY, L.A., and CHRISMAN, M.E., (1965b), Numerical values of the path integrals for LF and VLF, *Natl. Bur. Std. Tech. Note No. 319*.

BERRY, L.A., GONZALEZ, G., and LLOYD, J.L., (1969), Wave hop series for an anisotropic ionosphere, *Radio Science*, $\underline{4}$, 1025-1027,

BEZRODNY, V.G., NICKOLAENKO, A.P., AND SINITSIN, V.G., (1977), Radio propagation in natural waveguides, *J. Atmos. and Terr. Phys.*, 39, 661-688, (good review of recent Soviet work on the subject.)

BICKEL, J.E., (1967), VLF attenuation rates deduced from aircraft observations near the antipode of NPM, *Radio Science*, 2 (New Ser.), No. 6, 575-580.

BREMMER, H., (1949), *Terrestrial Radio Waves*, Elsevier, Amsterdam.

BREMMER, H., (1958), Applications of operational calculus to ground wave propagation, particularly for long waves, *IRE Trans.*, AP-6, No. 3, 267-272.

BUDDEN, K.G., (1961), *Radio Waves in the Ionosphere*, Cambridge Univ. Press, London and New York.

BUDDEN, K.G., and MARTIN, H.G., (1962), The ionosphere as a whispering gallery, *Proc. Roy. Soc.*, A265.

BURGESS, B., and JONES, T.B., (1967), Solar flare effects and VLF radio wave observations of the lower ionosphere, *Radio Science*, 2 (New Ser.), No. 6, 619-626.

CROMBIE, D.D., (1964a), The effects of a small local change in phase velocity on the propagation of a VLF radio signal., *J. Res. Natl. Bur. Std.*, D68, No. 6, 709-716.

CROMBIE, D.D., (1964b), Periodic fading of VLF signals received over long paths at sunrise and sunset, *J. Res. Natl. Bur. Std.*, D68, No. 1, 27-34.

CROMBIE, D.D., ed., (1967), Special issue on the propagation of long radio waves, *Radio Science*, 2, No. 6, 521-657.

DEEKS, D.G., (1966), D-region electron distributions in middle latitudes deduced from the reflection of long radio waves, *Proc. Roy. Soc.*, A291, No. 1426, 413-437.

FOCK, V.A., (1965), *Electromagnetic Diffraction and Propagation Problems*, Pergamon Press, Oxford.

GALEJS, J., (1964), Propagation of VLF waves below an inhomogeneous and stratified anisotropic ionosphere, *J. Geophys. Res.*, 69, No. 17, 3639-3650.

GALEJS, J., (1967), Propagation of VLF waves below an anisotropic stratified ionosphere with a transverse static magnetic field, *Radio Science*, 2(New Series), No. 6, 557-574.

GALEJS, J., (1972, *Terrestrial Propagation of Long Electromagnetic Waves*, Pergamon Press, Oxford.

HORNER, F., (1964), I. Terrestrial radio noise: Origins and importance, *Advan. Radio Res.*, 2, 122-204.

JONES, R.M., (1968), Application of the geometrical theory of diffraction to terrestrial LF radio wave propagation, *Mitteilungen aus dem Max Planck-Institut für Aeronomie*, (Lindau), Nr. 37(1), 3-26.

KAISER, A.B., (1968), Identification of a new type of mode interference in VLF propagation, *Radio Science*, 3, 545-549.

KRASNUSHKIN, P.E., (1962), On the propagation of long and very long radio waves around the earth, *Nuovo Cimento*, 26, 50-112.

KRASNUSHKIN, P.E., and FEDOROV, YE.N., (1973), Effects of the terrestrial magnetic field in the waveguide formed by the earth and the ionosphere, *Radio Eng. & Elect. Phys.*, 18, No. 1, 9-18.

LYNN, K.J.W., (1967), Anomalous sunrise effects observed on a long trans-equatorial VLF propagation path, *Radio Science*, 2(New Ser.), No. 6, 521-530.

MAHMOUD, S.F., and BEAL, J.C., (1971), VLF propagation parameters derived from observations of sunrise and sunset phenomena, *Proc. IEE*, Vol. 118, No. 10, 1351-1357.

MAKAROV, G.I., NOVIKOV, V.V., and ORLOV, A.B., (1970), Modern state of investigations of ultralong waves in the earth-ionosphere waveguide channel, *Radiofizika*, 13, No. 3, 321-355.

MAKAROV, G.I., NOVIKOV, V.V., and RYBACHEK, S.T., (1970), Propagation of EM waves in a spherical impedance waveguide, part III, *Diffraction and Radio Wave Propagation Problems*, 10, 3-64.

ORLOV, A.B., and AZARIN, (1970), The laws of propagation of VLF signals in the earth-ionosphere waveguide (survey of experimental results), *Diffraction and Radio Wave Propagation Problems*, 10, 3-107 (Leningrad State University).

PAPPERT, R.A., AND BICKEL, J.E., (1970), Vertical and horizontal VLF fields excited by dipoles of arbitrary orientation and elecation, *Radio Science*, 5, No. 12, 1445-1452.

PAPPERT, R.A., GOSSARD, E.E., and ROTHMULLER, I.J., (1967), A numerical study of classical approximations used in VLF propagation, *Radio Science*, 2(New Ser.), No. 4, 387-400.

PAPPERT, R.A., and MORFITT, D G., (1975), Theoretical and experimental sunrise mode conversion at V.L.F , *Radio Science*, 10, No. 5, 537-546.

PIERCE, J.A. and NATH, S.C., (1961), VLF propagation, *Ann. Progr. Rept.*, No. 60, pp. 1-6, Cruft Lab, Harvard Univ., Cambridge, Mass.

RAYLEIGH, LORD (J.W. Strutt), (1910), The problem of the whispering gallery, *Phil. Mag.*, 20, 1001-1004.

RHOADES, F.J., and GARNER, W.E., (1967), An investigation of the modal interference of VLF radio waves, *Radio Science*, 2(New Ser.), No. 6, 531-538.

RIES, G., (1967), Results concerning the sunrise effect of VLF signals propagated over long paths, *Radio Science*, 2(New Ser.), No. 6, 531-538.

RIKITAKE, T., (1966), *Electromagnetism and the Earth's Interior*, Chapter 12, Elsevier, Amsterdam.

ROUND, H.J.T., ECKERSLEY, T.L , TREMELLEN, K., and LUNNON, F.C., (1925), Report on measurements made on signal strength at great distances during 1922 and 1923 by an expedition sent to Australia, *J. IEE*, (London), 63, 933-1011.

RUGG, D.E., (1967), Theoretical investigations of the diurnal phase and amplitude variations of VLF signals, *Radio Science,* 2(New Ser.), No. 6, 551-556.

RYBACHEK, S.T., (1970), Applicability of Airy functions to the problem of the determination of the eigen-functions of a spherical waveguide, *Diffraction and Radio Wave Propagation Problems,* 10, 181-183.

SOLYMAR, L., (1959), Mode conversion in pyramidal-tapered waveguides, *Electron. Radio Engr.,* 36, No. 12, 461-463.

SOMMERFELD, A.N., (1949), *Partial Differential Equations,* Academic Press, New York.

SPIES, K.P. and WAIT, J.R., (1966), On the calculation of the ground wave attenuation factors at low frequencies, *IEEE Trans.,* AP-14, No. 4, 515-517. (correction in AP-27, No. 2, 286, 1979).

STEELE, F.K., and CHILTON, C.J., (1964), The measurement of the phase velocity of VLF propagation in the earth-ionosphere waveguide, *J. Res. Natl. Bur. Std.,* D68, No. 12, 1269-1274.

STEELE, F.K., and CROMBIE, D.D., (1967), Frequency dependence of VLF fading at sunrise, *Radio Science,* 2(New Ser.), No. 6, 547-549.

SWANSON, E.R., (1964), Rept. No. 1239, U.S. Navy Electron. Lab.

TAYLOR, W.L., (1960a), Daytime attenuation rates in the VLF band using atmospherics, *J. Res. Natl. Bur. Std.,* D64, No. 4, 349-355.

TAYLOR, W.L., (1960b), VLF attenuation for east-west and west-east propagation using atmospherics, *J. Geophys. Res,* 65, 1933-1938.

TAYLOR, W.L., and LANGE, L.J., (1958), Some characteristics of VLF propagation using atmospheric waveforms, *Proc. Conf. Atmospheric Elec.,* 2nd, pp. 609-617, Pergamon Press, Oxford.

TIBBALS, M.L., (1960), U.S. Navy Electron. Lab, San Kiego, Ca. Private Communication.

WAIT, J.R., (1951), The magnetic dipole over the horizontally stratified earth, *Can. J. Phys.,* 29, 577-592.

WAIT, J.R., (1953), Induction in a conducting sheet by a small current-carrying loop, *Appl. Sci. Res. Sect. B*, 3, 230-235.

WAIT, J.R., (1956), Radiation from a vertical antenna over a curved stratified ground, *J. Res. Nat. Bur. Std.*, 56, No. 4, 237-244.

WAIT, J.R., (1961a), A new approach to the mode theory of VLF propagation, *J. Res. Natl. Bur. Std.*, D65, No. 1, 37-46.

WAIT, J.R., (1961b) A diffraction theory for LF sky-wave propagation, *J. Geophys. Res.*, 66, No. 6, 1713-1724.

WAIT, J.R., (1962a), Model studies of the influence of ionosphere perturbations on VLF propagation, in *Tech. Sum. Rept.* by S.W. Maley and E. Bahar, Appendix I and II, Elec. Eng. Dept. Univ. of Colorado, Boulder, Colorado.

WAIT, J.R., (1962b), Mode conversion in the earth-ionosphere waveguide, *Natl. Bur. Std. Tech. Note No. 151*.

WAIT, J.R., (1963), The mode theory of VLF radio propagation for a spherical earth and a concentric anisotropic ionosphere, *Can. J. Phys.*, 41, 299-315, 819, 2267.

WAIT, J.R., (1964a), Electromagnetic surface waves, *Advan. Radio Res.*, 1, 157-217.

WAIT, J.R., (1964b), Two-dimensional treatment of mode theory of the propagation of VLF radio waves, *J. Res. Natl. Bur. Std.*, D68, No. 1, 81-93.

WAIT, J.R., (1964c), On phase changes in very low frequency propagation induced by ionospheric depression of finite extend, *J. Geophys. Res.*, 69, No. 3, 441-445.

WAIT, J.R., (1964d), Influence of a circular ionospheric depression on VLF propagation, *J. Res. Natl. Bur. Std.*, D68, No. 8, 907-914.

WAIT, J.R., (1967), The whispering gallery nature of the earth-ionosphere waveguide at VLF, *IEEE Trans.*, AP-15, No. 4, 580-581. (Correction AP-16, No. 1, 147(1968).

WAIT, J.R., (1968), Mode conversion and refraction effect in the earth-ionosphere waveguide for VLF radio waves, *J. Geophys. Res., Space Phys.*, 73, No. 11, 3537-3548.

WAIT, J.R., (1970), 2nd Ed.), *Electromagnetic Waves in Stratified Media*, Pergamon press, Oxford.

WAIT, J.R., Ed., (1974), Special issue on ELF Communications, *IEEE Trans. Comm.*

WAIT, J.R., (1978), Concise theory of radio transmission in the earth-ionosphere waveguide, *Reviews of Geophys. and Space Phys.*, 16, No. 3, 320-326 (gives a more general development of the wave-hop theory including a combined representation of rays and modes).

WAIT, J.R., and BAHAR, E., (1966), Simulation of curvature in a straight model waveguide, *Electron. Letrs.*, 2, No. 10, 358.

WAIT, J.R., and CONDA, A.M., (1961), A diffraction theory for LF sky-wave propagation - An additional note, *J. Geophys. Res.*, 66, No. 6, 1725-1729.

WAIT, J.R., and SPIES, K.P., (1964), Characteristics of the earth-ionosphere waveguide for VLF radio waves, *Natl. Bur. Std. Note No. 300* available from National Technical Information Service, Springfield, Va. under Accession No. PB168048.

WAIT, J.R., and SPIES, K.P., (1968), On the calculation of mode conversion at a graded height change in the earth-ionosphere waveguide at VLF, *Radio Science*, 3 (New Ser.), No. 8.

WAIT, J.R., and SPIES, K.P., (1970), VLF mode propagation across an elevated coastline, *Radio Science*, Vol. 5, Nos. 8-9, 1169-1173.

WAIT, J.R, and WALTERS, L.C., (1963), Reflection of VLF radio waves from an inhomogeneous ionosphere. P. I. Exponentially varying isotropic model, *J. Res. Natl. Bur. Std.*, D67, No. 3, 361-367; P. II. Perturbed exponential model, D67, No. 5, 519-523; P. III. Exponential model with hyperbolic transition, D67, No. 6, 747-752.

WATT, A.D., (1967), *VLF Radio Engineering*, Pergamon Press, Oxford.

WATT, A.D., and CROGHAN, R D., (1964), Comparison of observed VLF attenuation rates and excitation factors with theory, *J. Res. Natl. Bur. Std.*, D68, No. 1, 1-9.

WIEDER, B., (1967), Transmission of VLF radio waves through the ionosphere, *Radio Science*, 2(New Ser.), No. 6, 595-605.

ZABINA, I.N., (1969), Propagation of VLF waves, *Diffraction and Radio Wave Propagation Problems*, 9, 64-75 (published by Leningrad State University).

21
Guiding of Microwaves by an Elevated Tropospheric Layer

Introduction

There has been a great deal of discussion in the literature on the mechanism of long-distance propagation of microwaves in the earth's atmosphere. Although turbulence, no doubt, plays a major role, there is also evidence that stable layer formations will provide an excellent and efficient guiding mechanism. Various vertical profiles of the refractive index have been considered [*Pekeris*, 1946; *Langer*, 1951; *Brekhovskikh*, 1960; *Wait*, 1962; *Fock*, 1965; *Gerks and Anderson*, 1966]. It does seem surprising, however, that little attention has been given to the internal guiding properties of a single elevated layer. Only recently, *Nicolis* [1968] pointed out the striking focusing properties of a smooth concave interface between two homogeneous air masses, and the importance of multiple internal reflections has been demonstrated by *Hall* [1968]. A preliminary analysis [*Wait*, 1968] of this internal guiding mechanism has shown that the attenuation rate of the dominant modes is exceedingly small. The similarity of these modes with Lord Rayleigh's whispering-gallery acoustic modes was also noted.

In the present paper, we wish to expand on the calculation of these internally guided or whispering-gallery modes. In particular, we shall show that the low attenuation is not restricted to a sharply bounded curved layer.

Basic Analysis

The general problem of propagation of horizontally polarized waves in a uniformly stratified troposphere over a spherical earth can be reduced to an equivalent planar problem [*Pekeris*, 1946; *Wait*, 1962]. In this analog, the earth's surface is the plane $z=0$ and the equivalent dielectric susceptibility $\chi(z)$, as a function of height z, is given by

$$\chi(z) = (2z/a) + N^2(z) - N^2(0) \tag{1}$$

where a is the actual radius of the earth and $N(z)$ is the actual refractive index of the atmosphere at the height z. As indicated, the susceptibility is normalized to be zero at the earth's surface.

If we now consider propagation in the horizontal x direction, the electric field E_y of a given mode has the form

$$E_y = f(z)\exp[-ikSx] \tag{2}$$

where $k = N(0)\omega/c$ is the wave number *at the surface* of the earth and $f(z)$ satisfies

$$\partial^2 f/\partial z^2 + k^2[\chi(z) + C^2]f = 0 \tag{3}$$

where $C^2 = 1 - S^2$. The eigenvalues C are yet to be determined.

To facilitate the discussion, we assume that the equivalent susceptibility can be approximated over an interval of z by a linear function. Thus, for a restricted region of height,

$$\chi(z) = \chi + Kz \tag{4}$$

---

*Reprinted in part from: J.R. Wait and K.P. Spies, *Radio Science*, 4, 319–326, 1969.

where χ and K are constants. For example, for just above the earth's surface we can write

$$\chi(z) = K_0 z \tag{5}$$

Using (1), we see that

$$K_0 = \frac{2}{a} + 2N(0)N'(0) = \frac{2}{a_e} \tag{6}$$

where, by definition, a_e is the effective earth's radius.

We now introduce the normalized height variable y by

$$y = (K_0/k)^{1/3} kz = [2/(ka_e)]^{1/3} kz \tag{7}$$

Thus, for any individual layer, (3) is expressible in the form

$$\partial^2 f/\partial y^2 + (T - t + \alpha^3 y) f = 0 \tag{8}$$

where $T = (k/K_0)^{2/3} \chi$, $\alpha^3 = K/K_0$, and $-t = (k/K_0)^{2/3} C^2$.

The solutions of (8) are of the type

$$f = w[(t - T)\alpha^{-2} - (\alpha y)] \tag{9}$$

where $w(\tau)$ is an Airy function, which by definition satisfies

$$w''(\tau) - \tau w(\tau) = 0 \tag{10}$$

Thus, in general, f for a given layer is comprised of a linear combination of the Airy functions $w_1(\tau)$ and $w_2(\tau)$. In terms of the standard notation,

$$w_1(\tau) = -i\pi^{1/2}[Ai(\tau) + iBi(\tau)] \tag{11}$$

and

$$w_2(\tau) = +i\pi^{1/2}[Ai(\tau) - iBi(\tau)] \tag{12}$$

where Ai and Bi are functions defined and tabulated by *Miller* [1946].

To obtain the desired solutions of the problem, we must impose appropriate continuity conditions. Specifically, the tangential electric field E_y and the tangential magnetic field H_x must be continuous at the boundaries of the layers. Because $H_x = +(i\mu_0\omega)^{-1} \partial E_y/\partial z$, we see immediately from (2) that both f and $\partial f/\partial y$ are continuous at the interfaces between the layers. In addition, at the earth's surface ($z = 0$), we will impose the impedance boundary condition that states that

$$\eta H_x = \Delta E_y \tag{13}$$

where $\eta = \mu_0 \omega/k = 120\pi/N(0)$ is the characteristic impedance of the atmosphere at the earth's surface and Δ is the normalized surface impedance of the earth. Using (7), we readily find that our impedance boundary condition is equivalent to

$$\{\partial f/\partial y + qf = 0\}_{y=0} \tag{14}$$

where $q = -i(k/K_0)^{1/3}\Delta$. In dealing with tropospheric radio propagation, it turns out that q is effectively infinite, which means that the boundary condition at the earth's surface could be taken as simply $f = 0$; however, we will retain q as a finite parameter for the time being.

General Three-Region Problem

We now consider specifically a three-region problem as indicated in Figure 1. The equivalent susceptibility and the appropriate solutions are listed as follows:

For $0 < z < z_0$

$$\chi(z) = K_0 z \tag{15}$$

$$f = A[w_1(t-y) - \beta w_2(t-y)] \tag{16}$$

For $z_0 < z < z_1$

$$\chi(z) = \chi_1 + K_1 z \tag{17}$$

$$f = C w_1\left(\frac{t - t_b - \alpha^3 y}{\alpha^2}\right) + D w_2\left(\frac{t - t_b - \alpha^3 y}{\alpha^2}\right) \tag{18}$$

For $z_1 < z < \infty$

$$\chi(z) = \chi_0 + K_0 z \tag{19}$$

$$f = B w_1(t - t_a - y) \tag{20}$$

Here, A, B, C, D, and β are arbitrary constants and

$$t = -(k/K_0)^{2/3} c^2$$

$$t_a = (k/K_0)^{2/3} \chi_0$$

$$t_b = (k/K_0)^{2/3} \chi_1$$

By using (14), we see immediately that the constant β in (16) is given by

$$\beta = \frac{w_1'(t) - q w_1(t)}{w_2'(t) - q w_2(t)} \tag{21}$$

We then match f and $\partial f / \partial y$ at the interfaces $y = y_0 = (K_0/k)^{1/3} k z_0$ and $y = y_1 = (K_0/k)^{1/3} k z_1$ to obtain the linear system

$$A[w_1(t-y_0) - \beta w_2(t-y_0)] = C w_1 + D w_2 \tag{22}$$

$$A[w_1'(t-y_0) - \beta w_2'(t-y_0)] = C \alpha w_1' + D \alpha w_2' \tag{23}$$

$$B w_1(t - t_a - y_1) = C \hat{w}_1 + D \hat{w}_2 \tag{24}$$

$$B w_1'(t - t_a - y_1) = C \alpha \hat{w}_1' + D \alpha \hat{w}_2' \tag{25}$$

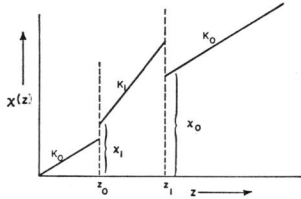

Fig. 1. The dielectric susceptibility profile used for the general analysis. For χ_1 read $\chi_1 + K_1 z_0$, and for χ_0 read $\chi_0 + K_0 z_0$.

Guiding of Microwaves by an Elevated Tropospheric Layer

where we have used the designations

$$w = w\left(\frac{t-t_b}{\alpha^2} - \alpha y_0\right)$$

$$\hat{w} = w\left(\frac{t-t_b}{\alpha^2} - \alpha y_1\right)$$

For a nontrivial solution, (22), (23), (24), and (25) require that

$$(Pw_1 + \alpha w_1')(Q\hat{w}_2 + \alpha \hat{w}_2') = (Q\hat{w}_1 + \alpha \hat{w}_1')(Pw_2 + \alpha w_2') \qquad (26)$$

where

$$P = -\frac{w_1'(t-y_0) - \beta w_2'(t-y_0)}{w_1(t-y_0) - \beta w_2(t-y_0)} \qquad (27)$$

and

$$Q = -\frac{w_1'(t-t_a-y_1)}{w_1(t-t_a-y_1)} \qquad (28)$$

Equation 26 yields the eigenvalues $t = t_e$ of the problem.

Continuous Linear Segmented Profile

Keeping in mind the objectives of the present paper, we now specialize the profile shown in Figure 1 to a form where there is no discontinuity in $\chi(z)$. Obviously, this requires that

$$\chi_0 = (K_1 - K_0)(z_1 - z_0)$$

and

$$\chi_1 = -(K_1 - K_0)z_0$$

If the upper region (i.e., $z > z_1$) is to have a refractive index Δn less than the lowest region (i.e., $0 < z < z_0$), it is evident from (15) and (19) that $\chi_0 = -2\Delta n$, bearing in mind that $|\Delta n| \ll 1$. Also, for convenience, we designate the thickness $z_1 - z_0$ by 2δ and the mean height $(z_1 + z_0)/2$ by h. Thus, we find

$$y_0 = (K_0/k)^{1/3} k(h - \delta)$$

$$y_1 = (K_0/k)^{1/3} k(h + \delta)$$

$$K_1 - K_0 = -\Delta n/\delta$$

$$t_a = -(k/K_0)^{2/3} 2\Delta n$$

$$t_b = (k/K_0)^{2/3}(\Delta n/\delta)(h - \delta)$$

$$\alpha = [1 - (\Delta n/K_0\delta)]^{1/3}$$

The corresponding profile is sketched in Figure 2a for $2\delta > 0$ and the limiting form is shown in Figure 2b for $2\delta \to 0$.

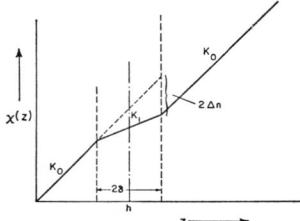
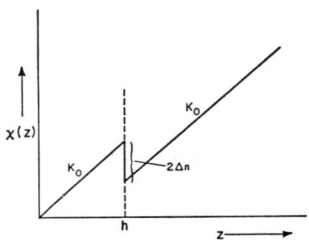

Fig. 2a. The linear segmented profile which is continuous at all heights.

Fig. 2b. The limiting case of the linear segmented profile when the transition width 2δ approaches zero.

After a little algebra, the relevant arguments of the Airy functions occurring in (26) can be written concisely as follows:

$$t - y_0 = \hat{t}$$

$$t - t_a - y_1 = \hat{t} + A$$

where

$$A = -2k\delta\alpha^3(K_0/k)^{1/3}$$

$$\frac{t-t_b}{\alpha^2} - \alpha y_0 = \frac{\hat{t}}{\alpha^2} = \tau_0$$

$$\frac{t-t_b}{\alpha^2} - \alpha y_1 = \frac{\hat{t}+A}{\alpha^2} = \tau_1$$

The mode equation (26) for the continuous profile shown in Figure 2a may thus be written compactly as

$$F(\hat{t}) = 0 \tag{29}$$

where

$$F(\hat{t}) = PQf_1 + \alpha Pf_2 + \alpha Qf_3 + \alpha^2 f_4 \tag{30}$$

where

$$f_1 = Bi(\tau_0)Ai(\tau_1) - Bi(\tau_1)Ai(\tau_0)$$

$$f_2 = Bi(\tau_0)Ai'(\tau_1) - Bi'(\tau_1)Ai(\tau_0)$$

$$f_3 = Bi'(\tau_0)Ai(\tau_1) - Bi(\tau_1)Ai'(\tau_0)$$

$$f_4 = Bi'(\tau_0)Ai'(\tau_1) - Bi'(\tau_1)Ai'(\tau_0)$$

$$P = -\frac{w_1'(\hat{t}) - \beta w_2'(\hat{t})}{w_1(\hat{t}) - \beta w_2(\hat{t})}$$

$$Q = -\frac{w_1'(\hat{t}+A)}{w_1(\hat{t}+A)}$$

As usual, the primes indicate differentiation with respect to the argument of the function.

Guiding of Microwaves by an Elevated Tropospheric Layer

For the calculations that follow, we set $\beta=1$. This is justified since, for the whispering-gallery or earth-detached type modes of interest, $|t|\gg 1$ with $|\arg t|<\pi/3$ and, consequently, (21) tells us that β is very close to 1. It should be stressed, however, that solutions of (26) and (29) exist where β differs from 1. These solutions correspond to modes that are reflected from the ground surface. They are not considered further here.

To solve the mode equation $F(\hat{t})=0$, we may employ Newton's method. Thus, a correction $\Delta\hat{t}_n$ to an approximate root $\hat{t}_{a,n}$ is

$$\Delta t = -F(\hat{t}_{a,n})/F'(\hat{t}_{a,n})$$

The process may be iterated any number of times until the desired accuracy is obtained. The method is facilitated by having a good starting solution. In the present case, the limiting sharply bounded model (i.e., $2\delta=0$) turned out to be useful. The mode equation (26) now reduces to the simple form $P=Q$, where $A=2(\Delta n)(k/K_0)^{2/3}=2(\Delta n)(ka_e/2)^{2/3}$. The corresponding explicit form for (29) is

$$Ai'(\hat{t}) - \frac{w_1'(\hat{t}+A)}{w_1(\hat{t}+A)} Ai(\hat{t}) = 0$$

Some results based on this equation were quoted previously [*Wait*, 1968].

Discussion of Calculated Data

The physical quantities of interest are the attenuation rate and the phase velocity. Conventionally, these quantities are referred to the earth's surface. Thus, the attenuation rate α_n, for the nth-order mode, is obtained from

$$\alpha_n = -k \operatorname{Im} S_n \approx k \operatorname{Im}(C_n^2/2) \quad \text{nepers per unit distance}$$

where C_n has the usual interpretation *Wait*, 1962] as the cosine of the (complex) angle of incidence at the ground of the nth mode. In a similar fashion, we find that the phase velocity v_n of the nth-order mode expressed as a ratio to the plane-wave velocity v_s at the earth's surface is

$$t_n/v_s = 1/\operatorname{Re} S_n \approx 1 + \operatorname{Re}(C_n^2/2)$$

Here, we note that $v_a = cN(0)$, where c is the velocity of light in free space and $N(0)$ is the refractive index of the air at the earth's surface.

The solutions of the mode equation (30) yield the discrete set of eigenvalues $t_n = \hat{t}_n + y_0$. Then, we obtain the required C_n values via the relation

$$-t_n = (ka_e/2)^{2/3} C_n^2$$

In actually displaying the calculated data, it is more convenient and more meaningful to *refer* the phase velocity to the level $z=h$, which is the center of the layer. It is a very simple matter to show that

$$v_n^h/v_s \approx v_n/v_s + h/a_e$$

where the superscript h reminds us that the velocity is being referred to the level $z=h$.

Without going into any details of the calculations, we first show the calculated attenuation rate α_n in Figure 3 as a function of the effective earth's radius a_e where $2\delta=0$ and the wavelength at the earth's surface is 10 cm, and where $\Delta n = 10^{-5} = 10$ N units. Here, for convenience, we choose $h=1$ km, but actually the results do not depend significantly on this value for the whispering-gallery or earth-detached modes being considered. Only the first four modes, which are the modes of lowest attenuation, are shown here. The corresponding phase velocity increments $(v^h/v_s)-1$ are shown in Figure 4 for the same first four modes.

316 Wave Propagation Theory

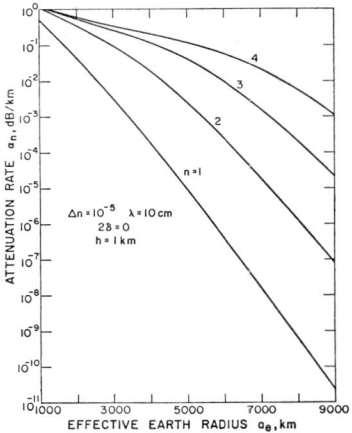

Fig. 3. Attenuation rate as a function of the effective earth radius a_e.

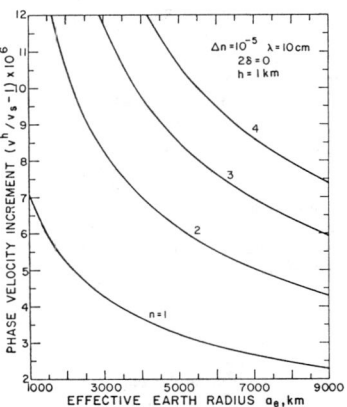

Fig. 4. Phase velocity as a function of the effective earth radius a_e.

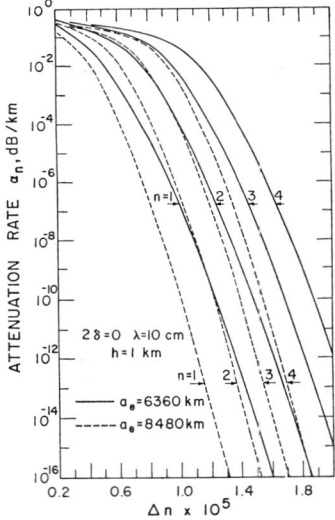

Fig. 5. Attenuation rate as a function of the refractive index contrast Δ_n.

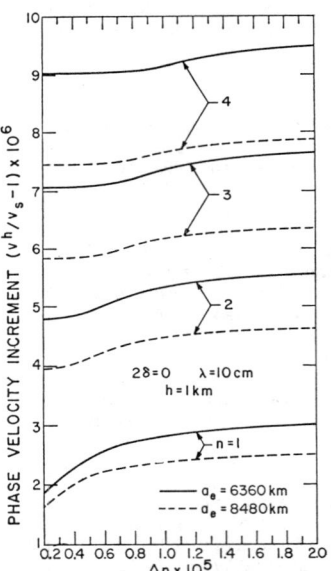

Fig. 6. Phase velocity as a function of the refractive index contrast Δ_n.

Guiding of Microwaves by an Elevated Tropospheric Layer

The remarkable feature of the calculations shown in Figure 3 is the smallness of the attenuation. These first two modes, for a_e greater than about 5000 km, lead to attenuations less than a decibel for a 100-km path. Unfortunately, the energy in these almost unattenuated modes is confined to a region just below the interface. In fact, the 'track width' w of the modes is readily shown to be given by [*Wait*, 1962, 1968]

$$w \simeq \left[(v^h/v_s) - 1\right] a_e$$

For example, using the calculated data in Figure 4 for $a_e = 5000$, we see that w for the first modes are, respectively, 16.2, 30.7, 42.5, and 53.7 meters. Outside the 'track width,' the field is evanescent and heavily damped with vertical distance.

In the previous example, Δn was 10 N units, which is a relatively large but not unusual change of refractive index at a subsidence inversion in the troposphere [*Saxton et al.*, 1964; *Lane*, 1965]. To illustrate the dependence of the modal characteristics on the refractive index jump at a sharp inversion, we plot the attenuation rate and the phase-velocity increment as a function of Δn. The results are shown in Figures 5 and 6 for the first four modes for a wavelength $\lambda = 10$ cm, and where $2\delta = 0$ corresponds to the sharply bounded situation. The effective earth radius a_e is assigned the value $a_e = 6360$ and 8480 km. The 6360 value represents the case of almost no atmosphere refraction in the lowermost region, and the 8480 value corresponds to more or less standard atmosphere refraction [*Bean*, 1966]. As indicated in Figure 5, the magnitude of the attenuation rates is strongly dependent on Δn, although from a practical standpoint the modes are almost unattenuated if Δn is greater than about 4 or 5 N units.

The wavelength dependence of the modal characteristics is illustrated in Figures 7 and 8. Here, we take $\Delta n = 10$ N units and $a_e = 6360$ km. Both a sharply bounded layer (i.e. $2\delta = 0$) and the case of a finite layer width (i.e. $2\delta = 10$ meters) are shown. The attenuation results shown in Figure 7 illustrate the extremely low attenuations associated with the shorter wavelengths. It also appears that the guiding properties of the concave layer are not appreciably degraded by choosing a finite value of the transition distance 2δ.

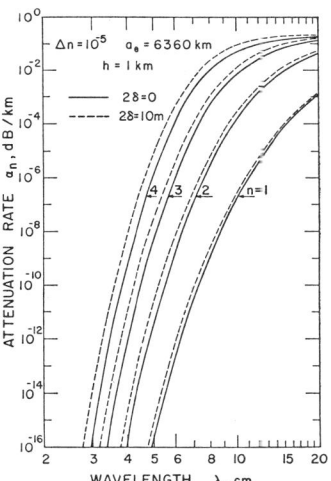

Fig. 7. Attenuation rate as a function of the operating wavelength λ.

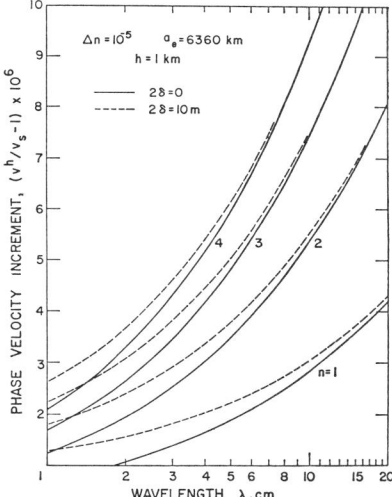

Fig. 8. Phase velocity as a function of the operating wavelength λ.

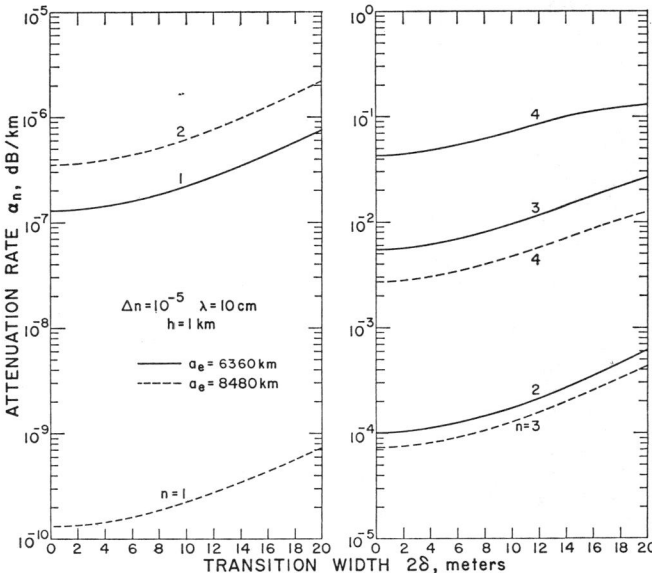

Fig. 9. Attenuation rate as a function of the transition width 2δ.

The dependence on the transition width 2δ is illustrated in Figure 9 for the conditions indicated. Although the attenuation rate of the first-order mode is exceptionally small, it is evident that the attenuation increases monotonically with 2δ. The effect is less noticeable for the higher-order modes. In general, it appears that the elevated layer with a linear transition zone behaves as a sharp boundary if the track width w of the mode is somewhat greater than the transition width 2δ.

Concluding Remarks

The results of the present analysis indicate that an elevated tropospheric inversion layer has a remarkable ability to guide electromagnetic waves along its concave side. For the smooth uniform structure considered, the most critical factors considered are the refractive index contrast Δn and the effective radius of curvature a_e of the layer. Less important is the degree of abruptness of the layer as described by the transition width 2δ. Also, it appears that the shorter centimetric wavelengths are guided most effectively.

The role of these whispering-gallery or earth-detached modes appears to be a neglected aspect of the subject. However, it is important to stress that the results obtained here refer only to a smooth and uniform layer structure. Whether conclusions based on such a model have important practical consequences remains to be investigated. Such factors as nonuniformity of the layer structure, mode conversion phenomena, excitation of the modes from ground-based terminals, and relationships to conventional surface ducted modes are fertile areas for further investigations. Hopefully, we have unturned a few stones that had been gathering moss since the early days of radar.

References

BEAN, B.R., (1966), *Radio Meterology*, Chapt. 3, National Bureau of Standards Monograph 92, Washington, D.C.
BOOKER, H.G., and WALKINSHAW, W., (1946), The mode theory of tropospheric refraction and its relation to waveguides and diffraction, in *Meteorlogical Factors in Radio-Wave Propagation*, Physical Society, London.
BREKHOVSKIKH, L., (1960), *Waves in Layered Media*, Academic Press, London.
CHANG, H.T., (1971), The effect of tropospheric layer structure on long range VHF radio propagation, *IEEE Trans.*, AP-19, 751–756.
CHO, S.H., and WAIT, J.R., (1978), Analysis of microwave ducting in an inhomogeneous troposphere, *Pure and Appl. Geophysics*, 116, 1118–1142.
FOCK, V., (1965), *Propagation and Diffraction Problems*, Pergamon Press, Oxford.
GERKS, I., and ANDERSON, R.M., (1966), Diffraction of radio waves in a stratified troposphere, *Radio Science*, 1(8), 897–912.
HALL, M.P.M., (1968), VHF propagation by double-hop reflection from a tropospheric layer, *Proc. IEE*, 115(4), 503–506.
LANE, J.A., (1965), Some investigations of the structure of elevated layers in the troposphere, *J. Atmos. Terrest. Phys.*, 27, 969–978.
LANGER, R.E., (1951), Asymptotic solutions of a differential equation in the theory of microwave propagation, in *Symposium on the Theory of Electromagnetic Waves*, pp. 73–84, Interscience Publishers, New York.
MILLER, J.C.P., (1946), *The Airy Integral*, Cambridge University Press, Cambridge.
NICOLIS, J.S., (1968), The role of curvature of a smooth tropospheric boundary in beyond-the-horizon radio paths, *IEEE Trans. Antennas Propagation*, 16(2), 271–272. (See also 'Corrections,' *IEEE Trans. Antennas Propagation*, 16(6), 1968.
PAPPERT, R.A., and GOODHART, C.L., (1977), Case studies of beyond-the-horizon propagation in tropospheric ducting environments, *Radio Science*, 12, 75–88.
PEKERIS, C.L., (1946), Asymptotic solutions for the normal modes in the theory of microwave propagation, *J. Appl. Phys.*, 17(12), 1108–1124.
SAXTON, J.A., LANE, J.A., MEADOWS, R.W., and MATHEWS, P.A., (1964), Layer structure of the troposphere, *Proc. IEE*, 111(2), 275–283.
WAIT, J.R., (1962), *Electromagnetic Waves in Stratified Media*, Pergamon Press, Oxford.
WAIT, J.R., (1968), Whispering-gallery modes in a tropospheric layer, *Elec. Lettrs.*, 4(18), 377–378.
WAIT, J.R., (1969), Guiding of microwaves by an elevated tropospheric layer, *Radio Science*, 4, 319–326.

22
Scattering from an Isolated Irregularity in a Tropospheric Duct

INTRODUCTION

A number of formalisms exist for the transmission of electromagnetic waves in a uniform tropospheric duct [e.g., Chang, 1971; Grikurov and Salikov, 1978]. But, of course, nothing in nature is uniform and a good example is an isolated change of refractive index that is limited in both vertical and lateral extent. In this note we provide a simple two-dimensional analysis of such a configuration. The method is much simpler to apply than the mode matching procedure [e.g., Cho and Wait, 1978].

The starting point is the cylindrical model where the atmosphere is taken to be radially inhomogeneous such that the effective wave number $k(r)$ is only a function of the radial coordinate r. In terms of cylindrical coordinates (r,θ,z), the earth's surface is at $r = a_o$ and we imagine θ to be an infinitely extended angular domain such that a mode in the absence of lateral inhomogeneities would propagate in the positive direction, as $\exp(-i\nu_m\theta)$, for a time factor $\exp(i\omega t)$, where ν_m is the appropriate factor for a mode of order m. When the boundary condition at $r = a_o$ is of the impedance type, the modal spectrum is entirely discrete, so the question of the continuous spectrum does not arise. To be specific, we choose a line current source of strength I_o amperes at $\theta = \theta_o$ and at $r = r_o$ where $r_o > a_o$. We can now assert that the resultant field $E_z(r,\theta)$ for $r \geq a_o$ can be expressed in the form

$$E_z = \sum_m A_m \phi_m(r) \exp[\mp i\nu_m(\theta - \theta_c)] \tag{1}$$

where the $-(+)$ sign in the exponent is to be used for $\theta > \theta_o (\theta < \theta_o)$. Here $\phi(r)$ is the appropriate radial wave function that satisfies

Scattering from an Isolated Irregularity

$$\left[\frac{1}{r}\frac{\partial}{\partial r}\left(r\frac{\partial}{\partial r}\right) - \frac{\nu^2}{r^2} + k^2(r)\right]\phi(r) = 0 \qquad (2)$$

The boundary condition, that yields the set $\nu = \nu_m$, is

$$[d\phi(r)/dr - ik_o\Delta_a\phi(r)]_{r = a_o} = 0 \qquad (3)$$

where $k_o = k(a_o)$ and Δ_a is a normalized wave admittance. The coefficients A_m are determined by the source conditions as we shall indicate below.

From Maxwell's equations we have that

$$i\mu_o\omega H_r = -\frac{\partial E_z}{r\partial\theta} = \pm i\sum_m A_m\nu_m\frac{\phi_m(r)}{r}\exp[\mp i\nu_m(\theta - \theta_o)] \qquad (4)$$

where μ_o is the magnetic permeability assumed to be constant everywhere. As indicated in (1) and (4), the summations are to include **all** discrete modes of order m [Bahar, 1975].

The required condition at the source is that

$$H_r\big]_{\theta_o + 0} - H_r\big]_{\theta_o - 0} = -I_o\delta(r - r_o) \qquad (5)$$

where $\delta(r - r_o)$ is the unit impulse function. This, in effect, says that the horizontal magnetic fields above and below the source are discontinuous by the amount of the impressed current. From (4) and (5), it is immediately required that

$$\frac{2}{\mu_o\omega}\sum_m A_m\nu_m\frac{\phi_m(r)}{r} = -I_o\delta(r - r_o) \qquad (6)$$

In the usual way, we now invoke the orthogonality property

$$\int_{r = a_o}^{\infty} \phi_m(r)\phi_n(r)r^{-1}dr = 0 \text{ for } m \neq n \qquad (7)$$
$$= \bar{M}_n \text{ for } m = n$$

Then clearly

$$A_m = -\frac{\mu_o\omega I_o}{2}\frac{\phi_m(r_o)}{\bar{M}_m\nu_m} \qquad (8)$$

which is the desired form of the coefficient for the expansion given by (1).

The field produced in the uniform duct for this source is called the primary field E_z^p. From (1) and (8) we see that

$$E_z^p(r,\theta) = -\frac{\mu_0 \omega I_0}{2} \sum_{m=1}^{\infty} \frac{\phi_m(r_0)}{\nu_m \bar{M}_m} \phi_m(r) e^{-i\nu_m \theta} \qquad (9)$$

for $\theta > 0$ and a line source current at $(r_0, 0)$. The summation index here is indicated to range from $m = 1$ through integers to infinity. These terms are ordered such that $-\text{Im}.\nu_1 > -\text{Im}.\nu_2 > -\text{Im}.\nu_3 \ldots\ldots\ldots$ so that the attenuation increases progressively with the mode number except in the case of degenerate roots of the mode equation; this case can be dealt with separately if not circumvented by slightly changing the parameters of the problem.

ELEMENTARY FORMULATION FOR ISOLATED SCATTERER

Now we consider an elementary scatterer, at $r = r_s$ and $\theta = \theta_s$, located in the duct which is otherwise uniform. To be definite we imagine this to be a thin filamental dielectric cylinder of cross-sectional area δA with permittivity $(1 + \Delta\varepsilon)$ relative to the background $\varepsilon(r_s)$. Since the problem is entirely two dimensional we can argue that the excess current I induced in the filament is

$$\delta I = E_z^p(r_s, \theta_s) i\omega\varepsilon(r_s) \cdot \Delta\varepsilon \cdot \delta A$$

when $E_z^p(r_s, \theta_s)$ is the primary electric field that drives the current along the filament. Of course, actually the total E_z field drives this current, but the Born-type approximation indicated is well justified since $|\Delta\varepsilon| \ll 1$.

To be explicit

$$\delta I = -ik_0^2 \frac{I_0 \Delta\varepsilon}{2} \sum_{m=1}^{\infty} \frac{\phi_m(r_0)}{\nu_m \bar{M}_m} \phi_m(r_s) e^{-i\nu_m \theta_s} \cdot \delta A \qquad (10)$$

where $k_0^2 = \varepsilon_0 \mu_0 \omega^2$. Here the wave number $k(r_0)$ is replaced by k_0 since it occurs as a multiplier so the resulting error is inconsequential.

We now consider the secondary field E_z^s that is produced by δI. Evidently

Scattering from an Isolated Irregularity

this is given by

$$E_z^S = -\frac{\mu_0 \omega \delta I}{2} \sum_{n=1}^{\infty} \frac{\phi_n(r_s)}{\nu_n \bar{M}_n} \phi_n(r) \exp[-i\nu_n(\theta - \theta_s)] \tag{11}$$

for $\theta > \theta_s$. Equivalently

$$E_z^S = \frac{i\mu_0 \omega k_0^2}{4} \sum_{m=1}^{\infty} \sum_{n=1}^{\infty} \frac{\phi_m(r_o)}{\nu_m \bar{M}_m} \phi_m(r_s) \frac{\phi_n(r_s)\phi_n(r)}{\nu_n \bar{M}_n} I_0 \Delta \varepsilon \delta A \tag{12}$$

$$\times \exp[-i\nu_m \theta_s - i\nu_n(\theta - \theta_s)] I_0 \Delta \varepsilon$$

also for $\theta > \theta_s$ which is the forward scatter case. In the case of backscatter, when $\theta < \theta_s$, we replace $-\nu_n(\theta - \theta_s)$ by $-\nu_n(\theta_s - \theta)$ in (11) and (12).

To express this secondary field in a more physically meaningful form, it is desirable to change notation. To this end we note first that the primary field is expressible in the form

$$E_z^P = -\frac{\mu_0 \omega I_0}{2} \sum_{m=1}^{\infty} \Lambda_m^2 G_m(r_o) G_m(r) e^{-i\nu_m \theta} \tag{13}$$

where

$$\Lambda_m^2 = \frac{1}{\nu_m \int_{a_o}^{\infty} r^{-1} [G_m(r)]^2 dr} = \frac{[\phi_m(a)]^2}{\nu_m \bar{M}_m} \tag{14}$$

and

$$G_m(r) = \phi_m(r)/\phi_m(a)$$

Clearly Λ_m is an "excitation factor" for a mode of order m while $G_m(r)$ is the corresponding height gain function.

The corresponding form for the secondary field for $\theta > \theta_s$, is found to be

$$E_z^S = -\frac{\mu_0 \omega I_0}{2} \sum_{m=1}^{\infty} \sum_{n=1}^{\infty} \Lambda_m G_m(r_o) \Delta S_{n,m} G_n(r) \Lambda_n \tag{15}$$

$$\times \exp[-i\nu_m \theta_s - i\nu_n(\theta - \theta_s)]$$

where

$$\Delta S_{n,m} = \frac{ik_0^2 \cdot \Delta \varepsilon \cdot \delta A}{2} \Lambda_m \Lambda_n G_m(r_s) G_n(r_s) \tag{16}$$

Here $\Delta S_{n,m}$ can be interpreted as a mode conversion coefficient that is related to scattering from an incident mode of order m to a mode of order n. The excitation factors Λ_m and Λ_n and height-gain function G_m and G_n are included in the definition for $\Delta S_{n,m}$ since they are a measure of how well the localized irregularity, of given strength, will produce secondary scatter.

The total field E_z of course is the sum $E_z^p + E_z^s$. An explicit expression for this quantity, for $\theta > \theta_s$, is

$$E_z = -\frac{\mu_0 \omega I_0}{2} \sum_{m=1}^{\infty} \sum_{n=1}^{\infty} \Lambda_m G_m(r_o) S_{n,m} G_n(r) \\ \times \Lambda_n \exp[-i\nu_m \theta_s - i\nu_n(\theta - \theta_s)] \quad (17)$$

where the (total) conversion coefficient is given by

$$S_{n,m} = \delta_{n,m} + \Delta S_{n,m} \quad (18)$$

where

$$\delta_{n,m} = 1 \text{ if } m = n \\ 0 \text{ if } m \neq n$$

As indicated above, we can extend the results given by (15) and (17) to the backscatter case by merely changing the sign of $\theta - \theta_s$ in the exponents. This tells us that an electrically small column scatters in an isotropic sense. To extend the present formulation to an irregularity of finite lateral and vertical extent, we may integrate over both the angular coordinate θ_s and the vertical coordinate r_s with an appropriate distribution of permittivity with the volume. This idea of merely superimposing the fields of individual scatterers is really the basis of the so-called Born approximation. In effect, this neglects the multiple scattering that could be important. Such effects could be considered by extending the above formulation to allow for interaction between the secondary sources, but the procedure would be cumbersome. A more fruitful approach is to employ a coupled mode formalism.

Scattering from an Isolated Irregularity

MODE COUPLING FORMALISM

The approach, now considered, is to use the ideal modes of the structure. For example

$$E_z = \sum_m [{}_oA_m e^{-i\nu_m\theta} + {}_oB_m e^{i\nu_m\theta}]\phi_m(r) \qquad (19)$$

is a general representation for the uniform structure where the wave number $k(r)$ is <u>only</u> a function of r. The coefficients ${}_oA_m$ and ${}_oB_m$ are not dependent on r and θ.

We are now interested in the case where the wave number is $k(r,\theta)$ being a function of both r and θ. Following the ideas of Schelkunoff [1963], we will now demonstrate a mode coupling formalism that uses the ideal or uniform modes as the basis functions. This approach enables us to use a perturbation scheme that exhibits the mode conversion, due to both single and multiple scattering, in a clear cut fashion.

Now again, for a uniform structure,

$$H_r = \sum_m \frac{\nu_m}{r\mu_o\omega} [{}_oA_m e^{-i\nu_m\theta} - {}_oB_m e^{i\nu_m\theta}]\phi_m(r) \qquad (20)$$

which is merely a consequence of Maxwell's equations.

We now postulate the following representations for the field in the modified system (i.e., where $k(r)$ is replaced by $k(r,\theta)$ and the boundary condition at $r = a$ remains unchanged):

$$E_z = \sum_m [A_m(\theta) + B_m(\theta)]\phi_m(r) \qquad (21)$$

and

$$H_r = \sum_m \frac{\nu_m}{r\mu_o\omega} [A_m(\theta) - B_m(\theta)]\phi_m(r) \qquad (22)$$

Here $A_m(\theta)$ and $B_m(\theta)$ will be found from the requirement that the resultant fields must satisfy Maxwell's equations in the modified systems that are

$$i\mu\omega \frac{\partial H_r}{r\partial\theta} - \left[k^2(r,\theta) + \frac{1}{r}\frac{\partial}{\partial r}\left(r\frac{\partial}{\partial r}\right)\right]E_z = 0 \qquad (23)$$

$$\frac{\partial E_z}{\partial \theta} + i\mu\omega r H_r = 0 \qquad (24)$$

Here (23) is obtained by simple eliminating H_θ from the pair $\partial E_z/\partial r = i\mu\omega H_\theta$ and $i\varepsilon\omega r E_z = -\partial H_r/\partial \theta + \partial(rH_\theta)/\partial r$.

On applying (23) and (24) to (21) and (22) it is found that

$$\sum_m \frac{i\nu_m}{r^2}(A'_m - B'_m)\phi_m(r) - \left[k^2(r,\theta) + \frac{1}{r}\frac{\partial}{\partial r} r \frac{\partial}{\partial r}\right](A_m + B_m)\phi_m(r) = 0 \qquad (25)$$

and

$$\sum_m (A'_m + B'_m)\phi_m(r) + i\nu_m(A_m - B_m)\phi_m(r) = 0 \qquad (26)$$

where $A_m = A_m(\theta)$, $B_m = B_m(\theta)$, $A'_m = \partial A_m(\theta)/\partial \theta$ and $B'_m = \partial B_m(\theta)/\partial \theta$.

Of course, (26) immediately tells us that

$$A'_n + B'_n + i\nu_n(A_n - B_n) = 0 \qquad (27)$$

for any order n. To deal with (25) we note that

$$\left(\frac{1}{r}\frac{\partial}{\partial r} r \frac{\partial}{\partial r}\right)\phi_m(r) = \left[\frac{\nu_m^2}{r^2} - k^2(r)\right]\phi_m(r) \qquad (28)$$

Thus,

$$\sum_m \left\{\frac{i\nu_m}{r^2}(A'_m - B'_m) - \left[k^2(r,\theta) - k^2(r) + \frac{\nu_m^2}{r^2}\right](A_m + B_m)\right\}\phi_m(r) = 0 \qquad (29)$$

Now individual terms cannot be set equal to zero here because of the θ dependent $k(r,\theta)$ term. We now invoke the orthogonality property that

$$\int_{a_o}^{\infty} \phi_m(r)\phi_n(r)\frac{dr}{r} = \bar{M}_n \delta_{m,n} \qquad (30)$$

where $\delta_{m,n} = 0$ if $m \neq n$
$= 1$ if $m = n$.

This suggests we multiply (29) by $r\phi_n(r)$ and integrate r from a_o to ∞.

This leads to

$$A'_n - B'_n + i\nu_n(A_n + B_n) + \frac{i}{\nu_n \bar{M}_n}\sum_m (A_m + B_m)\int_{a_o}^{\infty}[k^2(r,\theta) - k^2(r)]\phi_m(r)\phi_n(r) r dr = 0 \qquad (31)$$

Scattering from an Isolated Irregularity

Combining (27) and (31) leads to the coupled pair

$$A_n' + i\nu_n A_n = \sum_m C_{n,m}(A_m + B_m) \qquad (32)$$

and

$$B_n' - i\nu_n B_n = -\sum_m C_{n,m}(A_m + B_m) \qquad (33)$$

where the coupling coefficient $C_{n,m}$ is defined by

$$C_{n,m} = \frac{1}{2i\nu_n \bar{M}_n} \int_{a_o}^{\infty} [k^2(r,\theta) - k^2(r)]\phi_m(r)\phi_n(r) r\, dr . \qquad (34)$$

Now we can actually rewrite (32) and (33) in the form

$$\frac{d}{d\theta}[A_n(\theta) e^{i\nu_n\theta}] = e^{i\nu_n\theta}\sum_m C_{n,m}(A_m + B_m) \qquad (35)$$

$$\frac{d}{d\theta}[B_n(\theta) e^{-i\nu_n\theta}] = -e^{-i\nu_n\theta}\sum_m C_{n,m}(A_m + B_m) \qquad (36)$$

The θ dependence of $C_{n,m}$, A_m, and B_m is here understood. These coupled equations can be solved by using a perturbation scheme that, in principle, works well when the coupling is small and the modes in the system are not degenerate. To this end we write

$$A_n(\theta) = A_n^{(o)} + A_n^{(1)}(\theta) + A_n^{(2)}(\theta) + \ldots \qquad (37)$$

and

$$B_n(\theta) = B_n^{(o)} + B_n^{(1)}(\theta) + B_n^{(2)}(\theta) + \ldots \qquad (38)$$

where $A_n^{(o)}$ and $B_n^{(o)}$ are the zero order θ dependent forms. Succeeding terms are θ dependent corrections of first order, second order, and so on. In this scheme, we treat $C_{n,m}$ as a quantity of first order of smallness. Thus, for example,

$$\frac{d}{d\theta}[A_n^{(1)} e^{i\nu_n\theta}] = e^{i\nu_n\theta}\sum_m C_{n,m}[A_m^{(0)} + B_m^{(0)}] \qquad (39)$$

and

$$\frac{d}{d\theta}[B_n^{(1)} e^{-i\nu_n\theta}] = -e^{-i\nu_n\theta}\sum_m C_{n,m}[A_m^{(0)} + B_m^{(0)}] \qquad (40)$$

Now we integrate the first of the above pair over the range from $-\infty$ to θ and the second over the range from θ to ∞. Thus we readily find that

$$A_n^{(1)} e^{i\nu_n \theta} = \sum_m \int_{-\infty}^{\theta} [A_m^{(0)} + B_m^{(0)}] e^{i\nu_n \theta'} C_{n,m} d\theta' \qquad (41)$$

and

$$B_n^{(1)} e^{-i\nu_n \theta} = \sum_m \int_{\theta}^{\infty} [A_m^{(0)} + B_m^{(0)}] e^{-i\nu_n \theta'} C_{n,m} d\theta' \qquad (42)$$

where $C_{n,m}$ is defined by (34) where θ is replaced by θ' the dummy variable in (41) and (42). Formally the interative process can be continued and we would have, in general, that

$$A_n^{(j)} e^{i\nu_n \theta} = \sum_m \int_{-\infty}^{\theta} [A_m^{(j-1)}(\theta') + B_m^{(j-1)}(\theta')] e^{i\nu_n \theta'} C_{n,m}(\theta') d\theta' \qquad (43)$$

and

$$B_n^{(j)} e^{-i\nu_n \theta} = \sum_m \int_{\theta}^{\infty} [A_m^{(j-1)}(\theta') + B_m^{(j-1)}(\theta')] e^{-i\nu_n \theta'} C_{n,m}(\theta') d\theta' \qquad (44)$$

where $j = 1, 2, 3, \ldots$ is the perturbation order.

A word might be said about the limits of integration over θ in (41), (42), (43) and (44). In fact they are chosen on the basis that $A_n^{(j)}$, for $j > 0$, is zero as $\theta \to -\infty$. Certainly this is obvious if $j = 1$ because no mode conversion is generated in the forward wave until θ is non-zero. Also, we assume that $B_n^{(j)}$ for $j > 0$ is zero for $\theta \to \infty$.

If we now just consider modes incident from the left (i.e., propagating in the positive θ direction it is clear $B_m^{(0)} = 0$, but of course that does not mean we neglect $B_m^{(j)}$ in general. In fact, backward mode conversion may be important in some cases.

At this stage, it is useful to apply this coupled mode formalism to the isolated irregularity that was treated earlier. To this end we can write

$$A_n^{(1)} = e^{-i\nu_n \theta} \sum_m \int_{-\infty}^{\theta} A_m^{(0)} e^{i\nu_n \theta'} C_{n,m} d\theta' \qquad (45)$$

and

$$B_n^{(1)} = e^{i\nu_n \theta} \sum_m \int_{\theta}^{\infty} A_m^{(0)} e^{-i\nu_n \theta'} C_{n,m} d\theta' \qquad (46)$$

where

$$A_m^{(0)} = -\frac{\mu_0 \omega I}{2} \frac{\phi_m(r_0)}{\nu_m \bar{M}_m} e^{-i\nu_m \theta}$$

for the line source excitation at $\theta = 0$ and $r = r_0$. Now the defining integral for $C_{n,m}$ given by (34) can be approximated by

$$C_{n,m} \simeq \frac{1}{2i\nu_n \bar{M}_n} k_0^2 \Delta\epsilon \phi_m(r_s) \phi_n(r_s) r_s \delta r_s \tag{47}$$

where δr_s is the vertical extent of the irregularity. Now we further note that

$$\int_{-\infty}^{\theta} e^{i(\nu_n - \nu_m)\theta'} C_{n,m}(\theta') d\theta' \simeq e^{i(\nu_n - \nu_m)\theta_s} C_{n,m}(\theta_s) \delta\theta_s \quad \text{for } \theta > \theta_s$$
$$= 0 \quad \text{for } \theta < \theta_s \tag{48}$$

where $\delta\theta_s$ is the angular extent of the irregularity. Similarly

$$\int_{-\infty}^{\theta} e^{-i(\nu_n + \nu_m)\theta'} C_{n,m}(\theta') d\theta' \simeq e^{-i(\nu_n + \nu_m)\theta_s} C_{n,m}(\theta_s) \delta\theta_s \quad \text{for } \theta < \theta_s$$
$$= 0 \quad \text{for } \theta > \theta_s \tag{49}$$

Noting that $\delta A = r_s \delta r_s \delta\theta$ is the elemental area we are able to write

$$A_n^{(1)} = \frac{I_0 i \mu_0 \omega k_0^2 \Delta\epsilon}{4} \sum_n \sum_m \frac{\phi_m(r_0)}{\nu_m \bar{M}_m} \phi_m(r_s) \frac{\phi_n(r_s)}{\nu_n \bar{M}_n} \delta A \tag{50}$$

$$\times \exp[-i\nu_m \theta_s - i\nu_n(\theta - \theta_s)]$$

for $\theta > \theta_s$ and, of course, $A_n^{(1)} = 0$ for $\theta < \theta_s$. The corresponding expression for $B_n^{(1)}$ for $\theta < \theta_s$ is identical in form to (50) if $-i\nu_n(\theta - \theta_s)$ is replaced by $+i\nu_n(\theta - \theta_s)$. Then, of course, $B_n^{(1)} = 0$ for $\theta > \theta_s$.

CONCLUDING REMARKS

The equations developed here can be employed most easily when the scattering irregularity is limited in lateral extent. The approach is definitely not suitable for dealing with a gradual or tapered change of the vertical profile. In that case we should employ a coupled wave formalism that uses range dependent modes.

REFERENCES

BAHAR, E., (1975), Field transforms for multi-layered cylindrical and spherical structures of finite conductivity, *Can. Jour. Phys.*, 53, No. 11, 1078-1087.

CHANG, H.T., (1971), The effect of tropospheric layer structure on long range VHF radio propagation, *IEEE Trans.*, AP-19, No. 6, 751-756.

CHO, S.H., and WAIT, J.R., (1978), Analysis of microwave ducting in an inhomogeneous troposphere, *Pure and Applied Geophys.*, 116, No. 6, 1118-1142.

GRIKUROV, V.E., and SALIKOV, S.P., (1978), Numerical comparison of the ray method and the normal wave method for a tropospheric waveguide, *Radiotekhnika i Elektronika*, 23, (8), 1578-1587.

SCHELKUNOFF, S.A., (1963), *Electromagnetic Fields*, Chapt. 9, pp. 285-303, Blaisdell Publishing Co.

Scattering from an Isolated Irregularity

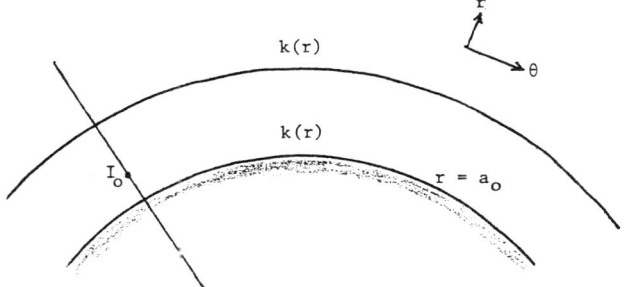

Fig. 1. 2D Model (z axis into paper).

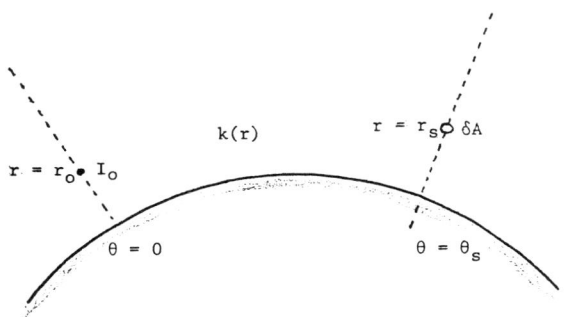

Fig. 2. Filamental scatterer at (r_s, θ_s) for source at $(r_o, 0)$.

23
Coupled Mode Analysis for a Non-Uniform Tropospheric Waveguide

INTRODUCTION

The waveguide model has been very useful in analytical investigations of radiowave transmission in naturally occurring ducts in the atmosphere [Bremmer, 1958; Wait, 1970]. A complicated feature is that the ducts are often non-uniform in the sense that the vertical profile of the refractive index changes along the path. When this change is sufficiently slow, the modes do not interact significantly and, thus, each mode can be tracked separately. This so-called adiabatic approximation has been used extensively in underwater sound [Deavenport, 1966; Nagl, et al, 1978] and atmospheric acoustic waves [Pierce, 1965]. Only quite recently has the importance of mode conversion been appreciated when dealing with non-uniform natural waveguides [Bahar and Wait, 1963; Wait, 1968; Bahar, 1971; Wait, 1974; Cho and Wait, 1978; Rutherford and Hawker, 1979].

In this paper we present a self-contained analysis for propagation in a two dimensional curved model of the earth where the refractive index is a function of both height and range. The method of dealing with the mode coupling is similar to that used in propagation in stratified media where correction terms to the WKB approximation are generated by successive iteration [Bremmer, 1958; Wait, 1970]. An analogous problem arises when dealing with the photon propagator in general electromagnetic cavities [Ledinegg, 1979].

FORMULATION

The following cylindrical model is assumed. The earth's surface is at $r = a_o$ with respect to cylindrical coordinates (r,θ,ϕ). The permittivity $\varepsilon(r,\theta)$, for $r > a_o$, is assumed to be independent of z. The magnetic permeability

Coupled Mode Analysis

is μ_o everywhere and does not vary. Thus, for a harmonic time factor $\exp(i\omega t)$ the "wave number" is $k(r,\theta) = i[\varepsilon(r,\theta)\mu_o]^{1/2}\omega$. To simplify our discussions we also assume the fields do not vary with z so that $\partial/\partial z = 0$ throughout. Furthermore, we take the excitation to be a uniform electric line source at $r = r_o$ and $\theta = 0$. Thus, the only non-vanishing field components are E_z, H_r and H_θ. The boundary condition at the earth's surface is taken to be

$$\left[\frac{\partial E_z}{\partial r} = i\mu_o\omega Y E_z\right]_{r=a_o} \tag{1}$$

where Y is the assumed constant surface admittance of the ground.

Maxwell's equations are

$$H_r = -\frac{1}{i\mu_o\omega}\frac{\partial E_z}{r\partial\theta} \tag{2}$$

$$H_\theta = \frac{1}{i\mu_o\omega}\frac{\partial E_z}{\partial r} \tag{3}$$

$$ri\varepsilon(r,\theta)\omega E_z = \frac{\partial}{\partial r}(rH_\theta) - \frac{\partial H_r}{\partial\theta} \tag{4}$$

On eliminating H_θ we obtain the basic pair

$$\frac{\partial E_z}{\partial \theta} = -i\mu_o\omega r H_r \tag{5}$$

and

$$\frac{\partial H_r}{r\partial\theta} = \frac{-}{i\mu_o\omega}\left(k^2(r,\theta) + \frac{1}{r}\frac{\partial}{\partial r}r\frac{\partial}{\partial r}\right)E_z \tag{6}$$

These latter two equations must be satisfied for the total transverse fields $E_z(r,\theta)$ and $H_r(r,\theta)$ for $r > a_o$. If we further eliminate H_r from (5) and (6) we see that

$$\left[\frac{1}{r}\frac{\partial}{\partial r}r\frac{\partial}{\partial r} + \frac{\partial^2}{r^2\partial\theta^2} + k^2(r,\theta)\right]E_z = 0 \tag{7}$$

except at the source.

UNIFORM MODEL

As an iterim step we momentarily consider that $k^2(r,\theta)$ is replaced by an θ independent form $k^2(r)$. A standard separation of variables solution then applies. The solutions for E_z are clearly of the form

$$\exp(\mp i\nu\theta)\phi(r)$$

where the radial functions of order ν satisfy

$$\left[\frac{1}{r}\frac{\partial}{\partial r} r \frac{\partial}{\partial r} - \frac{\nu^2}{r^2} + k^2(r)\right]\phi(r) = 0 \tag{8}$$

For fields propagating in the positive θ direction the general solution, for $r > a_o$, would be of the form

$$E_z = \sum_m A_m e^{-i\nu_m\theta} \phi_m(r) \tag{9}$$

where ν_m are the discrete solutions of the boundary condition at $r = a_o$ given by (1) and A_m is a coefficient that does not depend on θ. The corresponding radial magnetic field is given by

$$H_r = \frac{1}{r\mu_o\omega} \sum_m A_m \nu_m e^{-i\nu_m\theta} \phi_m(r) \tag{10}$$

In both (9) and (10), $\phi_m(r)$ satisfies (8) when ν is replaced by ν_m. If we were now dealing with waves propagating in the negative θ direction, the forms for E_z and H_r would be

$$E_z = \sum_m B_m e^{i\nu_m\theta} \phi_m(r) \tag{11}$$

$$H_r = -\frac{1}{r\mu_o\omega} \sum_m B_m \nu_m e^{i\nu_m\theta} \phi_m(r) \tag{12}$$

COUPLED MODE FORMALISM

We now return to the θ dependent problem. A solution of the following forms is postulated

$$E_z = \sum_m [A_m(\theta) + B_m(\theta)]\phi_m(r,\theta) \tag{13}$$

Coupled Mode Analysis

and

$$H_r = \frac{1}{r\mu_0\omega} \sum_m [A_m(\theta) - B_m(\theta)]\nu_m(\theta)\phi_m(r,\theta) \tag{14}$$

where the allowed θ dependence on ν_m, A_m and B_m and ϕ_m is indicated explicitly. We choose the ϕ_m to satisfy the Helmholz equation in a local sense, i.e.,

$$\left[\frac{1}{r}\frac{\partial}{\partial r} r \frac{\partial}{\partial r} - \frac{\nu_m^2(\theta)}{r^2} + k^2(r,\theta)\right]\phi_m(r,\theta) = 0 \tag{15}$$

The boundary condition that determines $\nu_m(\theta)$ follows from (1) and is written

$$\left[\frac{\partial\phi_m(r,\theta)}{\partial r} - i\mu_0\omega Y\phi_m(r,\theta)\right]_{r=a} = 0 \tag{16}$$

The next step, which is the crux of the procedure, is to insert the mode expansions (13) and (14) into the Maxwellian equations (5) and (6). Using a rather abbreviated notation, this leads readily to

$$\sum_m \{(A_m' + B_m')\phi_m + (A_m + B_m)\phi_m' + i\nu_m(A_m - B_m)\phi_m\} = 0 \tag{17}$$

and

$$\sum_m \nu_m\{(A_m - B_m)\phi_m' + \frac{\nu_m'}{\nu_m}(A_m - B_m)\phi_m + (A_m' - B_m')\phi_m$$

$$+ i\nu_m(A_m + B_m)\phi_m\} = 0 \tag{18}$$

where $A_m = A_m(\theta)$, $B_m = B_m(\theta)$, $A_m' = \partial A_m(\theta)/\partial\theta$, $B_m' = \partial B_m(\theta)/\partial\theta$, $\phi_m = \phi_m(r,\theta)$, $\phi_m' = \partial\phi_m(r,\theta)/\partial\theta$, $\nu_m = \nu_m(\theta)$ and $\nu_m' = \partial\nu_m/\partial\theta$.

We now utilize the orthogonality of the radial functions [Wait, 1974]. This property is stated succinctly as follows

$$\int_{a_c}^{\infty} \phi_m\phi_n (1/r) dr = M_n\delta_{m,n} \tag{19}$$

where M_n is the normalization factor and

$$\delta_{m,n} = \begin{matrix} = 1 \text{ if } m = n \\ = 0 \text{ if } m \neq n \end{matrix}$$

Bearing in mind that ϕ_m satisfies (15) and a similar equation holds for ϕ_n, one can easily show that

$$M_n = -\frac{a_o}{2\nu_n}\phi_n(a_o,\theta)\left[\frac{\partial}{\partial\nu}\left\{\left(\frac{\partial\phi_\nu(r,\theta)}{\partial r}\right)_{r=a_o} - i\mu_o\omega Y\phi_\nu(a_o,\theta)\right\}\right]_{\nu=\nu_n} \quad (20)$$

where the θ dependence has been indicated explicitly. In writing (20) we have also exploited the boundary condition (16). One might remark at this stage that the mode spectrum is entirely discrete. This is a consequence of the assumed boundary condition (i.e., Y is a constant independent of the mode number). Also, because of the cylindrical geometry, only modes are employed that satisfy individually the radiation condition at $r \to \infty$.

We now multiply both (17) and (18) by ϕ_n/r and integrate, with respect to r, from a_o to ∞. Using (19), this leads to

$$A_n' + i\nu_n A_n + B_n' - i\nu_n B_n + \sum_m (A_m + B_m)c_{n,m} = 0 \quad (21)$$

and

$$A_n' + i\nu_n A_n - B_n' + i\nu_n B_n + \frac{\nu_n'}{\nu_n}(A_n - B_n)$$
$$+ \sum_m \frac{\nu_m}{\nu_n}(A_m - B_m)c_{n,m} = 0 \quad (22)$$

where the prime indicates differentiation with respect to θ. On adding and subtracting the latter pair, we get the alternative coupled forms

$$A_n' + i\nu_n A_n + \frac{1}{2}\frac{\nu_n'}{\nu_n}(A_n - B_n) + \sum_m \frac{1}{2}\left[A_m\left(1 + \frac{\nu_m}{\nu_n}\right) + B_m\left(1 - \frac{\nu_m}{\nu_n}\right)\right]c_{n,m} = 0 \quad (23)$$

and

$$B_n' - i\nu_n B_n + \frac{1}{2}\frac{\nu_n'}{\nu_n}(B_n - A_n) + \sum_m \frac{1}{2}\left[B_m\left(1 + \frac{\nu_m}{\nu_n}\right) + A_m\left(1 - \frac{\nu_m}{\nu_n}\right)\right]c_{n,m} = 0 \quad (24)$$

where

$$c_{n,m} = \frac{1}{M_n}\int_{a_o}^{\infty}\frac{\phi_m'\phi_n}{r}dr \quad (25)$$

is the coupling coefficient.

A useful observation is that the factor $(1 - \nu_m/\nu_n)$ will be small for modes of nearly the same order. As a consequence, coupling from forward to backward modes is small in such cases. This fact will be evident in what follows below.

WKB UNCOUPLED FORM

The first step in dealing with the coupled equations (23) and (24) is to ignore all coupling. Then they reduce to

$$A_n' + i\nu_n A_n + (1/2)(\nu_n'/\nu_n)A_n = 0 \tag{26}$$

and

$$B_n' - i\nu_n B_n + (1/2)(\nu_n'/\nu_n)B_n = 0 \tag{27}$$

It is easily verified that solutions are

$$A_n = (\nu_n)^{-1/2} \exp(-i\int^\theta \nu_n d\theta) \tag{28}$$

and

$$B_n = (\nu_n)^{-1/2} \exp(+i\int^\theta \nu_n d\theta) \tag{29}$$

The perceptive reader will note these have the expected WKB forms for wave transmission in a slowly varying duct. In the absence of any dissipation, the exponential factors account for the accumulated phase while the inverse square root factor preserves the power flow in the θ direction.

MODIFIED COUPLED FORM

Taking a hint from the WKB limiting forms, we are led to choose our basic solutions of the coupled equations (23) and (24) to have the form

$$A_n = a_n \nu_n^{-1/2} \exp(-i\int_0^\theta \nu_n d\theta) \tag{30}$$

and

$$B_n = b_n \nu_n^{-1/2} \exp(+i\int_0^\theta \nu_n d\theta) \tag{31}$$

where a_n and b_n are expected to be slowly varying functions of θ when the coupling effects are not strong. We now see that

$$A_n' = \frac{a_n'}{\nu_n^{1/2}} \exp(-i\int_0^\theta \nu_n d\theta) - \frac{1}{2}\frac{\nu_n'}{\nu_n} A_n - i\nu_n A_n \tag{32}$$

and
$$B_n' = \frac{b_n'}{\nu_n^{1/2}} \exp\left(i\int_0^\theta \nu_n d\theta\right) - \frac{1}{2}\frac{\nu_n'}{\nu_n} B_n + i\nu_n B_n \tag{33}$$

Using (30), (31), (32) and (33), we see that (23) and (24) are transformed to

$$a_n' = \frac{\nu_n'}{2\nu_n} b_n \exp\left(2i\int_0^\theta \nu_n d\theta\right) - \sum_m \frac{\nu_n^{1/2}}{2\nu_m^{1/2}} c_{n,m}\left[\left(1 + \frac{\nu_m}{\nu_n}\right) a_m \exp\left(-i\int_0^\theta (\nu_m - \nu_n)d\theta\right)\right.$$

$$\left. + \left(1 - \frac{\nu_m}{\nu_n}\right) b_m \left(\exp i\int_0^\theta (\nu_m + \nu_n)d\theta\right)\right] \tag{34}$$

and

$$b_n' = \frac{\nu_n'}{2\nu_n} a_n \exp\left(-2i\int_0^\theta \nu_n d\theta\right) - \sum_m \frac{\nu_n^{1/2}}{2\nu_m^{1/2}} c_{n,m}\left[\left(1 + \frac{\nu_m}{\nu_n}\right) b_m \left(\exp i\int_0^\theta (\nu_m - \nu_n)d\theta\right)\right.$$

$$\left. + \left(1 - \frac{\nu_m}{\nu_n}\right) a_m \exp\left(-i\int_0^\theta (\nu_m + \nu_n)d\theta\right)\right] \tag{35}$$

These are the most useful forms of the coupled equations, in spite of their complicated appearance. A convenient approach to the solution is to use the formalism of perturbation theory. To this end we write

$$a_n(\theta) = {}_0a_n + {}_1a_n(\theta) + {}_2a_n(\theta) + \ldots \tag{36}$$

$$b_n(\theta) = {}_0b_n + {}_1b_n(\theta) + {}_2b_n(\theta) + \ldots \tag{37}$$

Where the θ-dependent coefficients are expanded in a series of ordered terms. The leading or zero-order terms are θ-independent, while the succeeding terms of order one, two, ... etc., are θ-dependent. For this approach to be meaningful we regard the coupling coefficient $c_{n,m}$ and ν_n'/ν_n to be quantities of first order. On examining (34) and (35) this tells us immediately that if a_n and b_n are j'th order quantities, then the derivatives a_n' and b_n' are of $(j + 1)$th order or smaller. Of course, this rather tenuous argument breaks down completely if a mode degeneracy is approached wherein the coupling coefficient $c_{n,m}$ becomes indefinitely large; this would violate the basic premise of the perturbation theory.

AN APPLICATION

To illustrate the application of the perturbation theory we imagine that a mode of order n is incident from the left. Thus

$$E_z^{(inc.)} = \sum_n {}_0a_n \frac{\phi_n(r,\theta)}{[\nu_n(\theta)]^{1/2}} \exp\left(-i\int_0^\theta \nu_n(\theta)d\theta\right) \tag{38}$$

Now according to the coupled mode theory, the fields corrected to first order would be given by

$$E_z = E_z^{(inc.)} + \sum_n \left[{}_1a_n \exp\left(-i\int_0^\theta \nu_n(\theta)d\theta\right) \right.$$
$$\left. + {}_1b_n \exp\left(i\int_0^\theta \nu_n(\theta)d\theta\right)\right][\nu_n(\theta)]^{-1/2} \phi_n(r,\theta) \tag{39}$$

where

$$_1b_n' \simeq (\nu_n'/2\nu_n)_0 a_n \exp(-2i\int_0^\theta \nu_n d\theta)$$

and

$$_1a_n' \simeq -\sum_m {}_0a_m c_{n,m} \exp\left(-i\int_0^\theta (\nu_m - \nu_n)d\theta\right) \tag{40}$$

In writing (39) and (40) we have assumed that for the important modes ν_n/ν_m can be replaced by one where it appears. We now integrate both sides of these two equations with respect to θ. Noting that $_1b_n$ vanishes for $\theta \to \infty$, we choose the limits in (39) from ∞ to θ. Similarly, because $_1a_n$ vanishes for $\theta \to -\infty$ we choose the limits in (40) from $-\infty$ to θ. Thus we deduce that

$$_1b_n(\theta) \simeq -{}_0a_n \int_\theta^\infty \frac{\nu_n'(\theta)}{2\nu_n(\theta)} \exp\left(-2i\int_0^\theta \nu_n(\hat{\theta})d\hat{\theta}\right)d\theta \tag{41}$$

and

$$_1a_n(\theta) \simeq -\sum_m {}_0a_m \int_{-\infty}^\theta c_{n,m}(\theta) \exp\left(-i\int_0^\theta [\nu_m(\hat{\theta}) - \nu_n(\hat{\theta})]d\hat{\theta}\right)d\theta \tag{42}$$

where the θ dependencies are indicated explicitly.

The physical meaning of (41) is clear. For example, the first order reflected wave is a mode of the same order and it represents the integrated effect of the reflection from the local gradient ν_n' in the wave number. For any extended region, the rapidly varying phase factor exponential term in (41) will produce cancellation so the net effect of the total reflected wave is small. Furthermore, the consideration of second order quantities shows that the backward mode conversion (i.e., $_2b_n$, $_3b_n$, etc.) in the present problem is even of less consequence.

The first order expression for the forward mode conversion given by (42) is more interesting. Here the incident mode is converted to modes of different order and the resulting mode conversion can be very significant.

THE COUPLING COEFFICIENT

As indicated above, the coupling coefficient $c_{n,m}$, as defined by (25), plays an important role. To evaluate this integral we first differentiate (15) with respect to θ to yield

$$\left(\frac{\partial}{\partial r} r \frac{\partial}{\partial r} - \frac{\nu_m^2}{r} + k^2 r\right)\phi_m' - \frac{2\nu_m \nu_m'}{r}\phi_m + 2kk'r\phi_m = 0 \qquad (43)$$

We also have

$$\left(\frac{\partial}{\partial r} r \frac{\partial}{\partial r} - \frac{\nu_n^2}{r} + k^2 r\right)\phi_n = 0 \qquad (44)$$

Now we multiply (43) by ϕ_n and (44) by ϕ_m' and subtract the resulting equations to yield

$$\frac{\partial}{\partial r}\left[r\phi_n \frac{\partial \phi_m'}{\partial r} - r\phi_m' \frac{\partial \phi_n}{\partial r}\right] - \frac{2\nu_m \nu_m'}{r}\phi_m \phi_n$$
$$+ 2kk'r\phi_m\phi_n = (\nu_m^2 - \nu_n^2)r^{-1}\phi_m'\phi_n \qquad (45)$$

Both sides are integrated with respect to r from a_o to ∞ to give

$$(\nu_m^2 - \nu_n^2) \int_{a_o}^{\infty} \frac{\phi_m' \phi_n}{r} dr = \left[r\phi_n \frac{\partial \phi_m'}{\partial r} - r\phi_m' \frac{\partial \phi_n}{\partial r} \right]_{r=a_o} \qquad (46)$$

$$- 2\nu_r \nu_n' M_n \delta_{m,n} + 2\int_0^{\infty} kk' r\phi_m \phi_n dr.$$

The square bracket term on the right hand side of (46) vanishes at the upper limit because of the radiation condition. At the lower limit we utilize the boundary condition for ϕ_n at $r = a_o$ given by (16) and further note it also holds for the θ derivative ϕ_n'. Thus, the square bracket term vanishes identically. Now, provided $m \neq n$ we can deduce from (46) that

$$c_{n,m} = \frac{2}{(\nu_m^2 - \nu_n^2)M_n} \int_{a_o}^{\infty} kk' \phi_m \phi_n r dr \qquad (47)$$

The case $n = m$ can be treated separately. Thus, for example,

$$c_{n,n} = \frac{1}{M_n} \int_{a_o}^{\infty} \phi_n \phi_n' r^{-1} dr = \frac{1}{2M_n} \frac{\partial}{\partial \theta} \int_{a_o}^{\infty} (\phi_n)^2 r^{-1} dr \qquad (48)$$

or, more simply

$$c_{n,n} = (2M_n)^{-1} M_n' \qquad (49)$$

The expression for the coupling coefficient given by (47) is very convenient. The integration over the radial (i.e., height) coordinate need only extend over the range where $\partial k/\partial \theta$ is non-vanishing. It is also interesting to note that the expression for self-mode coupling given by (49) does not depend on the specific form of radial profile of the wave number.

AN EXAMPLE

A concrete example is useful to illustrate some of the points we have been trying to make. We choose a model such that

$$k^2(r,\theta) = k^2(r) + \Delta k^2 u(\theta - \theta_o) \qquad (50)$$

for $a_2 > r > a_1$ where $u(\theta - \theta_o)$ is the unit step function at $\theta = \theta_o$.

Outside this range $k(r,\theta) = k(r)$, i.e., where $r > a_2$ and $a_1 > r > a_0$. Physically we imagine this as an elevated layer of enhanced refractivity that abruptly begins at $\theta = \theta_0$ and is uniform thereafter. Thus, for the height interval $a_2 > r > a_1$, we have

$$2kk' = \partial k^2/\partial \theta = \Delta k^2 \delta(\theta - \theta_0)$$

where $\delta(\theta - \theta_0)$ is the unit impulse function at $\theta = \theta_0$. For this case we see that the first order mode conversion coefficient given by (42) simplifies to

$$_1a_n(\theta) = 0 \quad \text{for} \quad \theta < \theta_0$$

$$= -\sum_m a_m \bar{c}_{n,m} \exp[-i\nu_m \theta_0 - i\hat{\nu}_n(\theta - \theta_0)] \quad \text{for} \quad \theta > \theta_0. \tag{51}$$

Here, in the case $m \neq n$,

$$\bar{c}_{n,m} = \int_{-\infty}^{\theta} c_{n,n}(\theta)d\theta = \int_{a_1}^{a_2} \frac{1}{M_n(\nu_m^2 - \hat{\nu}_n^2)} \int_{a_1}^{a_2} (\Delta k^2)\phi_m \hat{\phi}_n r dr \tag{52}$$

In the present example $\hat{\nu}_n$ is the appropriate wave number for the uniform structure for $\theta > \theta_0$ and the corresponding wave function is $\hat{\phi}_n$. Then, of course

$$M_n(\theta) = M_n \quad \text{for} \quad \theta < \theta_0$$
$$= \hat{M}_n \quad \text{for} \quad \theta > \theta_0 \tag{53}$$

so that

$$c_{n,n} \triangleq 2^{-1}(M_n\hat{M}_n)^{-1/2}(\hat{M}_n - M_n)\delta(\theta - \theta_0) \tag{54}$$

and thus

$$\bar{c}_{n,n} = 2^{-1}(M_n\hat{M}_n)^{-1/2}(\hat{M}_n - M_n) \tag{55}$$

In many cases we are only interested in approximate estimates of the mode conversion so $\bar{c}_{n,n}$ is of little consequence. Furthermore, in such cases we replace $\hat{\nu}_n$ and $\hat{\phi}_n$ in (52) by ν_n and ϕ_n.

Coupled Mode Analysis

COMPARISON WITH MODE MATCHING

It is useful to compare the specific example above with a simple mode matching procedure [Wait, 1968, 1974] where reflected modes are neglected at the outset. To that end we would write

$$E_z = \sum_m \frac{\alpha_m}{\nu_m^{1/2}} e^{-i\nu_m\theta} \phi_m(r) \quad \text{for} \quad \theta < \theta_o \tag{56}$$

and

$$E_z = \sum_m \sum_n \frac{\alpha_m}{\hat{\nu}_n^{1/2}} e^{-i\nu_m\theta_o} S_{n,m} e^{-i\hat{\nu}_n(\theta-\theta_o)} \hat{\phi}_n(r) \quad \text{for} \quad \theta > \theta_o \tag{57}$$

where α_m is to be specified by the source and $S_{n,m}$ is a mode conversion defined as indicated. Following the Kirchhoff procedure we match E_z at the aperture plane $\theta = \theta_o$. Thus

$$\frac{\phi_m(r)}{\nu_m^{1/2}} = \sum_n S_{n,m} \frac{\hat{\phi}_n(r)}{\hat{\nu}_n^{1/2}} \tag{58}$$

must hold for $a_o < r < \infty$. Then in the time honored tradition we multiply both sides of (58) by $\hat{\phi}_{n'}(r)/r$ and integrate with respect to r from a_o to ∞. Using the orthogonality property given by (19), it follows easily that

$$S_{n,m} = \left(\frac{\hat{\nu}_n}{\nu_m}\right)^{1/2} \frac{1}{\hat{M}_n} \int_{a_o}^{\infty} \frac{\phi_m(r)\hat{\phi}_n(r)}{r} dr \tag{59}$$

where

$$\hat{M}_n = \int_{a_o}^{\infty} [\hat{\phi}_n(r)]^2 \, r^{-1} \, dr \tag{60}$$

Now we note that

$$\left(\frac{\partial}{\partial r} r \frac{\partial}{\partial r} - \frac{\nu_m^2}{r} + rk^2\right) \phi_m = 0 \tag{61}$$

and

$$\left(\frac{\partial}{\partial r} r \frac{\partial}{\partial r} - \frac{\hat{\nu}_n^2}{r} + r\hat{k}^2\right) \hat{\phi}_n = 0 \tag{62}$$

The first of these is multiplied by $\hat{\phi}_n$ and the second by ϕ_m; subtracting leads to

$$\frac{\partial}{\partial r}\left[\hat{\phi}_n r \frac{\partial \phi_n}{\partial r} - \phi_m r \frac{\partial \hat{\phi}_n}{\partial r}\right] - \frac{\nu_m^2 - \hat{\nu}_n^2}{r} \phi_m \hat{\phi}_n + (k^2 - \hat{k}^2) r \phi_m \hat{\phi}_n = 0 \qquad (63)$$

On integrating this result from a_0 to ∞ leads readily to

$$S_{n,m} = \left(\frac{\hat{\nu}_n}{\nu_m}\right)^{1/2} \frac{-1}{\hat{M}_n (\nu_m^2 - \hat{\nu}_n^2)} \int_{a_0}^{\infty} r \phi_m \hat{\phi}_n \Delta k^2 \, dr \qquad (64)$$

where $\Delta k^2 = \hat{k}^2 - k^2$. This is essentially consistent with the coupled mode formalism, at least to the extent that factors such as $(\hat{\nu}_n/\nu_m)$ can be replaced by one.

Actually, if $k(r)$ is real and Y is effectively infinite, there may be some merit in numerical work if we replace $\phi_n(r)$ in (59) by it's complex conjugate. In this case M_n, as given by (60), now contains $\phi_n(r)^2$. This type of normalization of the modes is commonly used when dealing with loss free structures such as optical fibers (Marcuse, 1974; Spurleder and Unger, 1979).

FINAL REMARKS

We have attempted here to present a self-consistent approach to the mode coupling phenomena in non-uniform ducts. While the discussion was in the context of microwave transmission in the troposphere, the general method if certainly applicable to underwater sound channels and acoustic wave ducting in the atmosphere. The main shortcoming of the present analysis is the difficulty when mode degeneracy occurs. In such cases, the coupling coefficient becomes indefinitely large. An approach to treat problems like this has been given recently by Budden [1975]. This would appear to be a fruitful area for further work.

REFERENCES

BAHAR, E., (1971), Diffraction of electromagnetic waves by cylindrical structures characterized by variable curvature and surface impedance, *J. Math. Phys.*, 12, No. 2, 186-196.

BAHAR, E., and WAIT, J.R., (1963), Propagation in a model terrestrial waveguide of non-uniform height: theory and experiment, *Jour. Res. Nat'l. Bur. of Stand.*, D69, 1445-1463.

BREMMER, H., (1958), Propagation of electromagnetic waves, *Handbuch der Physik*, 16, 423-639.

BUDDEN, K.G., (1975), The critical coupling of modes in a tapered earth-ionosphere waveguide, *Math. Proc. Camb. Phil. Soc.*, 77, 567-580.

CHO, S.H., and WAIT, J.R., (1978), Analysis of microwave ducting in an inhomogeneous troposphere, *Pure and Appl. Geophys.*, 116, 1118-1142.

DEAVENPORT, R.L., (1966), A normal mode theory of an underwater acoustic duct by means of Green's function, *Radio Science*, 1, No. 6, 709-724.

MARCUSE, D., (1974), *Theory of Dielectric Optical Waveguides*, Academic Press, New York.

LEDINEGG, E., PAPOUSEK, W., and BSCHNIZER, B., (1975), Anamalous attenuation of VLF waves passing over Greenland ice sheet.

MCDANIEL, S.T., (1977), Mode conversion in "shallow water sound propagation", *Jour. Acoust. Soc. Amer.*, 62, 320-335.

NAGL, A., ÜBERALL, H., ZARUR, G.L., and HAUG, A.J., (1978), Adiabatic mode theory of underwater sound propagation in a range dependent environment, *J. Acous. Soc. of Amer.*, 63, 739-749.

PIERCE, A.D., (1965), Extension of the method of normal modes to sound propagation in an almost stratified medium, *J. Acous. Soc. Amer.*, 37, No. 1, 19-27.

RUTHERFORD, S.R., and HAWKER, K.E. (1979) An examination of the range dependence of the ocean bottom on the adiabatic approximation, *Jour. Acoust. Soc. Amer.*, 66, 1145-1151.

SPURLEDER, F., and UNGER, H.G., (1979), *Waveguide Tapers, Transitions and Couplers*, (IEE Electromagnetic Wave Series), Peter Peregrinus Ltd., Hitchin, Herts. England.

WAIT, J.R., (1968), On the theory of VLF propagation for a step model of the non-uniform earth-ionosphere waveguide, *Can. Jour. Phys.*, 46, 1979-1982.

WAIT, J.R., (1970), *Electromagnetic Waves in Stratified Media*, Pergamon Press, Oxford, 2nd Edition.

WAIT, J.R., (1974), Theory for the excitation and propagating of electromagnetic waves guided by an elevated refractive index discontinuity in the troposphere, *Can. J. Phys.*, 52, No. 19, 1852-1862.

Index

Airy Function Representations, 85, 192, 246, 275
Anisotropic ionosphere model, 234, 263
Asymptotic evaluation of fields, 117, 125
Anisotropic half-space, 96, 107, 110

Bremmer's WKB generalization, 186
Brewster's Angle, 15

Coupled mode analyses, 325, 332, 337

Day-to-Night transition, 276
Dipole integral formula, 17, 97

Earth crust waveguide, 153
Earth detached modes, 253
Earth flattening concept, 255
Earth ionosphere waveguide, 244
Excitation of surface waves, 69, 86
Exponential boundary transition, 165, 174

Flat earth limit, 252

Geometrical optics, 188
Green's function, 61, 67
Ground wave propagation, 83, 90, 101, 123, 244

Height Gain functions, 88, 251, 258
Hertz potentials, 44
Hyperbolic transition model, 226

Impedance matching, 15
Ionospheric reflection, 212

Lateral waves 156
Laterally varying waveguide model, 269
Linearly segmented atmosphere, 310
Loop-loop mutual coupling, 200

Magnetic line source, 68
Magneto-telluric fields, 18
Maxwell's Equations, 2, 97, 236
Microwave model for VLF, 257
Mode conversion 270, 273, 320, 325
Mode formulations 86, 157, 244, 310
Mode matching, 343
Modified saddle point method, 118

Natural Oscillations, 14
Non-uniform transmission line theory, 77
Norton surface wave, 101, 122
Numerical distance, 124

Ohm's Law, 1

Perturbed exponential model, 221
Phase integral, 82, 182, 185, 191
Plane wave reflection, 7, 14

Quasi static fields, 98, 104, 142

Rapid boundary transition, 195
Residue series, 154

Sky waves, 248
Sommerfeld attenuation function, 71, 114
Sommerfeld integral, 16
Source location, 140
Special conductivity profiles, 164, 182, 212
Spherically stratified conductor, 76
Surface impedance, 10, 18, 20, 22, 44, 51, 59, 84, 126
Surface impedance tensor, 47
Surface wave of Zenneck, 15, 56

Transmission line analogy, 9, 14
Trapped miners, 129
Tropospheric inversion, 318
Tropospheric propagation, 310, 320
Tropospheric reflection, 187
Tropospheric scattering, 318

Wave hops, 249
Wave tilt, 10, 11, 15, 46, 91
WKB method, 184
Wire loop source, 129, 204
Whispering gallery modes, 252

About the Author

JAMES R. WAIT is Professor of Electrical Engineering and Geosciences at the University of Arizona. Among the many distinguished awards he has won is the 1973 NOAA Scientific Research and Achievement Award for "his outstanding achievement as a scientist and research leader in theoretical studies of electromagnetic wave propagation in the earth and its atmosphere."